About this Publication

Title:

Mike Becker's Helicopter Handbook

Volume 1: Basic Training for Helicopter Pilots and Instructors

Series:

For Helicopter Pilots

Edition:

First published 1986, 2nd Release 1997, 3rd Release 1997, 4th Release 2000

Revised Edition 2023.

Principal Author:

Mike Becker, ATPL(H), FIR, FER, Diploma (Training and Assessment)

Editor:

Bev Austen, BTech(CompSt), MEd(DTL)

Copyright

Copyright © 2023 Becker Helicopter Services Pty Ltd

Photos and Illustrations

Most photos and illustrations in this document have been sourced from Becker Helicopter Services Pty Ltd. The remainder is taken from the internet from various sources; Every effort has been made to ensure images with Creative Commons Licences have been used and/or appropriate attribution provided.

Disclaimer

Nothing in this text supersedes any operational documents issued by any civil aviation authority or regulatory body, aircraft, engine, and avionics manufacturers or the operators of aircraft throughout the world. No responsibility is taken for interpreting and applying the information in this document. Managing the safety of the aircraft is the sole responsibility of the pilot-in-command.

Every possible effort has been made to establish the accuracy of the information contained in this book; however, the author, Becker Helicopter Services Pty Ltd, accepts no responsibility for errors or omissions.

The Publisher and the Author make no representations or warranties for the accuracy or completeness of the contents of this work and expressly disclaim all warranties, including warranties of fitness for a particular purpose, without limitation. No warranty may be created or extended by sales or promotional materials. The advice and strategies contained herein may not be suitable for every situation. This work is sold with the understanding that the author is not engaged in rendering legal, accounting, or other professional services. If professional assistance is required, the services of a competent professional person should be sought. Neither the Publisher nor the Author shall be liable for damages arising therefrom.

The fact that an organisation or website is referred to in this work as a citation and/or a potential source of further information does not mean that the author or the publisher endorses the information the organisation or website may provide or recommendations it may make. Further, readers should be aware that internet websites listed in this work may have changed or disappeared between when it was written and when it was read.

Feedback and comments

As this edition involved significant revisions since the last release, there may be some inadvertent errors, omissions or even differences in opinion on technique.

We welcome your feedback via this email: shop@beckerhelicopters.com.

Contents

Pre-Flight, Start-up, Run-up and Shutdown 1-1

Effects of Controls 2-1

Elementary Handling 3-1

Hover 4-1

Lift-off, Transition and Landing 5-1

The Circuit 6-1

Basic Autorotation 7-1

Manoeuvring during Autorotation 8-1

Flight Control Emergencies (FCEs) 9-1

Limited Power Operations 10-1

Confined Areas 11-1

Low Level Flying 12-1

Weight and Balance 13-1

Detailed Table of Contents

About this Publication	i
Contents	ii
Detailed Table of Contents	iii
About the Author	ix
Author's Dedication	ix
About this Book	x
Acknowledgements	x

Pre-Flight, Start-up, Run-up and Shutdown 1-1

    Preparation: Pre-Flight Inspections 1-2
        Documents to be Carried 1-2
        Blade Sailing 1-2
        Approaching and Departing a Helicopter 1-5
        Fuel Policy 1-7
    Air Exercises: Pre-Flight Inspections 1-8
        Air Exercise 1-1: Pre-Flight Inspection 1-8
        Air Exercise 1-2: Start and Run-up 1-11
        Air Exercise 1-3: Engine Shutdown 1-12
        Air Exercise 1-4: Refuelling 1-12

Effects of Controls 2-1

    Preparation: Effects of Controls 2-2
        Cyclic 2-2
            Swashplate Assembly 2-3
        Collective 2-7
        Throttle 2-11
        Anti-torque pedals 2-15
            What is Yaw? 2-19
            What is Balance? 2-19
        Summary of Helicopter Controls 2-22
        Magnitude and Rate of Control 2-23
        Flapback 2-24
        Stability 2-25
        Limits for Forward Flight 2-27
        Secondary Controls 2-30
            Carburettor Icing 2-30
    Air Exercises: Effects of Controls 2-34
        Handing Over and Taking Over (HOTO) Control 2-34
        The Work Cycle 2-35
        Air Exercise 2-1: Using the Cyclic 2-40
        Air Exercise 2-2: Using the Pedals 2-43
        Air Exercise 2-3: Using the Collective 2-45
        Air Exercise 2-4: Using the Throttle 2-46
        Demonstration 2-1: Further Effects of Controls 2-47
        Demonstration 2-2: Effects of Controls at the Hover 2-47
            Cyclic 2-47
            Collective 2-48
            Anti-torque Pedals 2-49
            Throttle 2-49

Elementary Handling 3-1

- Preparation: Elementary Handling 3-2
  - Performance Equation 3-2
  - Forces in Forward Flight 3-5
  - Forces in the Climb 3-5
  - Forces in a Descent 3-6
  - Forces in a Turn 3-6
  - Power Changes 3-8
- Air Exercises: Elementary Handling 3-12
  - Air Exercise 3-1: Power Changes 3-12
  - Air Exercise 3-2: Turns 3-14
  - Air Exercise 3-3: Climbing and Descending Turns 3-15

Hover 4-1

- Preparation: Hover 4-2
  - The Work Cycle 4-2
  - Summary 4-7
  - Ground Effect at the Hover 4-8
  - The Effect of Wind at the Hover 4-10
  - Engine Failure at the Hover and Hover Taxi 4-14
- Air Exercises: Hover 4-17
  - Air Exercise 4-1: Effect of controls at the Hover 4-18
  - Air Exercise 4-2: Manoeuvring at the Hover 4-20
    - Turns about the mast 4-20
    - Hover Taxi 4-21
    - Sideways flight 4-22
    - Backwards flight 4-23
    - Turns about the Nose 4-24
    - Turns about the tail 4-25
    - Pattern Hover 4-26
    - Summary 4-26
  - Air Exercise 4-3: Engine Failure at the Hover and Hover Taxi 4-27

Lift-off, Transition and Landing 5-1

- Preparation: Lift-off and Landing 5-2
  - Translating Tendency (Tail Rotor Drift) 5-2
  - Dynamic Rollover 5-4
    - Static Rollover 5-4
    - Dynamic Rollover 5-6
    - Cyclic Limits and Dynamic Rollover 5-9
    - Scenarios that May Contribute to Dynamic Rollover 5-10
    - Preventing Dynamic Rollover 5-11
    - Recover from Dynamic Rollover 5-12
  - Ground Resonance 5-12
  - Wind 5-13
  - Centre of Gravity Effects on the Cyclic 5-14
  - Slope 5-16
  - HEFFR Check: Pre-Liftoff and Pre-Landing 5-21
  - Hover Checks 5-22
  - PWPTEM Check: Pre Departure Brief 5-23
  - Short Finals ARP Check: Pre-Landing 5-24
  - Transitions 5-26
  - Transverse Flow Effect or Inflow Roll 5-28
- Air Exercises: Lift-off, Transition and Landing 5-29
  - Air Exercise 5-1: Lift-off and Landing 5-30

	Lift-off	5-31
	Landing	5-32
Air Exercise 5-2: Transition		5-33
	Transition into forward flight	5-33
	Transition back to the hover	5-34
Air Exercise 5-3: Sloping Ground Take-offs and Landings		5-35
	Sloping Ground Landing	5-35
	Sloping Ground Take-Off	5-36

The Circuit 6-1

Preparation: The Circuit 6-2
- Height Velocity Graph or Dead Man's Curve 6-2
- Types of Circuits 6-4
- Circuit Height 6-6
- Circuit Direction 6-7
- Legs to a Circuit 6-9
- Effects of Wind 6-11
- Lookout in the Circuit 6-13
- Short Finals Check 6-15

Air Exercises: The Circuit 6-19
- Air Exercise 6-1: The Circuit 6-19
- Air Exercise 6-2: The Racetrack Circuit 6-23

Basic Autorotation 7-1

Preparation: Autorotation 7-2
- Practising Autorotations 7-2
- HASEL Check: Before Practising Autorotation 7-3
- Symptoms of an Engine Failure 7-4
- Clutch vs Freewheeling Unit 7-5
- Energy Management 7-7
- Controlling Rotor RPM 7-9
- Aerodynamics on the Blades during Autorotation 7-12
- Flares during Autorotation 7-14

Air Exercises: Autorotation 7-18
- Air Exercise 7-1: The Entry, Descent and Go-Around procedure 7-19
  - Entry and Descent 7-19
  - Decision and Go-Around 7-20
- Air Exercise 7-2: Power Termination procedure 7-21
  - Decision and Power Termination 7-21
- Air Exercise 7-3: Full Touch Down Autorotation 7-23
  - Full Touch Down 7-23
- Summary Diagram: Autorotation to Power Recover 7-24

Manoeuvring during Autorotation 8-1

Preparation: Manoeuvring during Autorotation 8-2
- Descent Configurations 8-2
- Turns during Autorotation 8-5
- Adjusting Speed during Autorotation 8-9
- Precision Autorotation 8-11

Air Exercises: Manoeuvring during Autorotation 8-15
- Air Exercise 8-1: Range, Endurance, Constant Attitude 8-16
  - Range 8-17
  - Endurance 8-20
  - Constant Attitude 8-23

Air Exercise 8-2: Turns during Autorotation 8-26
Air Exercise 8-3: Adjusting Speed during Autorotation 8-27
Air Exercise 8-4: Precision Autorotation 8-28

Flight Control Emergencies (FCEs) 9-1

Preparation: Flight Control Emergencies 9-2
    Collective and Throttle Steering 9-4
    Jammed Controls at the Hover 9-5
        Jammed Collective at the Hover 9-5
        Jammed Throttle at the Hover 9-6
        Jammed Cyclic at the Hover 9-8
        Jammed Pedals at the Hover 9-8
    Jammed Controls while in Forward Flight 9-13
        Jammed Collective from Forward Flight 9-13
        Jammed Throttle from Forward Flight 9-18
        Jammed Cyclic from Forward Flight 9-18
        Jammed Pedals from Forward Flight 9-19
        Left Pedal Jam in Forward Flight 9-22
        Right Pedal Jam in forward flight 9-23
    Yaw Control Emergencies (YCE) 9-24
        Tail Rotor Driveshaft Failure 9-24
        Tail Rotor Component Failure 9-24
        LTE at the Hover and Slow Forward Flight 9-25
        YCE Summary 9-26
    Hydraulics Failure 9-27

Air Exercises: Flight Control Emergency (FCE) 9-29
    Air Exercise 9-1: Jammed Controls at the Hover 9-30
        Jammed Collective at the Hover 9-30
        Jammed Throttle at the Hover 9-31
        Discussion Only: Jammed Cyclic at the Hover 9-32
        Pedals Jam in the Hover Position at the Hover 9-32
        Left Forward Pedal Jam at the Hover 9-33
        Right Forward Pedal Jam at the Hover 9-33
    Air Exercise 9-2: Jammed Controls in Forward Flight 9-34
        Collective Jams in Forward Flight 9-34
        Throttle Jams in Forward Flight 9-36
        Discussion Only: Jammed Cyclic from Forward Flight 9-37
        Pedals Jam in the Left Position in Forward Flight 9-37
        Pedals Jam in the Right Position in Forward Flight 9-39
    Air Exercise 9-3: Yaw Control Emergencies 9-41
        Discussion Only: Tail Rotor Driveshaft Failure at the Hover 9-41
        Discussion Only: Tail Rotor Driveshaft Failure in Forward Flight 9-41
        Discussion Only: Tail Rotor Component Failure at the Hover 9-42
        Discussion Only: Tail Rotor Component Failure in Forward Flight 9-43
        Discussion Only: LTE at the Hover and Slow Forward Flight 9-44
    Air Exercise 9-4: Hydraulic Failure 9-45
        Discussion Only: Cyclic Hard-over 9-45
        Hydraulic Failure from Forward Flight 9-46
        Hydraulic Failure at the Hover 9-47

Limited Power Operations 10-1

Preparation: Limited Power Operations 10-2
    Pitch and Power 10-3
    Performance Charts 10-3

Piston Engine Performance Check	10-5
Turbine Engine Power Assurance Check	10-6
Using a Performance Chart	10-8
Power Required and Power Available	10-10
Power Margin and Power Curve	10-12
Power Margin and Power Categories	10-14
Transients and Power limits	10-15
Power, Atmospheric Conditions and Altitude	10-17
Wind	10-25
Ground Cushion and Surfaces	10-28
Take-off and Approach Profiles	10-31
Normal Take-off and Approach	10-31
Constant Angle Take-off and Approach	10-32
Steep Take-off and Approach	10-33
Vertical Take-off and Approach	10-34
Very Limited Power Take-offs and Landings	10-34
Shallow Take-off and Approach Profile	10-35
Running Take-offs and Landings	10-36
Air Exercises: Limited Power Operations	10-37
Air Exercise 10-1: Determining Power and Wind	10-38
Nominate a Power Category at the Hover	10-38
Nominate a Power Category while in Forward Flight	10-39
Estimating the Wind at the Hover	10-40
Estimating the Wind while in Forward Flight	10-41
Air Exercise 10-2: Cushion Creep Take-off and Landings	10-42
Air Exercise 10-3: Running Take-off and Landings	10-44

Confined Areas ... 11-1

Preparation: Confined Areas	11-2
Reconnaissance	11-3
High Recce	11-4
Low Recce	11-5
Power Settling and Vortex Ring State	11-7
Forces on the Blade	11-9
Forces on the Blade: Hovering OGE in Nil Wind	11-9
Forces on the Blade: Shallow Forward Descent in Nil Wind	11-11
Forces on the Blade: Steep Forward Descent in Nil Wind	11-13
Effect on the Tail Rotor	11-14
The Effect of a Tail Wind	11-17
Effect of Collective during VRS	11-18
Conditions Which Produce Power Settling and VRS	11-19
Recovery from Vortex Ring State	11-19
How Does Wind Affect VRS	11-21
Maneuvering in a Confined Area	11-23
Recirculation	11-25
The PSWATP	11-26
P POWER, PILOT, PAYLOAD	11-27
S SIZE, SHAPE, SLOPE, SURFACE, SUN/MOON, SHADOWS, SURROUNDS, STOCK	11-28
W WIND, WIRES-WIRES-WIRES, WAY IN, WAY OUT	11-32
A APPROVAL, APPROACH, ABORT	11-39
T TURNING and TERMINATION POINTS. THREATS	11-43
P PLAN	11-45
Air Exercises: Confined Areas	11-46
Air Exercise 11-1: Power Settling an VRS	11-47

    Air Exercise 11-2: Confined Area Operations 11-49

    Summary 11-51

Low Level Flying 12-1

  Risk Analysis 12-2

  Example Risk Analysis 12-4

  Height Velocity Diagram 12-6

  Wind 12-8

  Control Effectiveness 12-10

  Low Level Manoeuvres 12-13

    Low Level Turns 12-13

    Maximum Performance Takeoff and Zero-Zero Landings 12-21

    Low Level Circuit 12-23

    Decelerating Climb and High Hover 12-24

    Rapid Descents 12-26

  Low Level Emergencies 12-27

  Air Exercises: Low Flying 12-29

    Air Exercise 12-1: Introduction to Low Flying 12-31

      Airspeed vs Groundspeed 12-31

      S-Turns and Reversal Turns 12-32

      Figure of 8 Turns 12-33

    Air Exercise 12-2: Quick Stops 12-34

      Straight ahead Quick Stops 12-34

      90-degree Quick Stop 12-35

      180-Degree Quick Stop 12-36

    Air Exercise 12-3: Reversal Turns 12-37

      Flat 180-Degree Turn 12-37

      Cyclic 180-Degree Turn 12-38

      Pedal and Torque 180 degree Turns 12-38

    Air Exercise 12-4: Low Level Circuits with a Maximum Performance Takeoff and a Zero Zero Landing 12-40

      Maximum Performance Takeoffs 12-40

      Low Level Circuit 12-41

      Zero Zero landing 12-42

    Air Exercise 12-5: Decelerating Climb, High Hover and Rapid Descents 12-43

      Decelerating Climb, High Hover and Rapid Descent 12-43

      Rapid Descent 12-44

    Air Exercise 12-6: Low Level Emergencies 12-45

      Engine Failure After Takeoff 12-45

      Engine Failure on Approach 12-46

      Engine Failure Low Level High Speed 12-47

Weight and Balance 13-1

  Preparation: Weight and Balance 13-2

    Balance Terms 13-2

    Weight Terms 13-5

    Converting Fluid Volume to Weight 13-7

    Example: Calculating the CofG Position 13-7

About the Author

Mike Becker is one of Australia's most experienced helicopter instructors, with over 16,000 hours of rotary wing flight experience. His career has taken him from the mountains in New Zealand to the outback of Australia to the jungles of Papua New Guinea. He has also worked in the United States, Italy and Borneo.

He has flown a range of helicopter types, including Robinson R22, Robinson R44, Bell 47, Hughes 269, Hughes 500, Bell 206, Bell 407, Bell 427, Bell 212, EC120, Dragon Fly, Brantley B2B, Enstrom EF28, Sikorsky S62A, Hiller 12ET, Bolkov Bo105, Aerospatial AS350, Agusta 109 Power, Agusta 109S Grand, and the Agusta 119 Koala.

Mike is a Grade One Flight Instructor and Flight Examiner with an Australian Air Transport Pilots Licence (Helicopter) and an Australian Commercial Pilots Licence (Fixed Wing).

He is experienced in a comprehensive range of helicopter operations, including high altitude, remote area operations, mustering, firefighting, tourism, sling load operations, specialised long-line operations, search and rescue, night unaided (VFR), instrument flight (IFR) and Night Vision Goggles (NVG) operations.

Mike is the Chief Pilot and Head of Training for his business, Becker Helicopter Services, in Australia. He, and his wife Jan, established Becker Helicopters in 1997 with one Bell 47 and have grown the company through a love of helicopters, hard work, and determination. Mike is the recipient of many awards, including the "Captain John Ashton Award for Flight Standards and Aviation Safety" by the Guild of Air Pilots and Air Navigators of London, which was awarded in recognition of over 18,000 accident-free flight training hours at Becker Helicopters.

"Mike Becker's Helicopter Handbook" was first published in 1986, followed by a range of theory books and instructional videos. In 2021, he worked with the Helicopter Association International publishing three Training Reference Guides around avoiding Inadvertant Flight into IMC (https://rotor.org/education/). For access to more titles by Mike Becker, visit shop.beckerhelicopters.com.

Author's Dedication

> This book is dedicated to Nana D'Ott, who had patience as I messed up her lounge and wore out her little typewriter when I started this project. Her encouragement and quiet support were a special part of my life. Thank you, Nana. And to Poppa D'Ott, who taught me the value and spirit of adventure. I miss you both.

About this Book

Introduction

This manual has been written for students during flight training and for instructors.

It is not designed as a theory manual but as a practical learning tool to be used during flight training where the student and instructor can read about what exercises they will be doing and learn about the related theory behind the exercises.

It is important to note that this book is not specific to any one type of helicopter. Although the R22 and B206 are used as examples at times for airspeeds, power settings and illustrations, what is more important and relevant to ALL helicopters is the techniques and processes described.

Any type-specific information must be taken from the applicable Rotorcraft Flight Manual (RFM), and any specific technique or process should be validated by the Flight School. This book is supplementary and supports your path to learning.

Layout

Each chapter will usually cover the following points:

Aim	Gives the aim of the exercise.
Objective	States the elements or objectives that will be covered within the exercise and reflect the items you need to remember and apply
Motivation	The reason why we want to know the information
Preparation	Covers the concepts, theory and flight techniques relevant to the air exercise(s).
Air Exercises	Goes through the exercises you will be doing. The air exercise is laid out in the same format as the instructor would be demonstrating it with you in the helicopter. Some of the common faults demonstrated by students are also included.

Acknowledgements

Editor: Bev Austen

Bev Austen has been analysing, designing and developing documentation and training solutions for many years. We have collaborated on many publications through the years since working on the First Edition of this book. She assisted in putting together the format of this manual, making suggestions on content and preparing graphics.

1

Pre-Flight, Start-up, Run-up and Shutdown

Aim	To introduce the student to the helicopter and some fundamentals important to its operation.
Objectives	On completion of this lesson, the student will be able to: - identify documentation required prior to a flight - summarise a process for conducting a pre-flight inspection - state the disc protocols for safely approaching and departing a helicopter - outline a procedure for starting the helicopter and then conducting a run-up - outline a procedure for shutting down the helicopter, and - outline a procedure for refuelling a helicopter.
Motivation	Every helicopter you fly will have a different order and technique for pre-flight, start-up and shutdown within the Rotorcraft Flight Manual (RFM). In some cases, like the Airbus family of helicopters, you need to do an after-flight check and then a turnaround check, so their terminology does not include a pre-flight. This is all semantics, as you need to ensure your helicopter is inspected and safe to fly and has the required documentation before you start it, whether as a pre-flight or an after-flight. When you begin flying, the flight school will issue you with a checklist specific to the type of helicopter you will be flying and take you through the pre-flight. After doing a pre-flight (or "daily" as it is also commonly known) a few times, you will find that the checklist is no longer required as you will have it committed to memory. As you advance to larger, more sophisticated helicopters, the checklists will also become more complex, and you will need to refer to the checklists to avoid missing anything. This chapter outlines a typical checklist and pre-flight similar to what you may expect when you start flight training in a smaller helicopter. However, it is not particular to any one type of helicopter and is only an example. It is not to be used on a real helicopter. For further details, check with your flight school.

Preparation: Pre-Flight Inspections

Documents to be Carried

Documentation Before a flight, the following documentation shall be on board the helicopter:

- The Rotorcraft Flight Manual (RFM) which may also be referred to as the Helicopter Flight Manual (HFM) or Pilot Operating Handbook (POH).
- A current Maintenance Release (MR) signed for the day by a licensed and type-rated pilot or engineer.

Additionally, you may find on board the aircraft a:

- Certificate of Airworthiness
- Certificate licensing the radio installation, and a
- Certificate of Registration for the helicopter.

These items are usually found in the front of the RFM. If they are not there, consult with the helicopter operator.

The pilot must have in their possession:

- A valid licence and current Medical Certificate. Today these may be carried electronically; for example, a digital image on your mobile phone can suffice.
- Maps, documentation and other navigation equipment as required for the flight, which may also be electronic.

It is important to note that each country may vary in their requirements for what is required to be carried on a flight, so you may have additional items not listed above.

Blade Sailing

Introduction Blade sailing is the excessive flapping of the rotor blades, which can occur at low RPM with strong gusts of wind. Blade sailing can:

- damage the helicopter if the rotor blades strike the tail boom, or
- severely injure personnel approaching or departing the helicopter, as the tips of the blades can be at the same height as a person's head.

Therefore, understanding blade sailing early in your flight training is important.

When can it occur Blade sailing can occur whenever the Rotor RPM is low, such as:

- when the rotor is speeding up or slowing down, such as when the helicopter is being started or shut down
- in strong wind conditions, or
- if the disc experiences a brief gust.

With low Rotor RPM, the centrifugal force on the blades is less. If a gust of wind strikes the disc, flapback will occur. The advancing blade will flap up and the retreating blade down. If the helicopter is pointing into the wind, there is now an extreme danger of the blade striking the tail cone. At such low Rotor RPM, the cyclic has little to no effect.

Chapter 1 Pre-Flight, Start-up, Run-up and Shutdown

Two-bladed helicopters All helicopters are susceptible to blade sailing, but two-bladed helicopters with fully articulated rotor heads will experience it more because of the freedom of movement designed into the rotor head.

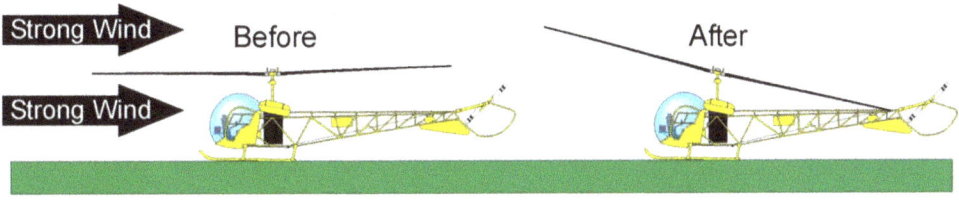

Any direction It is also important to remember that the blades will sail away from a gust of wind as the advancing blade climbs and the retreating blade descends. This means the blades may sail in any direction.

Consider a pilot departing the helicopter to the front while the blades are still turning. A gust of wind from behind causes the blades to sail forwards. There is the real possibility of the pilot being struck by the main rotor.

Blade sailing is most common when working around other helicopters. One helicopter may be parked and winding down while another is coming into land. The downwash from the arriving helicopter or even the prop wash from a taxiing aeroplane may cause a gust that produces a blade sail.

Park at 90° from the wind It may help to have the helicopter parked so the wind strikes the helicopter 90° from the side (from the right in anticlockwise turning rotor systems and from the left in clockwise turning rotor systems). With the wind at the side, as the blade passes over the tail, it climbs and is less likely to strike the tail cone.

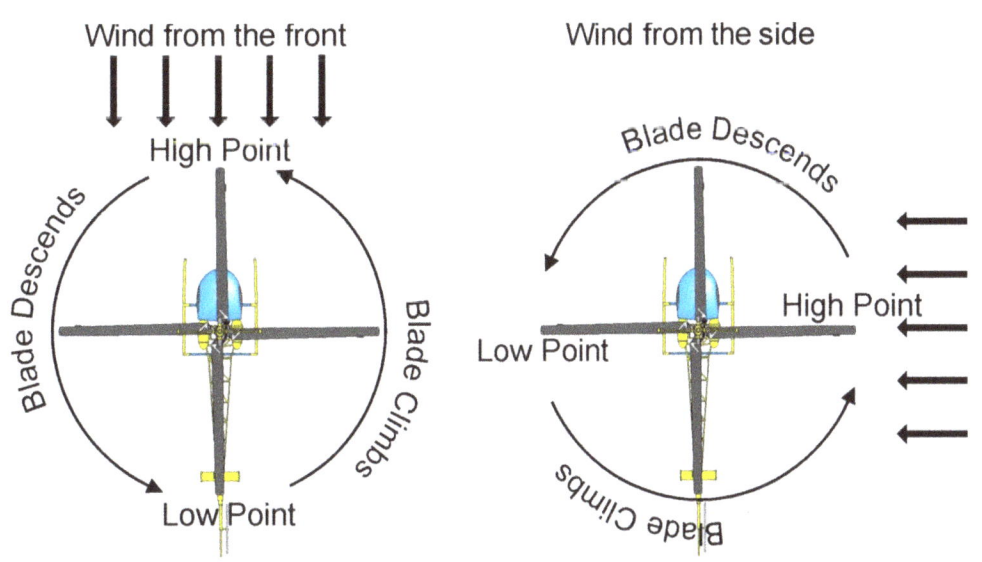

Flap restraints

Another method of reducing the effects of blade sailing is using flap restraints, sometimes referred to as "droop stops".

Flap restraints are designed to reduce the flapping amplitude at very low Rotor RPM, such as when the blades are winding down or just starting up. They are there to protect the main rotor head from mast bumping and protect the crew when working under the disc.

The flap restraints can either work:

- through centrifugal force (as in the case of the B47, B206, etc.), or
- as a floating ring that the blades rest against (as in the H300 or AS350).

When the flap restraints are engaged, they reduce the flapping a blade can experience by physically restricting its movement.

Flap restraints engaged

B47 flap restraints engaged   **Limited flapping**

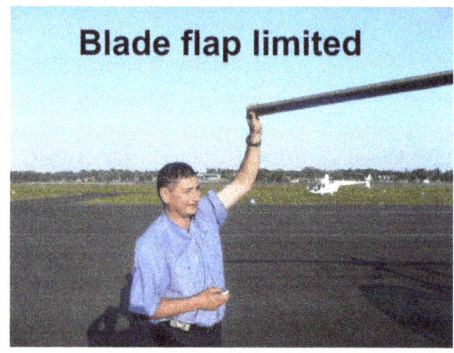

Flap restraints disengaged

The blade can flap over a larger distance when the restraints are disengaged.

B47 flap restraints disengaged   **Extended flapping**

Disc protocols

Most operators have specific disc protocols to reduce the risk of pilots and crew being inadvertently struck by the main rotor blade during operation.

If the blades are turning, the preference is to stay with the helicopter either in the cockpit or within close proximity of the fuselage (do not go near the blade tips) until the blades have stopped turning.

Upon landing, the pedals should be set neutral, the collective fully frictioned down, and locked, if available. The cyclic should be centred so the disc is parallel (level) with the ground. Do not place the cyclic into wind. When the disc is level and, if there are trims attached to the cyclic, set the trims so that the cyclic wants to remain in that position, then firmly apply the frictions and or locks.

Chapter 1 Pre-Flight, Start-up, Run-up and Shutdown

Using the cyclic to control blade sailing

A crewmember (pilot, co-pilot or student) should remain at the controls and ensure the security of the cyclic. If the blades, for whatever reason, want to sail, then use cyclic to try to control the disc. If a rotor brake is fitted, use it to rapidly reduce the Rotor RPM and minimise the sailing.

Maintain a level disc attitude

With the disc set parallel to the ground (level), there is less likelihood of anyone being struck by the main rotors. This is good.

Dangerous when disc is not level

If the disc is not set parallel to the ground (for example disc tilted forward into the wind when on the ground), there is a greater likelihood that the main rotors may strike someone. This is bad.

The pilot's professionalism can often be demonstrated by how the rotor disc is set when the helicopter is idling on the ground. A professional pilot maintains a level disc attitude; the amateur has no idea.

Impact of different helicopter designs

A Sikorsky S76 helicopter has the main transmission mounted at an angle so that the rotor disc with the cyclic set neutral is tilted forward. This tilt gives the aircraft a higher airspeed but also means that approaching and departing the helicopter from the front is forbidden.

Approaching and Departing a Helicopter

Signal

If a pilot or crew member wishes to depart or approach a helicopter when the blades are turning, they must first get permission from the crew member at the controls so that all concerned are aware and cautious in controlling the disc.

Operators will have a protocol for approaching and departing a helicopter, often with pre-arranged signals required before approaching or departing a helicopter. This signal may be in the form of a thumbs up or a nod of the head.

Note: It is essential that the pilot and operator are aware of any design features of a helicopter that requires approach and departure from specific directions.

| **Approaching a helicopter** | When approaching a helicopter:

- stop outside the rotor disc area
- make eye contact with the crewmember and look for a signal indicating it is safe to approach.
- If the signal is not forthcoming and you cannot make eye contact with the crewmember at the controls, **do not approach**.

Upon receiving a positive signal and with your helmet on, duck down and approach the helicopter.

As you are moving under the disc, try to keep an eye on what is happening and continuously be aware of your surroundings. If the disc sails, drop to the ground and either try to get closer to the fuselage or away from the blade tips.

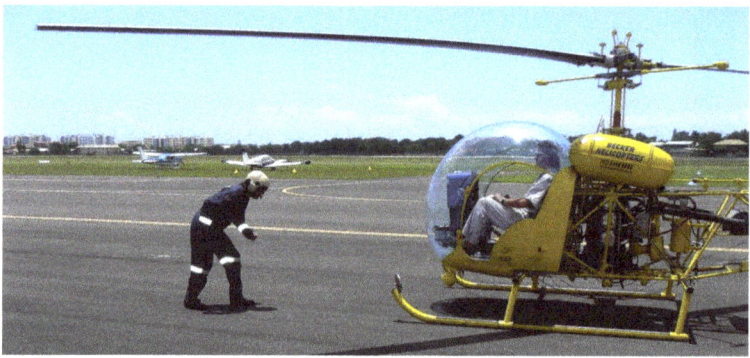

Many experienced pilots have been caught, with complacency being their downfall.

Ever heard of the term "Familiarity breeds contempt"?

**Older pilots (those with 1000s of hours)
with many years of working around helicopters
tend to get complacent with working under and around rotor blades.** |
|---|---|
| **Departing the helicopter** | If departing the helicopter, consult the crew member, ensuring they are in control of the disc and that it is level; then, with your helmet still on (if wearing one), duck down and depart towards the front or side. The best approach and departure angle is the 10 or 2 o'clock position. |

Fuel Policy

Fuel policy

When you become a commercial pilot, each company you fly with will have a fuel policy written into their Company Operations Manual or Standard Operating Procedures (SOPs). Following is an example of a fuel policy.

Introduction

The fuel policy aims to ensure that each flight has sufficient fuel on board so that 30 minutes of fuel remains in the helicopter's fuel tanks on landing. This fuel reserve acts as a safety buffer to prevent inadvertent fuel starvation in flight.

Before every flight

Before every flight, check the following:

- Fuel remaining in the tanks. If possible, this check is done visually with a calibrated dipstick. Only rely on the gauge if it is in a helicopter that is known to be reliable or it is virtually impossible to check visually (e.g. B206, AS350, etc.).
- If you have just conducted a refuel, then again dip the tank to double-check the amount of fuel or recalculate the fuel on board.
- After determining the amount of fuel, divide this amount by the helicopter's average consumption rate (usually stated in the Company Operations Manual or given to you by the flight school). This calculation will give you a "total endurance time" in minutes.
- Take away 30 minutes as a reserve from the "total endurance time", leaving you with a "total in the air time".
- Check the amount of oil remaining in the engine.

At start-up

Convert the "total in the air time" to your watch. This time is known as the FUEL ON GROUND TIME (FOG). Write this time down on your flight log, kneeboard, or some other way. It represents the time you must either:

- land and check your fuel remaining and either conduct a refuel or do another calibration based on what is still left in your tanks, or
- have completed the planned flight and have landed.

Example:

You dip the fuel tanks in a Bell 47 either with a dipstick or by visually looking and ascertaining the fuel level against the internal tank baffles. You determine there are 100 litres on board. The consumption figure in the company's Operations Manual is **60 litres per hour.** Therefore, 100 litres divided by 60 litres per hour equals an endurance of 1 hour and 40 minutes. Minus 30 minutes reserve from 1 hour 40 minutes "endurance time", and you are left with a "time in the air" of 1 hour and 10 minutes. If you start up at 0800 hours, add 1 hour and 10 minutes to give you a "Fuel on Ground" time of 0910.

Tip

When you land, determine how much fuel you should have remaining in your tanks by dividing the flight time by the known consumption rate. Then visually check to see how your calculations were. This check acts as a confirmation of your consumption rate to avoid nasty surprises on the next flight.

Air Exercises: Pre-Flight Inspections

Objectives

On successful completion of the air exercises, the student will be able to:

- identify the documents required on board the helicopter and carried by the pilot
- carry out a pre-flight inspection
- start, run-up and shutdown the helicopter, and
- refuel the helicopter.

Air Exercise 1-1: Pre-Flight Inspection

Introduction

The pre-flight inspection involves walking around the helicopter to check its general condition before a flight.

What to check for

During the pre-flight inspection, you need to check for any evidence of:

- damage to any part of the airframe, rotors, engine and transmission
- delamination (glue debonding from any surface)
- leakage of any fluids
- discolouration of paint due to heat
- dents, chafing, galling, nicks
- corrosion which looks like a bubbling of the paint
- cracks
- fretting of aluminium parts, which produces a fine black powder, and
- fretting of steel parts, which produces a reddish brown or black residue.

If you notice evidence of any of the above points, inform your instructor or contact an Aircraft Engineer.

Pre-flight inspection

The procedure below details an example of a pre-flight inspection similar to what you may expect to see when you start. Each item should be individually looked at and inspected for its overall condition and ability to function correctly during flight. Any evidence of damage should be rectified before departure.

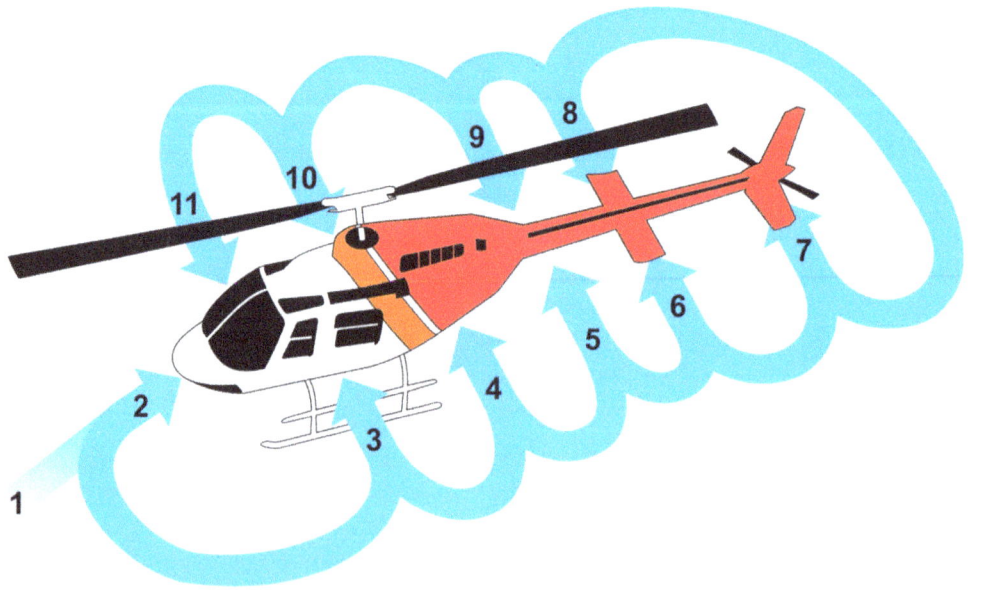

Chapter 1 Pre-Flight, Start-up, Run-up and Shutdown

Step	Section of Helicopter	Check
1	Walking to the helicopter	■ Look at how it is sitting. Does it look right? Is there anything that is catching your eye that looks unusual or different from last time?
2	Nose Section	■ Windshield condition and cleanliness ■ Fresh air vents ■ Landing lights ■ Pitot tube
3	Fuselage (Left Side)	■ Lift skid gear ■ Door hinges, latches and seals ■ Security of left seat, seat belts and controls
4	Engine (Left Side)	■ Rotor head and mast ■ Oil lines ■ Oil quantity ■ Steel tube joints ■ Left-hand fuel tank and fuel drain ■ Control rod ends ■ Inlet air duct ■ Carb air filter ■ Electrical terminals ■ Exhaust system
5	Tail Boom (Left Forward)	■ Attachment bolts ■ Forward drive shaft ■ Vee Belt condition ■ Tail rotor drive coupling ■ Cooling shroud ■ Cooling fan
6	Tail Boom (Left Side)	■ Driveshaft ■ Bearings ■ Joins, tail cone stem ■ Elevator ■ Control rods/cables
7	Tail Rotor	■ Tail rotor blades ■ Control rods ■ Gearbox oil ■ Main blades un-tied
8	Tail Boom (Right Side)	■ Driveshaft ■ Bearings ■ Joins, tail cone stem ■ Elevator ■ Control rods/cables

Step	Section of Helicopter	Check
9	Engine (Right Side)	<ul><li>Oil leaks</li><li>Vee Belt condition</li><li>Steel tube joints</li><li>Tail rotor drive coupling</li><li>Right-hand fuel tank and fuel drain</li><li>Control rod ends</li><li>Inlet air duct</li><li>Carb air filter</li><li>Oil quantity</li><li>Cooling shroud</li><li>Electrical terminals</li><li>Cooling fan</li><li>Exhaust system</li><li>Oil lines</li></ul>
10	Fuselage (Right Side)	<ul><li>Right skid gear</li><li>Door hinges, latches and seals</li></ul>
11	Interior	<ul><li>Flight Manual</li><li>Signed Maintenance Release</li><li>Fire extinguisher</li><li>Seat belts</li><li>Headsets</li><li>Instruments</li><li>Gauges</li><li>Radios</li><li>Switches</li><li>General security</li></ul>

Air Exercise 1-2: Start and Run-up

Starting a piston-engine helicopter The procedure below details an example of an engine start, rotor engagement and engine run-up similar to what you may expect when you do your first lesson with a flight school in a piston-engine helicopter.

Note: The run-up is a systems check carried out before take-off.

No	Stage	Description
1	Before starting the engine	Before starting the engine, you need to check: ■ seat belts are on ■ headsets are on ■ instruments are checked and set ■ circuit breakers are in ■ fuel shut-off value is on, and ■ flight controls for full and free movement.
2	Starting the engine	When starting the engine, you need to: ■ turn Carb Heat off ■ check the mixture is rich ■ check the master battery switch is on ■ check the clutch is disengaged ■ check the area around the helicopter is clear ■ check both magnetos are on ■ prime the engine ■ open a small amount of throttle ■ start the engine ■ set idle speed ■ switch on the alternator/generator ■ engage the clutch ■ check the clutch has engaged ■ check the oil pressure is rising, and ■ remain at the idle until the engine cylinder head temp is in the green.
3	Running up the engine	To run-up the engine, you need to: ■ increase throttle until RPM is at the top of the green range ■ check flat pitch MAP setting (should be between 13" - 16" MAP) ■ apply Carb Heat and note the drop in RPM and the rise in carb air temp ■ turn Carb Heat off ■ check the left and right magnetos ■ roll off the throttle to check the overriding clutch - RPM needles should split ■ check the low RPM warning horn, and ■ throttle back to idle.

Results Once you have completed the run-up procedures, the pilot can conduct pre-liftoff checks, and the helicopter is ready for take-off.

Air Exercise 1-3: Engine Shutdown

Engine shutdown The engine shutdown procedure is described in the table below.

Step	Action
1	Set RPM at idle.
2	Frictions on.
3	Idle for 2 to 3 minutes to cool the engine down (primarily the exhaust valves).
4	Disengage the clutch, if applicable.
5	Roll off the throttle.
6	Lean the mixture.
7	Turn magnetos off.
8	Turn the master battery and generator switch off.

Note: There are two ways to turn the engine off, by either:

- leaning the mixture, which will starve the engine of fuel, or
- turning off the magnetos which stop the spark to the spark plugs.

Air Exercise 1-4: Refuelling

Before flight The pilot must ensure enough fuel in the helicopter's fuel tanks to carry out the proposed flight, plus a reserve of fuel for contingencies.

During pre-flight The pilot, if possible, should visually check the fuel in the tanks. However, this may not always be possible depending on the helicopter's design. The amount of fuel can be estimated by:

- using the fuel gauge (which by itself can be inaccurate)
- using the fuel gauge and relating this to past fuel records indicating amounts put into the tanks against the amount of fuel burnt based on the flight time
- past experience of sighting the level of fuel in the tank against a known mark, such as a baffle, a rivet or similar, or
- using a calibrated dipstick, as shown in the picture.

Fuel consumption during the flight can now be monitored by relating the gauge movement to the known consumption rate against the amount of time flown.

Pilots need to be constantly aware of this and recalculate this in flight in order not to run out of fuel inadvertently.

Chapter 1 Pre-Flight, Start-up, Run-up and Shutdown

Refuelling equipment

Refuelling a helicopter can have many more traps compared to a fixed-wing because they are often refuelled in remote areas using:

- drums
- jerry cans
- mobile tankers, or
- airport refuelling bowsers.

In addition, helicopters are often kept running while refuelling, known as a 'hot refuel'.

The picture shows a typical refuelling set-up using a drum and pump with a manual rotary handle.

A good pump will have:

- a filter
- camlock fittings to seal the hose after use, and
- at least three (3) metres of hose, so the pilot does not have to land too close to the drum.

Refuelling

Because helicopters use hand pumps, drums, jerry cans, etc., fuel contamination is a significant concern when refuelling. Therefore, the following points must be covered.

1. Before putting the drum spike into the drum, check that it is clean and dry.
2. Before refuelling, always pump a minimal amount of fuel onto the ground or into a container; this ensures that any contamination in the system does not go into the fuel tank.
3. When refuelling, one side of the fuel will cross-flow into the second tank (in most helicopters), as shown below. Therefore, after refuelling the second tank and if you want maximum fuel on board, you need to go back and top up the first tank.

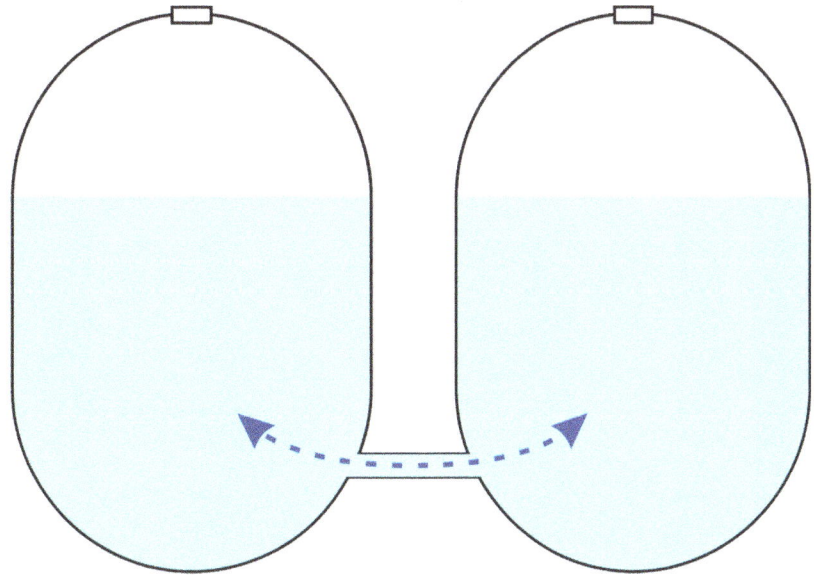

Earthing

Earthing is not a real problem with helicopters because they are not sitting on rubber wheels; therefore, the helicopter is already earthed. However, it is essential that the helicopter and the refuelling equipment are linked together to have the same electrical potential and, therefore, there is less likelihood of a static discharge. This is usually done by using a static line. If the refuelling pump has been manufactured correctly, the delivery hose will have small wires running through it that act as in-built static lines. This can be confirmed by reading the labelling on the hose, which should say *"fuel delivery hose"* or similar.

Fuel drain

After refuelling, the pilot should drain a small amount of fuel from each tank to check for contaminants.

If you find water or other contaminants, keep draining the tank until only clean fuel comes out.

Notes:

- Mogas is red
- Avgas is light blue or green
- Jet A1 is clear
- Water can be clear or muddy
- If you mix Mogas and Avgas, the combined fuel can change colour and appear clear, making you think it is water. Be careful!

Fuel contamination

The most common contaminant is water, which will either:

- form bubbles on the bottom of the testing bottle, or
- if there is enough water, it will form a layer on the bottom with fuel floating on top.

2

Effects of Controls

Aim	To introduce the flight controls.
Objectives	On completion of this lesson, the student will be able to: ■ identify and name the primary flight controls, and ■ describe their primary effect when moved while in forward flight and the hover.
Motivation	So you want to know how to fly a helicopter! The coordination of the flight controls required to fly a helicopter is gained by practice and, in time, anticipation. Good knowledge of what happens when each control is moved individually and then together is essential. As well as flying similarly to fixed-wing aircraft, the helicopter can: ■ hover ■ move horizontally in any direction ■ turn whilst hovering, and ■ climb and descend vertically. In this chapter, we will explore what happens primarily (first) to the helicopter when a control is moved and then observe the secondary effects (what happens next) if we do not coordinate the other controls with the first control movement. The following information will not be specific to any helicopter type but will instead look at various helicopters and various cyclic, collective, throttle and pedal designs.

Preparation: Effects of Controls

Primary controls The primary controls are those used to control the helicopter in flight. They include:

Primary Controls	Primary Effect in Forward Flight	Primary Effect at the Hover
Cyclic	Attitude inclusive of pitch and roll	Position and movement over the ground
Collective	Power inclusive of Engine power (when operating) and total rotor thrust	
Throttle	Primary Engine RPM Control	
Tail Rotor Pedals	Tail Rotor Thrust to manage Yaw (Balance)	Tail Rotor Thrust to control direction and rate of turn

Cyclic

Location The cyclic is positioned at the front centre of the pilot's seat and is moved using the right hand.

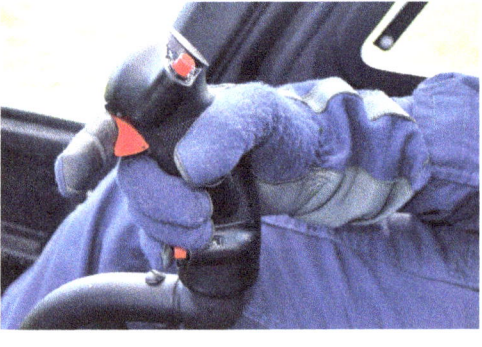

Different types of cyclic The cyclic is either a conventional cyclic coming up between the legs or an unconventional cyclic such as the R22/R44/R66 design, which has a control coming up between the two front seats and a T-bar pivot for each pilot. (Believe it or not, it was designed to allow the ladies to get in and out easily and has the added benefit of fewer moving parts to the system.)

On a cyclic hand grip, you may find radio and intercom transmit buttons, force trims and steerable landing light witches' hats, autopilot release buttons, cargo hook release buttons and possibly other ancillary equipment controls that are now easy to activate by the pilot.

Examples

Conventional EC120 cyclic   Unconventional R22 cyclic

Chapter 2 Effects of Controls

B47 Cyclic	H300 Cyclic
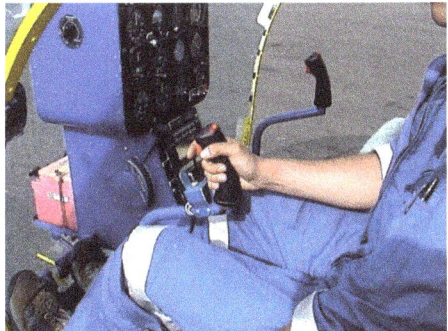	

What it does

The cyclic is used to tilt the rotor disc (plane of rotation) to control the helicopter's attitude and, therefore:

- move the helicopter horizontally in any direction, and
- maintain a hover.

In basic terms, the cyclic is the steering wheel for a helicopter.

Effect on rotor thrust

Moving the cyclic does not alter the amount of Total Rotor Thrust; it merely changes the direction of Total Rotor Thrust, which splits between the vertical and horizontal components.

Since Total Rotor Thrust acts at right angles to the plane of rotation, the horizontal component of this tilted Total Rotor Thrust acts to move the helicopter in the direction the disc is tilted.

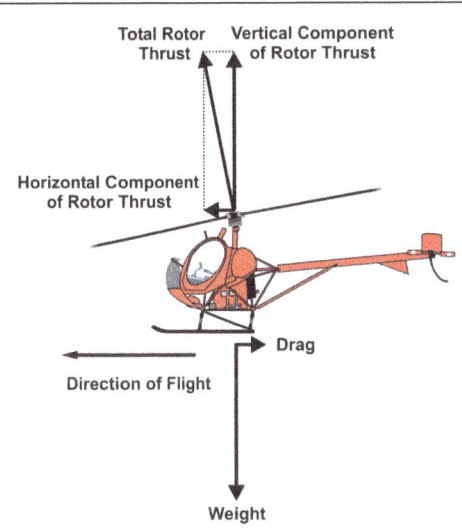

Swashplate Assembly

The swashplate assembly is the mechanical device that allows the cyclic and collective to communicate with the spinning rotor blades. It consists of two discs that are located between the rotor hub and the transmission system.

The tail rotor has a smaller form of the swashplate so that pedal movements can communicate with the spinning tail rotor blades.

Hughes 300 swashplate

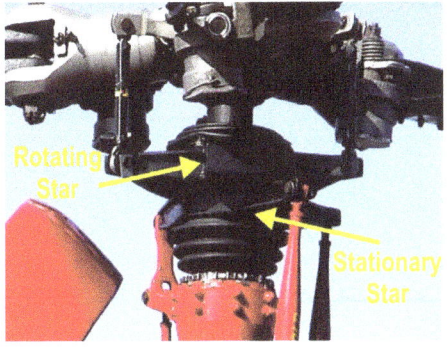

Stationary Star (Non-Rotating Disc)

The Stationary Star is the lower disc in the swashplate which **does not rotate** with the blades. It is linked to the cyclic and collective pitch control systems.

The Stationary Star can tilt in any direction and move up and down about the mast.

Rotating Star (Rotating Disc)	A bearing attaches the non-rotating disc (stationary star) to the upper disc, referred to as the "rotating star". The rotating star **rotates** and is mechanically linked to the rotor blades through control rods, pitch links and pitch horns and is how the pitch changes are introduced.
Example: Bell 47 swashplate	Below is an example of a Bell 47 swashplate.

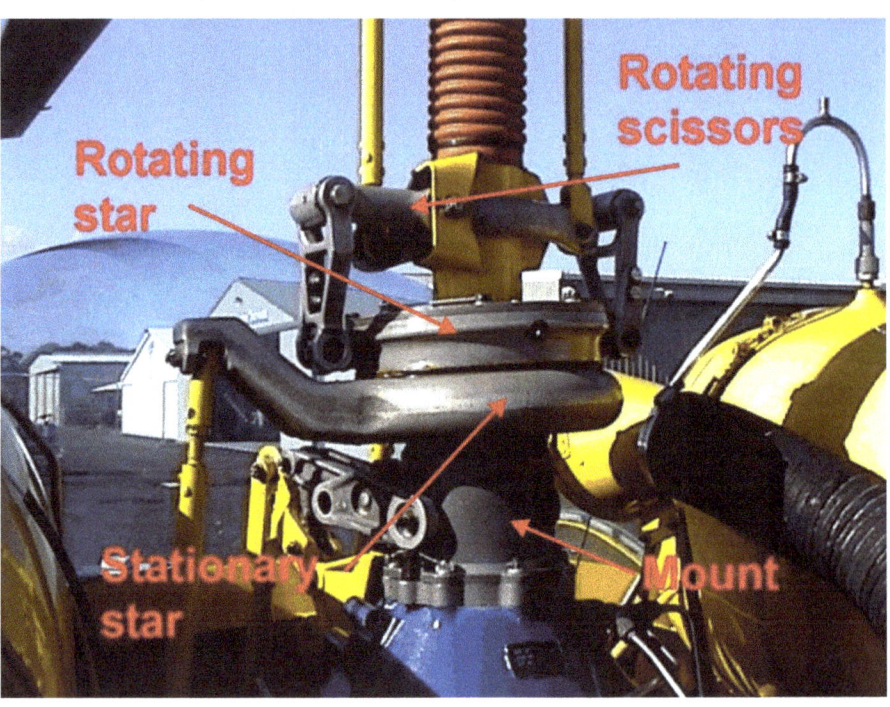

Rotating scissors	The rotating scissors (or lift links) are a hinged connection that are attached to a sleeve bolted to the main rotor mast. This connection makes the rotating star turn with the main rotor mast.
Changing Pitch	When the cyclic is moved, the swashplate tilts in the same direction.

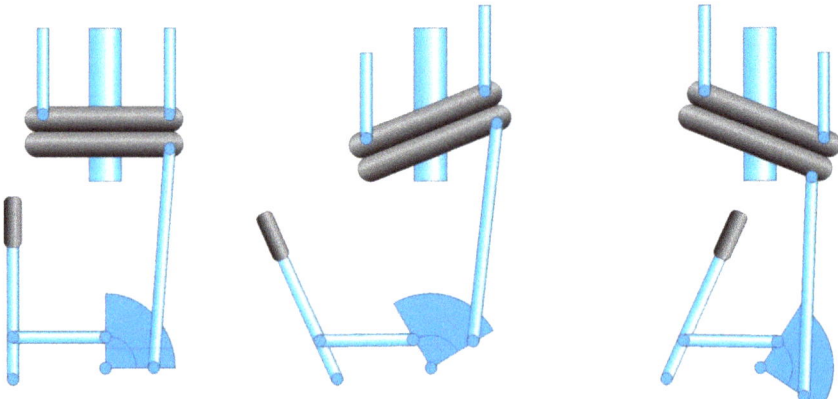

Because each blade is individually connected to the swashplate via control rods, pitch horns and pitch links, then each blade will have a continually changing pitch angle as it rotates through its 360° cycle of travel, depending on the position of the cyclic and cyclically achieving a tilted disc (hence the term cyclic).

Chapter 2 Effects of Controls

Forward cyclic

Forward cyclic will feather each of the blades. In an anti-clockwise rotating system:

- the advancing blade's pitch will decrease with a minimum pitch at 90 degrees to the pilot's right, and
- the retreating blade's pitch will increase with a maximum pitch at 90 degrees to the pilot's left.

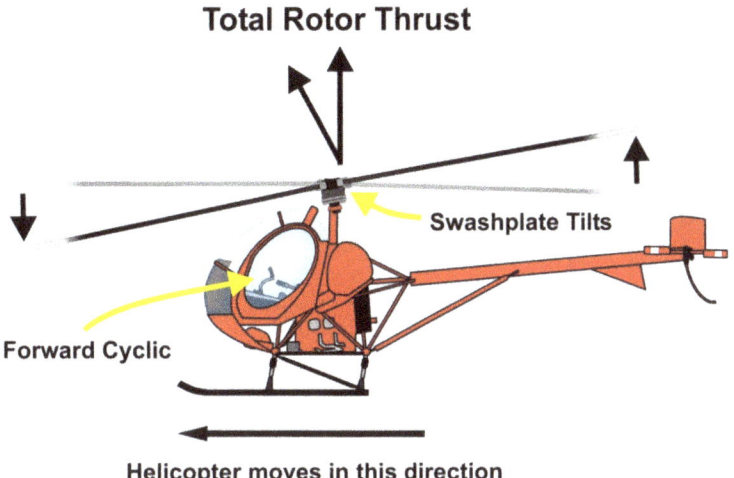

Effect of moving the cyclic forward

The diagram below depicts a helicopter with the cyclic in the forward position (swashplate tilted forward) and the disc tilted. This diagram ignores the effect of airflow from forward speed.

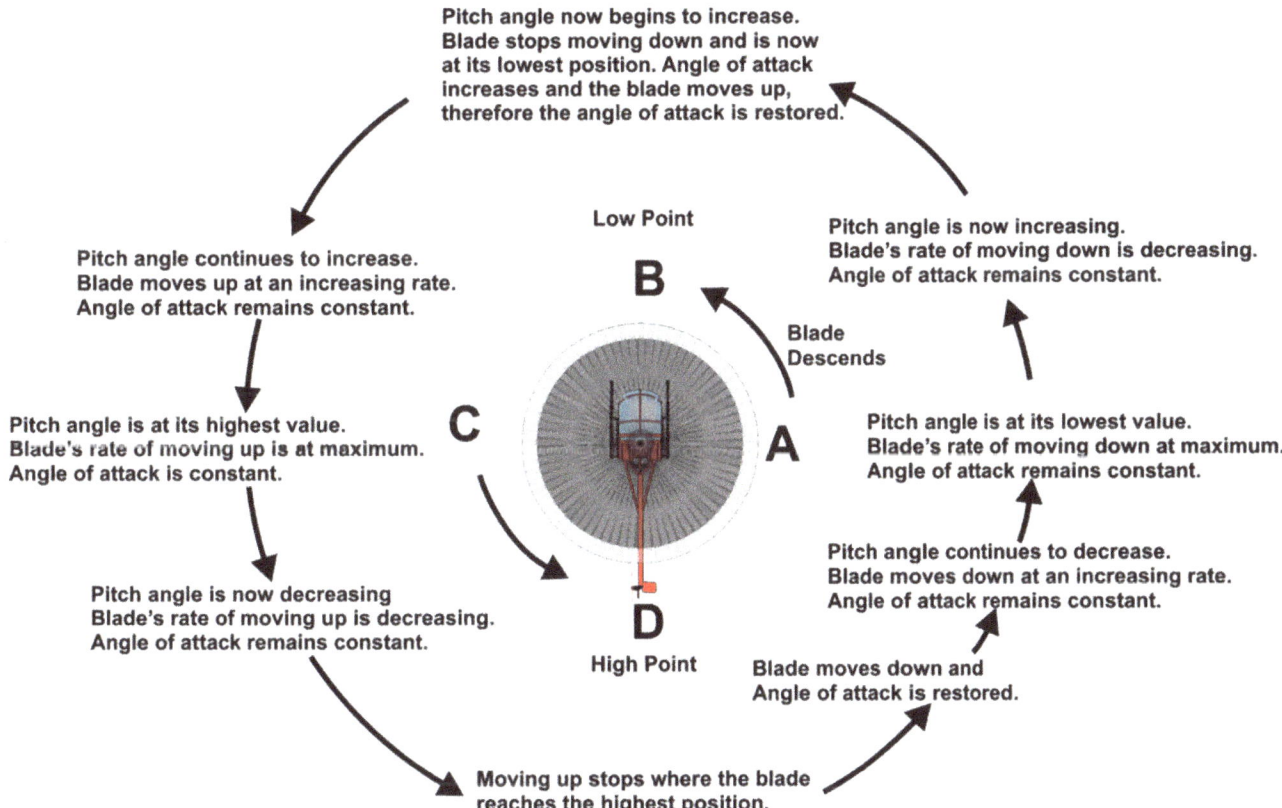

Summary   The table below summarises the effects of cyclic movements.

Cyclic movement	Nose	Altitude	IAS	G Force	RPM tendency	VSI
Cyclic Forward	↓	↓	↑	↓ Due to reduced coning angle	↓	↓
Cyclic Aft	↑	↑	↓	↑ Due to greater coning angle	↑	↑
Cyclic Right	Bank Right	Same	Same	↑	↑	Same
Cyclic Left	Bank Left	Same	Same	↑	↑	Same

Chapter 2 Effects of Controls

Feel of cyclic

When using the cyclic:

- it should be sensitive
- it does not require uniform control movement – for some exercises, the cyclic is extremely sensitive, yet for others requires large control movements
- if it is:
  - hydraulically boosted, will feel smooth with no feedback forces, and
  - not hydraulically boosted, will feel rough with stick shake and feedback forces.
- it does not centre itself, and
- you should never let it go.

Collective

Description

The collective lever is positioned on the left of the pilot's seat and is raised or lowered using the left hand. Collectives are pretty standard in every helicopter; the only differences will be in the amount of equipment a manufacturer wants to put on it and the amount of travel.

On a collective, you may find throttle/s, start buttons, landing light switches, governor beep switches, cargo hook release buttons or levers, or any other ancillary controls that may be required and easily accessible to the pilot.

Collective Friction

All collective levers have a friction device that can be tightened or loosened to make the collective a bit stiffer to move and help it hold its position when let go.

The pilot can adjust this at any time. It is not a lock and can take different forms, such as a lever (as shown) or a knob.

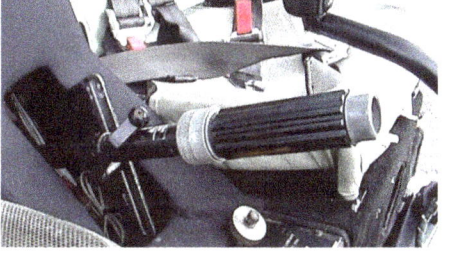

Examples

Bell 206 collective

H300 collective

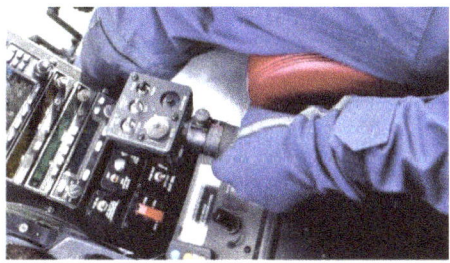
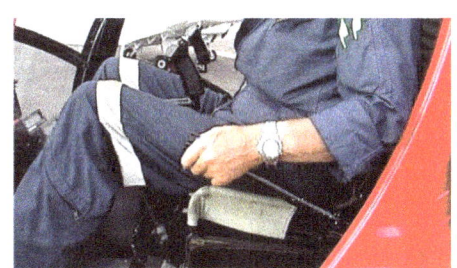

R44 collective

EC120 collective

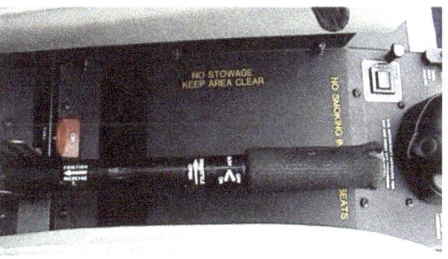
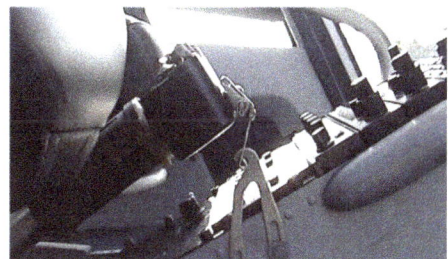

| **Feel of collective** | When using the collective, it should:

- feel slightly stiff
- hold position when released; however, this can vary depending on the friction set, "G" force, turbulence, helicopter type and rigging, and
- be smooth with no feedback forces. |

| **What it does** | Primarily, the collective increases or decreases the helicopter's total rotor thrust by simultaneously changing the pitch on all main rotor blades by an equal amount. |

| **Effect on rotor thrust** | The collective adjusts the pitch angle via the swashplate and, therefore, the angle of attack of all rotor blades equally and simultaneously, increasing and decreasing the Total Rotor Thrust.

To increase Total Rotor Thrust:

- you raise the collective, which
- increases the pitch angle and the angle of attack, and therefore
- increases Total Rotor Thrust.

To decrease Total Rotor Thrust:

- you lower the collective, which
- decreases the pitch angle and the angle of attack, and therefore
- decreases Total Rotor Thrust. |

| **Maintaining rotor RPM** | The diagram below shows that increasing the pitch angle by raising the collective will increase rotor thrust and rotor drag.

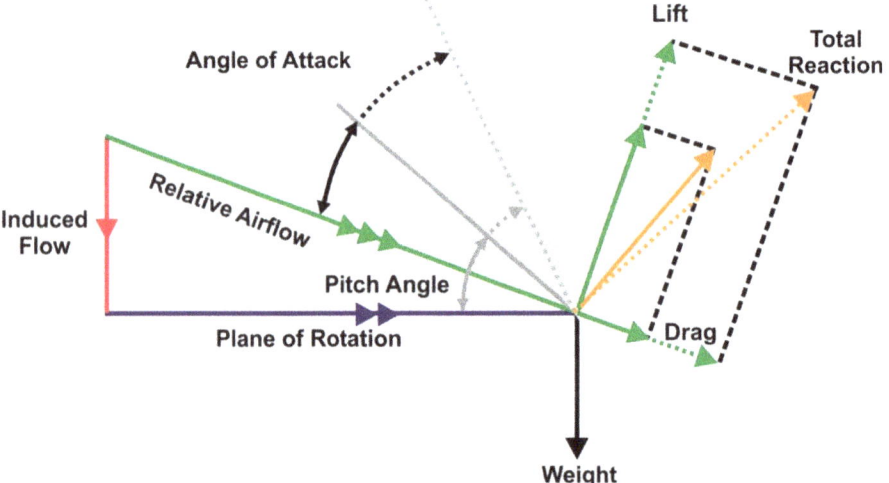

To maintain the Rotor RPM (rotational velocity), the increase in rotor drag must be overcome by increasing the power produced by the engine. The manufacturer knows this and has incorporated a system in the helicopter that will automatically increase and decrease engine power in response to collective inputs.

This system can either be mechanical, electrical or governed. |

Chapter 2 Effects of Controls

Mechanical correlator (piston)

The collective is connected to the engine through a correlation or cam box, which in theory, holds the Rotor RPM (rotational velocity) relatively constant during collective movements.

In a piston helicopter, as the collective is:

- **raised**, the correlator automatically opens the throttle butterfly in the carburettor, **increasing power**.
- **lowered**, the correlator automatically closes the throttle butterfly in the carburettor, **decreasing power**.

In practice, however, this only partially works, and the throttle must be used to fine-tune the Rotor RPM. The accuracy of the correlator box varies from helicopter to helicopter.

Mechanical correlator (turbine)

In a turbine, the mechanical correlator is more like an *anticipator,* which allows the governor to add or subtract fuel in anticipation of the Rotor RPM slowing or speeding up after a collective movement. The governor then fine-tunes the fuel flow to maintain the selected Rotor RPM.

Electrical governor in a piston engine helicopter

In some piston engine helicopters, an electrical governor monitors the RPM. In the R22/R44, a set of points in the magneto sends information to an electronic box, which operates an electric motor that moves the throttle; this system is known as a governor. When Rotor RPM wants to decay, for whatever reason, if the governor is on, then the throttle is automatically increased without pilot input so that it remains in a specified operating band; the opposite happens if the RPM wants to increase.

Electrical sensing governors are very accurate but may experience some lag if large quick collective inputs are made. They may also fail or become erratic if the information they receive is interrupted for any reason. Therefore, the electrical governor can be switched on or off by a switch in the cockpit (normally located at the end of the collective), and there is a warning light on the console to notify the pilot of its status.

R44 governor switch OFF   **R44 governor switch ON**

   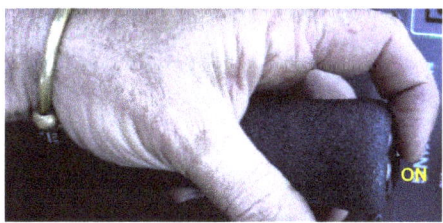

R44 governor warning light showing governor is OFF   **R44 governor warning light showing governor is ON**

Mechanical or electrical (computer) governor in a turbine engine helicopter

Turbine engines typically govern the RPM through a governing unit (governor) which is separate from the Fuel Control Unit (FCU). Like the electrical governor, they are very accurate but may experience some lag if large quick collective inputs are made.

These governors are permanently on and work as an independent unit attached to the accessories gearbox of the engine and determine the amount of fuel delivered to the engine dependent on pilot demand.

Like piston engines, an "anticipator" is directly connected between the collective and the governor to automatically add or reduce fuel flow in parallel to the electrical and pneumatic sensors controlling RPM.

An additional 'beep' switch can also be on the collective, allowing the pilot to activate a linear actuator (small electric motor). This switch is used to adjust the N2 RPM to a desired figure and is sometimes useful if the pilot is anticipating a large fast movement of the collective or is about to conduct a harsh manoeuvre.

Linear actuator activated by beep switch

Beep switch

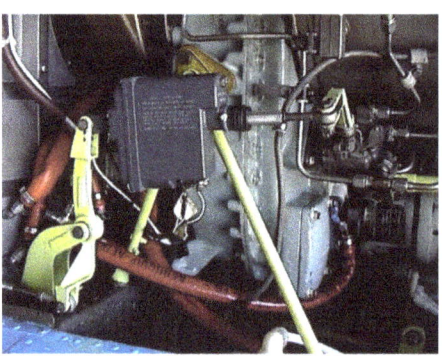
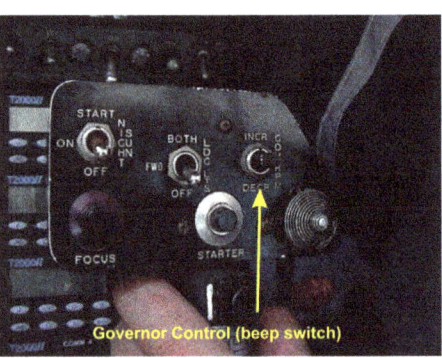

Summary

The table below summarises the effects of collective changes.

Collective movements	MAP/Tq	RPM tendency	YAW	Pedal required	Altitude/Power
Collective Up	↑	↓	→	Left	↑
Collective Down	↓	↑	←	Right	↓

Note: MAP = Manifold Air Pressure (piston). Tq = Torque (turbine).

Chapter 2 Effects of Controls

Throttle

The throttle is located on the end of the collective and is operated with the left hand in a similar fashion to the twist grip on a motorcycle but in the opposite direction.

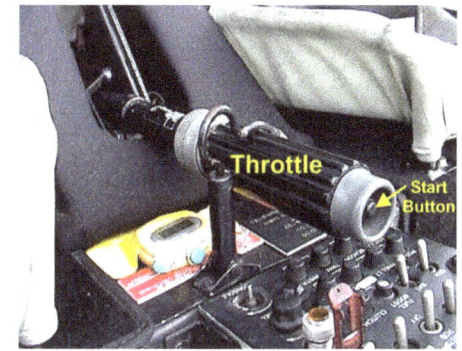

Some turbine helicopters have been designed with the throttle as a lever that is not attached to the collective. This lever can be located on the floor beside the collective or on the ceiling beside the pilot's head, as it is not designed to be used in flight.

B47 throttle

B47 throttle rolled off

B47 throttle rolled on

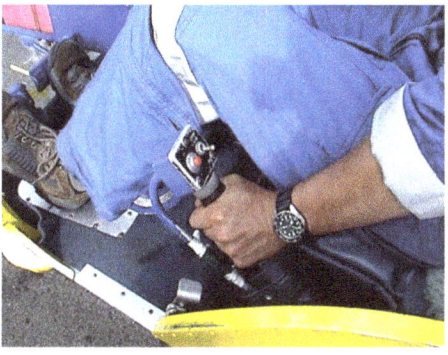
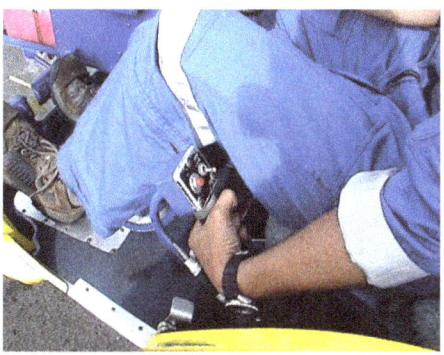

H300 throttle

H300 throttle rolled off

H300 throttle rolled on

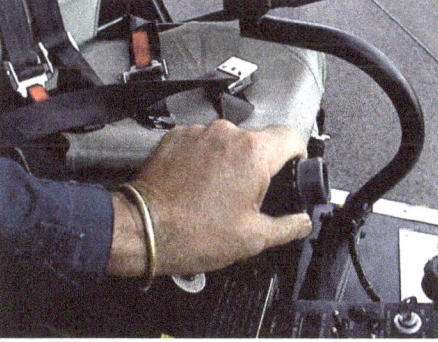

What it does

In a piston engine helicopter, the throttle is used for starting, stopping, setting an idle or flight RPM and during emergencies. Changes in collective (up or down) or harsh cyclic movements (disc loading) will require small corrective movements of the throttle by the pilot to maintain the RPM within the flight parameters. (This may not be required if the helicopter has a governing system.)

In a turbine helicopter, the throttle is used for starting, stopping, setting an idle or flight RPM and during emergencies. In flight, changes in collective (up or down) or harsh cyclic movements (disc loading) will be automatically corrected by the governing system, so during normal flight, the pilot should never have to manipulate the throttle.

Adjusting the throttle

Turning the throttle:

- outboard **increases** power and RPM, and
- inboard **decreases** power and RPM.

Feel of throttle

When using the throttle, it should:

- be very sensitive but firm, and
- hold position when released.

Note: *It is easy to overcontrol with the throttle. You must squeeze the throttle with gentle pressure to get the best result.*

Piston engine throttle positions

The table below illustrates the different throttle positions for piston engine helicopters.

Position	H300 throttle positions
Off override (piston) In the OFF override position, the throttle is usually held off hard against a spring or a detent for practice autorotations so that when the collective is raised or lowered, the correlator will not function.	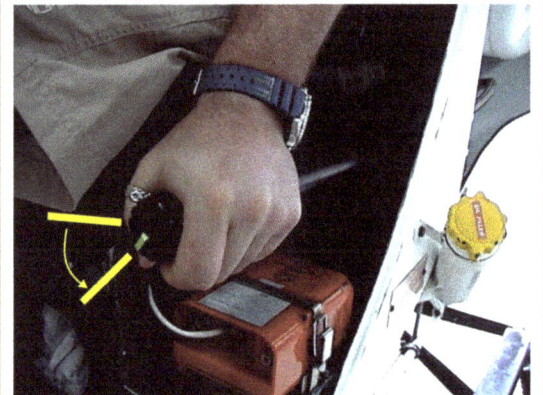
Starting position The starting position is usually a low idle position and can sometimes be felt as a small indent or detent, which is simply the throttle running over a small spring-loaded ball bearing.	
Ground idle position The ground idle position is where the engine and rotor RPM are running together to warm up.	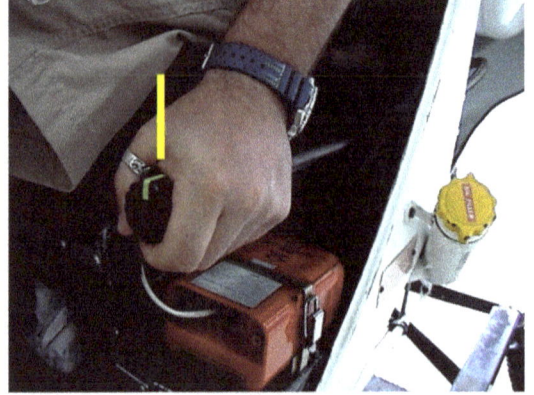

Chapter 2 Effects of Controls

Position	H300 throttle positions
Operating RPM The operating range is a range in which the throttle is moved to maintain RPM as the demands on power change.	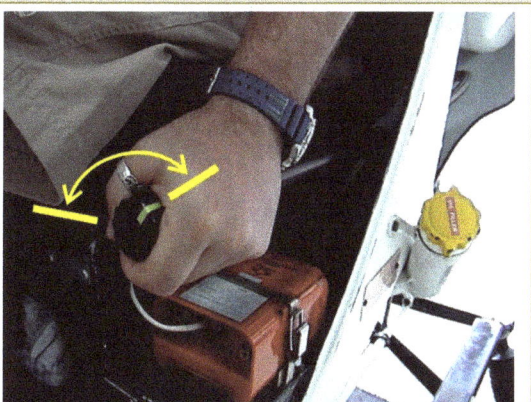
Full throttle The throttle is on full and can go no further.	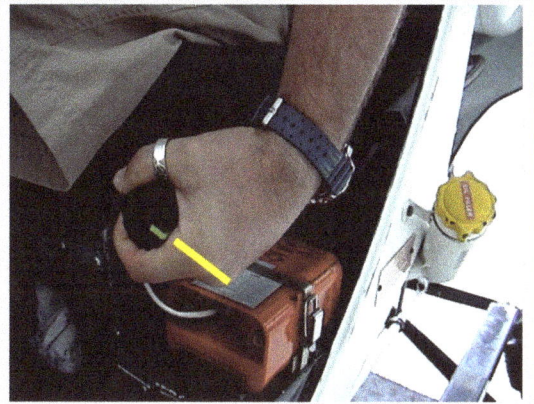

Turbine Engine Throttle positions   The table below illustrates the different throttle positions in turbine engine helicopters.

Position	EC120 throttle positions
Off	
Start This is usually a small indent or detent, which is simply the throttle running over a small spring-loaded ball bearing where the throttle position allows fuel to be provided to the engine.	

Page 2-13

Position	EC120 throttle positions
Ground idle position This is where the engine and rotor RPM run together to warm up, and the throttle is set.	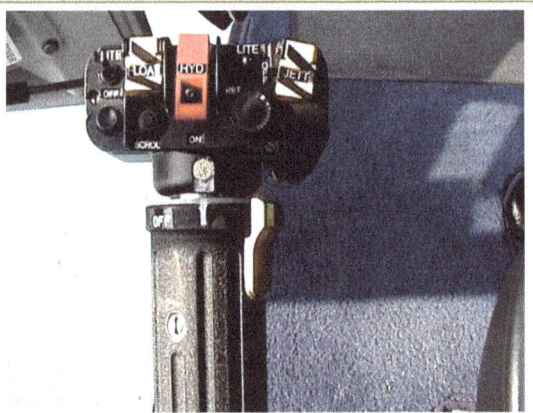
Full Throttle This is when the throttle is on full and can go no further, and the governor will now control the RPM automatically. It allows the FCU to use the maximum engine RPM to maintain rotor RPM, if needed.	

Effect on rotor thrust

When the engine is functioning properly, the Engine RPM and Rotor RPM are at a fixed ratio to each other, and the two needles on the RPM gauge are joined together.

A movement of the throttle is then designed to control Engine RPM but will have the effect of also controlling Rotor RPM. When describing RPM in relation to the throttle, we consider both Engine RPM and Rotor RPM together, which is described as RRPM. If the engine fails, the throttle will have no effect.

In helicopters with an electrical governor or governed engine (turbine), the throttle is not required to fine-tune RPM. Instead, it is used for starting, idling at low idle, bringing the RPM up to operating RPM and also for use in emergencies.

Because RPM relates to the V^2 portion of the lift formula, an increase in RPM will increase rotor thrust, and a decrease in RPM will decrease rotor thrust.

Important: *Rotor thrust must be maintained within the manufacturer's design limits as specified in the flight manual and displayed on the RPM gauge. RPM remains an approximate constant at 100% of design RRPM.*

Chapter 2 Effects of Controls

Effects of throttle summary

The table below summarises the effects of throttle changes in a conventional (anti-clockwise rotating main blades) helicopter:

Changes in throttle	MAP/Tq	RPM	YAW	Pedal required
Throttle Increase	↑	↑	→	Left
Throttle Decrease	↓	↓	←	Right

Anti-torque pedals

Anti-torque pedals are positioned at and operated by the pilot's feet.

Anti-torque pedals are also known as:

- Tail rotor pedals
- Pedals, and
- Yaw pedals

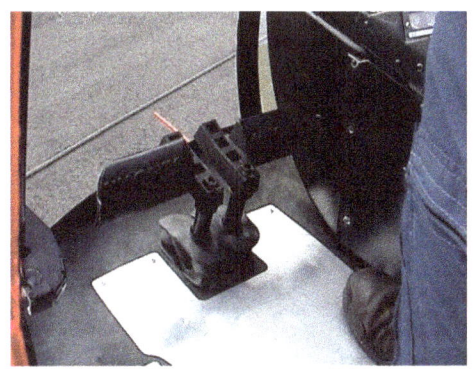

They are sometimes mistakenly referred to as rudder pedals. This would only be the case in non-conventional helicopters that do not have a tail rotor but do have a rudder.

B47 Pedals

B47 left forward pedals   B47 pedals neutral   B47 right forward pedals

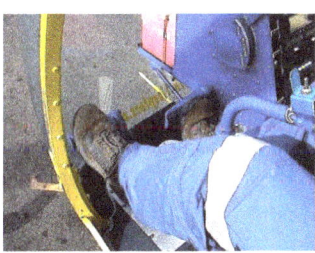  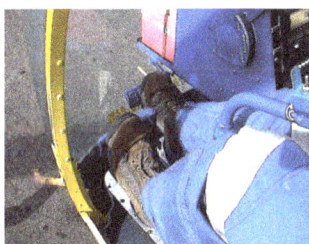

What they do	The tail rotor anti-torque pedals collectively adjust the pitch on the tail rotor blades to counter the torque effect between the engine and the main rotor blades. They are used for:

- Changing the direction of the fuselage (the way the nose points) while hovering. By operating the pedals to produce a tail rotor thrust greater or less than the torque reaction, the heading of the fuselage can be altered while the helicopter is hovering over a spot. The pedals operate in the correct sense, in that a turn to the left results from pushing on the left pedal and a turn to the right results from pushing the right pedal.

- Maintaining balance while in forward flight by controlling the yaw. Yaw (and therefore balance) is indicated by the balance ball, which is mounted on the console.

- Stopping the fuselage from rotating in power-off flight (autorotation). During autorotation, the fuselage will tend to turn in the direction of rotation of the main rotor due to friction in the drive shafts. The tail rotor blades are capable of positive and negative values of pitch. Negative pitch (right pedal) is used in autorotation to maintain directional control.

Feel of anti-torque pedals	When using the anti-torque pedals, they should be:

- light and sensitive, and
- at times twitchy.

They are not self-centring, meaning they will either remain in the position you left them or, as is more common, they take up a neutral position related to the minimum pitch on the tail rotor blades, which is not enough to counter the torque reaction. This means that, in most helicopters, when the engine is functioning, there is always a requirement for some positive pitch and pedal (left pedal in conventional helicopters). You must leave your feet on the pedals at all times.

Effect of anti-torque pedals summary

The table below summarises the effects of using the anti-torque pedals.

Use of pedals	RPM tends	Yaw or turn
Left Pedal	↓	←
Right Pedal	↑	→

To return to balance: Step on the Ball

Chapter 2 Effects of Controls

Effect on rotor thrust

Left pedal - SINK

When the tail rotor thrust is increased, it is achieved by increasing the pitch angle on the tail rotor blades, which will also increase drag and suck power from the system. For this reason, the RPM of the main rotor blades will decay, rotor thrust will decrease, and the helicopter will sink.

Right Pedal - CLIMB

When tail rotor thrust is decreased, it is achieved by decreasing the pitch angle on the tail rotor blades, which will also decrease drag and not use as much power as before. For this reason, the RPM of the main rotor blades will increase, rotor thrust will increase, and the helicopter will climb.

The piston pilot, therefore, is required to compensate for the changes in RRPM by using the throttle. A turbine engine does this automatically.

Understanding torque reaction

In a single-rotor helicopter, the spinning main rotor causes an action called *"torque reaction"*. Torque reaction is an example of Newton's third law of physics: "for every action, there is an equal and opposite reaction". Therefore, if the main rotor blades are turning under power in an anti-clockwise direction, the fuselage will want to turn to the right or in a clockwise direction. Torque reaction, therefore, is directly related to power.

If no power drives the blades, as in an engine failure, then there is no torque reaction trying to torque the fuselage in the opposite direction to the blades. Instead, the fuselage will try to turn in the same direction as the blades due to friction in the transmission and drive train components.

Anti-torque tail rotor

To counteract the torque reaction in a helicopter, an anti-torque device is installed. This device is the anti-torque tail rotor.

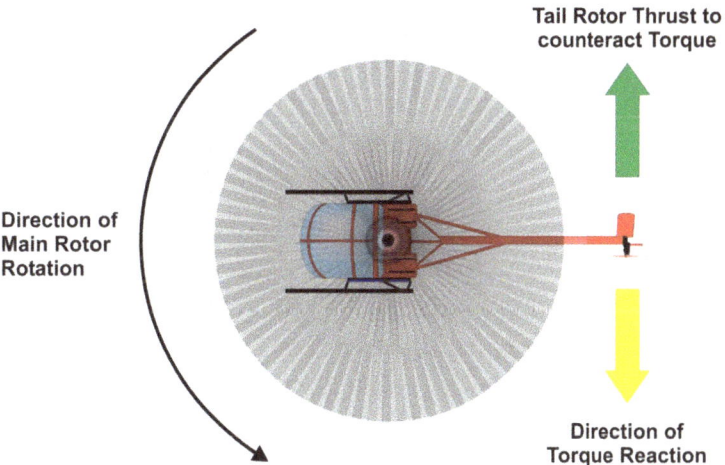

The tail rotor is driven by power taken from the main transmission of the main rotor. Since there is no method of shifting gears, the anti-torque tail rotor runs at a fixed ratio to the speed of the main rotor drive shaft (usually around 6:1, that is, six revolutions of the tail rotor to one revolution of the main rotor).

It absorbs approximately 10-15% of total engine power.

Balancing torque reaction

When engine power is varied, the amount of thrust the tail rotor generates must be varied to maintain a balance of the forces.

Changing the tail rotor pitch using the anti-torque pedals varies tail rotor thrust.

R44 changing pitch   **R44 changing pitch**

The diagram above shows that the tail rotor control mechanism moves in and out on a spline. This spline is connected to pitch links connected to the tail rotor blades. As the spline moves in and out, the pitch of the tail rotor blades changes.

Varying torque

The amount of torque produced varies when the engine power is varied:

When power is.	Then the fuselage.	Which is corrected using the	Which causes the RPM to
Increased	turns to the right	left pedal	⬇
Decreased	turns to the left	right pedal	⬆

Neutral position

When the pedals are in the neutral position (i.e. lined up together), the tail rotor will be in a "positive medium pitch" setting. More left pedal gives more positive pitch. More right pedal gives less positive pitch until you have negative pitch, such as when required in autorotations. In most helicopters, when cruising in forward flight, the pedals are in the neutral position providing a small positive medium pitch and allowing vertical fins (small wings) attached to the tail to do the work of the anti-torque tail rotor.

EC120 tail rotor pedals - neutral   **EC120 Vertical fin attached to Fenestron**

Chapter 2 Effects of Controls

What is Yaw?

What is Yaw?

Yaw is the horizontal movement of the helicopter about the normal axis (also referred to as the vertical axis), which in single-rotor conventional helicopters runs in line with the mast.

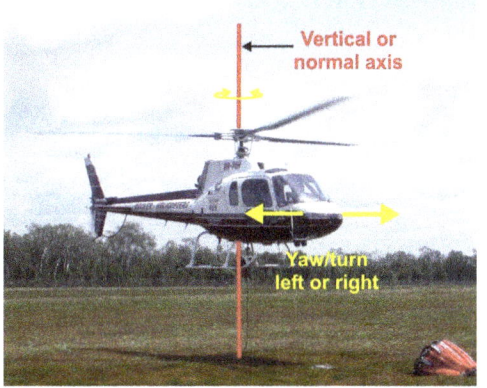

In forward flight, yaw is controlled by the tail rotor pedals and indicated by the balance ball.

At the hover, the balance ball is ineffective; however, the tail rotor pedals still control the turning about the normal (vertical) axis left or right.

Secondary effect in forward flight

In forward flight, the secondary effect of yaw is roll. When more of one side of the fuselage faces the oncoming wind due to the helicopter's forward movement (in effect, the headwind), this produces more drag or resistance on that side and causes the fuselage to roll.

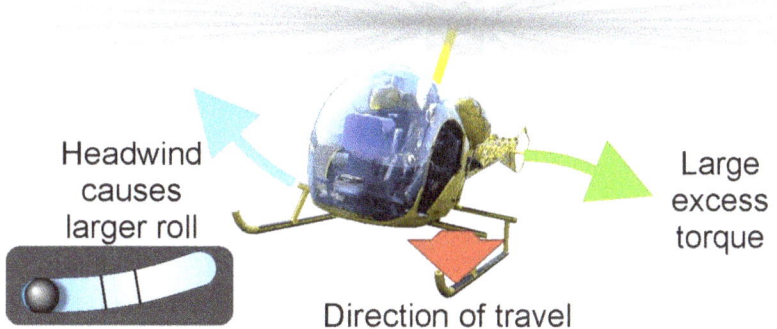

In forward flight, yaw can be measured by balance, and the pilot can make the appropriate changes.

Important

It is important to note several things here:

1. Balance is not considered at the hover. While taxiing a helicopter or on take-off and landing below about 300 ft AGL, the pilot is more inclined to keep the helicopter straight by aligning the skids with the direction of travel regardless of balance. This is done so that in the event of an engine failure, the skids are aligned with the direction of helicopter travel, and then on ground contact, the helicopter is less likely to roll over.
2. The balance ball may not be exactly in the middle for balanced flight. This is especially true in small helicopters when there is a change in the centre of gravity (C of G) with fuel burn or when the helicopter is flown solo, and the pilot's weight is pronounced. To find balanced flight, lift up to the hover and note where the balance ball is. This is now the position of the ball for balanced flight for this flight. But remember, as fuel is consumed, this position may change mid-flight, and as previously stated, this is quite noticeable in small training helicopters but not so noticeable in bigger machines.

What is Balance?

What is balance

When a helicopter is balanced, it means that the nose and the tail of the aircraft are following each other in the direction of travel with the floor of the helicopter level.

Example	Imagine sitting in a boat on the ocean, and you put a marble on the floor. If the boat rocked one way, the marble would roll towards the low side. If the boat rocked the other way, the marble would roll the other way. 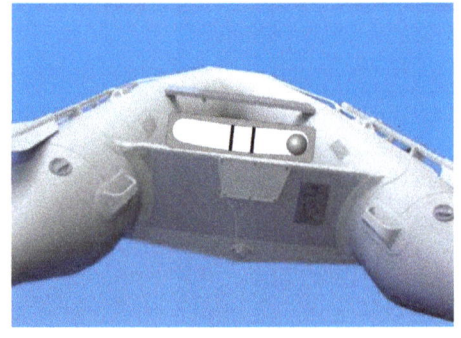 If the marble stayed in the middle of the boat and did not move, you would have to say the boat is in balance.
The balance ball	In aircraft, we treat balance the same way; however, we also have to deal with high forward speeds and the effects of G forces, including centrifugal and centripetal forces. We have a small 'marble' meter, called the balance ball, in the aircraft. It is a solid ball sealed inside a curved tube of liquid. 
In balance	Let's now take a helicopter that is in balanced forward flight. All forces would be balanced, and the ball would be in the middle.

Effect of increasing power	If we now increase power by raising the collective but do not increase left pedal, the nose of the helicopter would want to swing to the right and the tail to the left. If we continued to track straight ahead with cyclic, the rotor disc would pull us to the left while the nose tried to turn the helicopter to the right. The overall effect is that the helicopter would be slightly cocked off to the right while trying to fly left. It would feel very uncomfortable, and we would be 'out of balance', and the balance ball would have rolled to the low side of the cockpit (the left).

Chapter 2 Effects of Controls

Slips and Skids

A slip or skid will result if the helicopter is out of balance **during a turn**.

A **slip** occurs when the helicopter slides sideways **towards the centre of the turn**. It is caused by too little pedal pressure in the direction of the turn (or too much pedal in the opposite direction), effectively stopping the nose of the helicopter from following the turn. The rate of turn then is too slow for the bank angle being used. A slip would be indicated by the **balance ball moving toward the inside of the turn**.

A **skid** occurs when the helicopter slides sideways **away from the centre of the turn**. It is caused by too much pedal pressure in the direction of the turn (or too little pedal in the opposite direction), making the nose of the helicopter turn faster. The rate of turn then is too fast for the angle of bank being used. A skid would be indicated by the **balance ball moving toward the outside of the turn**.

Balance summary

If the ball is in the middle, you are in balance.

If the ball is out to the:

- **right**, you are out of balance to the right and need to apply more **right pedal** to bring it back in the middle.
- **left**, you are out of balance to the left and need to apply more **left pedal** to get it back in the middle.

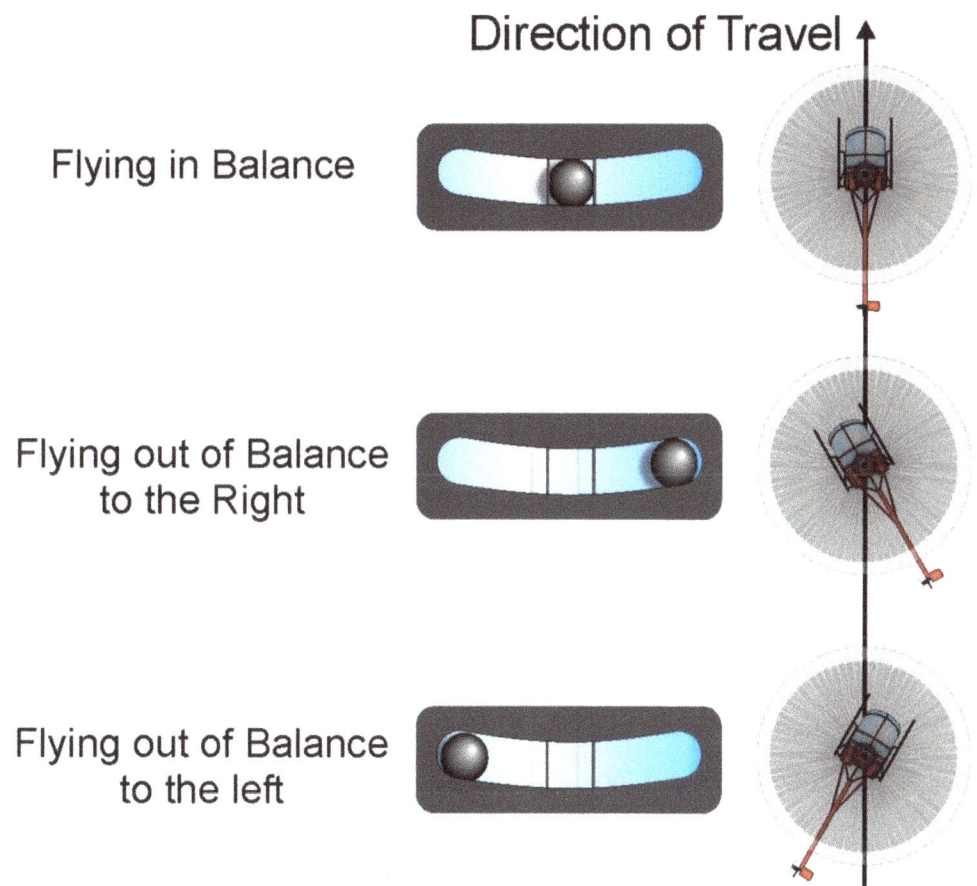

Mike Becker's Helicopter Handbook

Summary of Helicopter Controls

Cyclic

Direction — Attitude

Collective

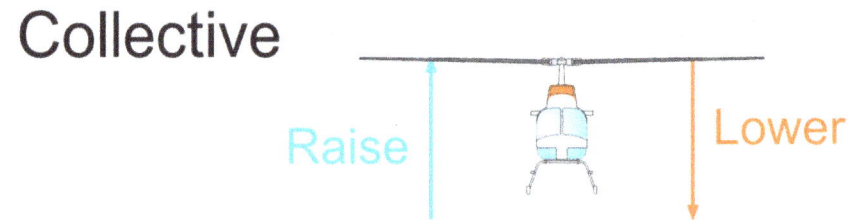

Raise — Lower

Pedals

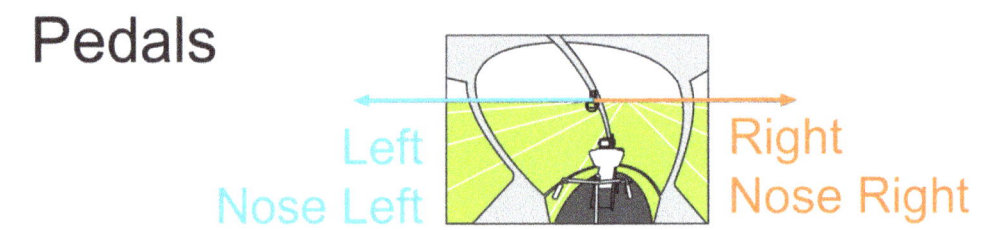

Left / Nose Left — Right / Nose Right

Throttle

Decrease / Roll In — Increase / Roll Out

Magnitude and Rate of Control

When moving the primary flight controls (cyclic, collective, pedals or throttle), it will take the student pilot time to understand and develop the feel required for managing the relationship between:

- **magnitude** (the amount of movement), and
- **rate** (the speed of the movement) of control movement.

The magnitude and rate of control movement can vary depending on the:

- aircraft's design
- manoeuvre
- aircraft's speed, and
- the situation.

Initially, it is common to overcontrol until this feel is developed.

Magnitude

The word magnitude describes the amount of control input required or given by the pilot to cause a change in attitude, power or yaw (direction). A large control input will cause a large change. A small control input will cause a small change.

Small cyclic input   **Large cyclic input**

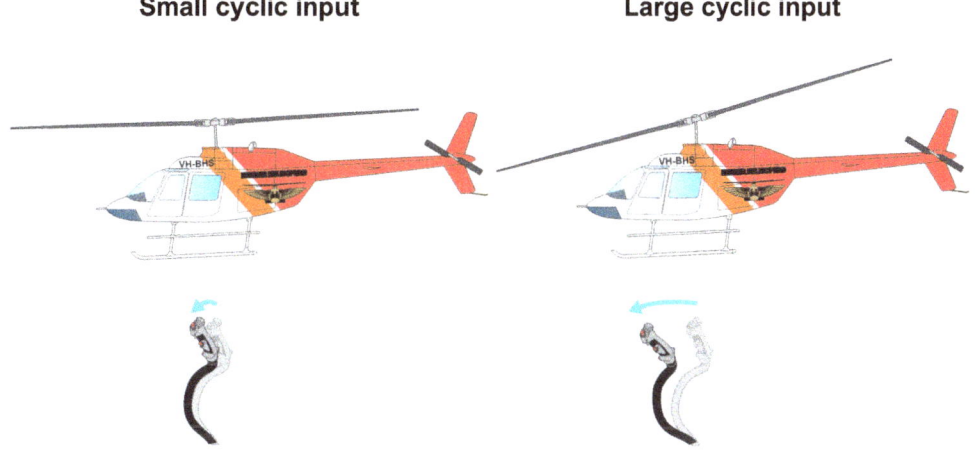

Rate

The word rate describes the speed at which the control input or change is made. A slow rate of control movement will cause a slow change. A fast rate of control movement will cause a fast change.

Momentum

It is important to note that large fast inputs cause the helicopter to move quickly. This leads to momentum in the fuselage, where it will take an even larger input to stop the one incorrectly entered. This can lead to the controls reaching their design limits and exceedances in power, RPM, attitude and yaw, placing the helicopter out of control. In the early stages of training, small slow inputs, closely mirrored by the instructor, will get you through this phase in time.

Flapback

Flapback and forward flight

In practical terms, when transitioning from hover to forward flight, as horizontal airflow, as a result of forward flight, takes effect on the blades, the blades will experience a dissymmetry of lift, and they will want to flap to equality, causing a change in disc attitude known as flapback (the disc will literally flap back).

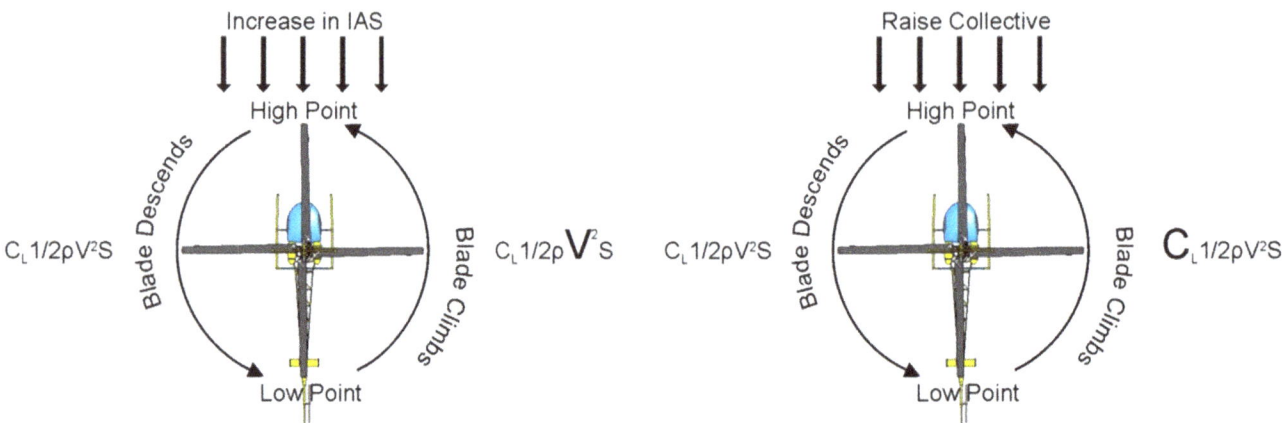

If no corrective action was taken, the helicopter would assume a nose-up attitude and probably lose forward speed.

When transitioning from hover to forward flight, to correct the effects of flapback, the cyclic would be moved forward, returning the disc to the desired attitude, and you would continue to gain forward speed.

Important:

It is important to note that if the helicopter is in forward flight or hovering into the wind and that wind dies out, there will be too much forward cyclic countering for the flapback that is no longer there. The helicopter would then nose over in response to the excessive cyclic.

This is commonly called flap forward, but in reality, there is **less flapback**.

Flapback and wind

Consider a helicopter hovering in still air conditions and then experiencing some wind. As the blades flap to equality in reaction to the new airflow, the disc tilts back, referred to as 'flapback'.

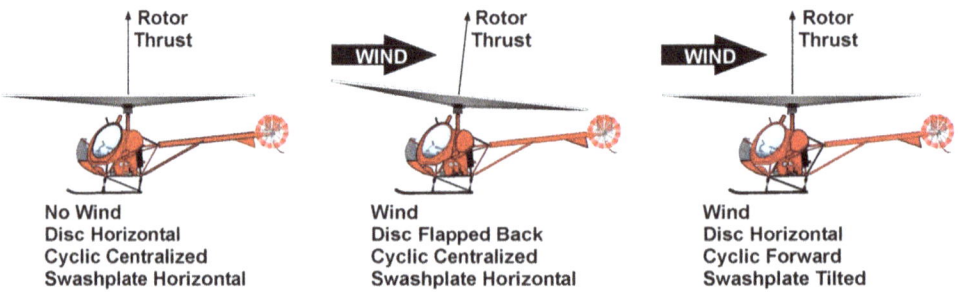

The effect of the wind at the hover is countered by changing cyclic pitch. Moving the cyclic forward tilts the swashplate forward, which results in a change of disc attitude and a change in the direction of rotor thrust, countering the effect of the wind.

Stability

Stability describes an object's characteristic in returning to its original state after it has been disturbed.

If a helicopter is disturbed from a given flight path by some outside influence, to be stable, it should return to its original state without any input from the pilot.

There are two aspects of stability: Static stability and Dynamic stability.

Static stability

Static stability deals with an object that is not moving.

If an object is disturbed from a given position and following this disturbance it	Then the object is said to be	Diagram example
returns to this position of its own accord	Statically Stable or Positively Statically Stable	
takes up a new position some distance away from the original position	Statically Neutrally Stable	
continues to move further and further away from its original position	Statically Unstable or Statically Negatively Stable	

Dynamic stability

An aircraft **must** have Static stability (whether built-in, or artificially provided by computers) before it can be considered for dynamic stability. If an object is Statically Stable, it will return to its original position, but in doing so, it may initially overshoot. How the object overshoots is described as the object's Dynamic stability. Dynamic stability deals with an object that is moving. For an aircraft, it is in a moving airflow, and energy can be extracted from this airflow, which affects the dynamic behaviour.

Types:

There are three types of Dynamic stability:

- Dynamically Stable
- Dynamically Neutrally Stable
- Dynamically Unstable

Dynamically Stable

If the amplitude of the oscillations	Then the object is said to be
Die out (damped down)	**Dynamically Stable** or **Dynamically Positively Stable**

Dynamically Neutrally Stable	If the amplitude of the oscillations	Then the object is said to be
	Continue at a constant amplitude	**Dynamically Neutrally Stable**

Dynamically Unstable	If the amplitude of the oscillations	Then the object is said to be
	Continue at an increasing amplitude (diverge)	**Dynamically Unstable** or **Dynamically Negatively Stable**

Helicopter stability

For a helicopter to have Static stability, it is assumed that the cyclic is held in a fixed position, referred to as a 'stick fixed' or 'stick held' position.

Hovering stability about the main rotor

If a gust of wind strikes the rotor disc, it will flap away from the wind. The helicopter will move away from the gust if no corrective action is taken.

However, when the gust dies off, the helicopter now experiences wind from the opposite side due to its sideways movement. The disc will flap away from this new airflow; the fuselage (hanging below like a pendulum) will also follow through, making the disc flapback even more. This continues for 3 or 4 cycles until it crashes. The movement of the helicopter will result in it experiencing continual sideways changes in the airflow affecting the disc.

The helicopter, therefore, is Statically Stable (tries to return to its original position), but Dynamically Unstable in the pitching and rolling planes at the hover.

Hovering stability in the yawing plane

If a gust of wind strikes the helicopter, it will affect the tail rotor. The helicopter will want to weathercock into the gust, thereby experiencing an airflow from the other side; therefore, a port/starboard swing begins with each oscillation getting worse.

The helicopter is Statically Stable, but Dynamically Unstable in the yawing plane at the hover.

A gust of wind alters the relative airflow on the tail rotor blade, thereby altering the tail rotor thrust. If the gust is such that the new relative airflow increases the angle of attack on the blade, the tail rotor thrust will increase and vice versa.

Chapter 2 Effects of Controls

Stability while in forward flight

If a gust of wind strikes the fuselage of a helicopter from the side while in forward flight, the effect is to yaw the helicopter, but the inertia of the helicopter will keep it on the original path.

Weathercock action will then return the fuselage to its original position. In forward flight, the helicopter is both Statically Stable and Dynamically Stable in the **yawing plane**.

If a gust of wind strikes the disc from ahead, the disc will flapback, and forward thrust will reduce. The aircraft will decelerate, and as it does so, the inertia of the fuselage will cause it to pitch up, taking the disc back further and thus decreasing speed even more. When the speed stabilises to a lower figure, the fuselage will start to pitch down, and at the same time, the disc will flap forward. As speed increases, the disc will flapback again, and the cycle will be repeated but with increasing amplitude.

The helicopter is Statically Stable but Dynamically Unstable in forward flight in the **pitching and rolling planes**.

Summary

A helicopter is statically stable and dynamically unstable, with the cyclic held in a fixed position, in all planes except the yawing plane in forward flight, where it is both statically and dynamically stable.

A helicopter is statically unstable and, therefore, dynamically unstable in all planes when the cyclic is not fixed (let go), sometimes referred to as a 'stick free' position.

Limits for Forward Flight

Design limits of the cyclic

To achieve forward flight, the cyclic is moved to tilt the disc. As the helicopter moves forward, the stick has to be moved further forward to prevent the disc from flapping back; therefore, a speed could be reached where the cyclic cannot go any further forward.

Airflow reversal

The speed of the retreating blade is high at the tip and low at the root, but the airflow from forward flight will be the same over the entire blade; therefore, the value of the relative airflow will vary along the blade until the airflow from forward flight is greater than that of the blade's rotation at the root end.

The airflow will reverse and be from the trailing edge and cause a loss of rotor thrust.

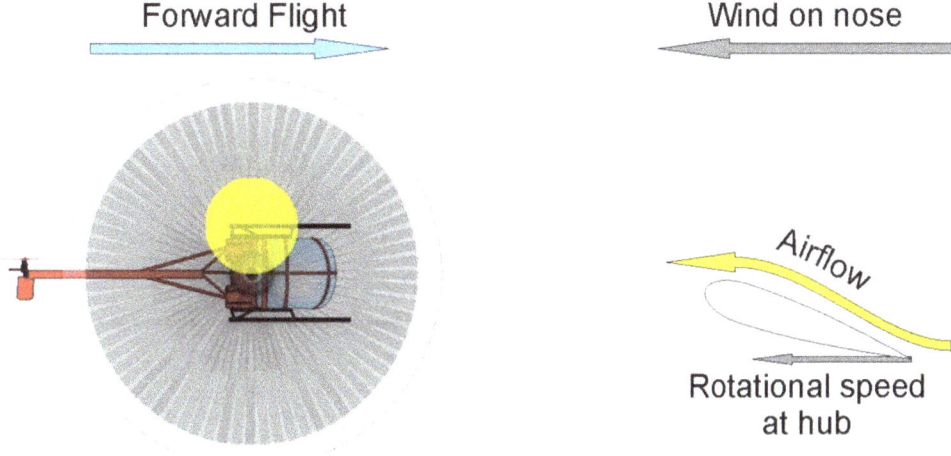

Note: *At this point, this section of the blade has **not** stalled; it is merely not producing any lift and is not acting like an aerofoil.*

Compressibility effect

Compressibility occurs when high forward speeds, along with a high RPM combined with the amount of curvature of the aerofoil, create a relative airflow approaching the speed of sound at the tip of the advancing blade.

The helicopter may shudder or vibrate, and the nose may pitch down and roll to the advancing blade side (right in anti-clockwise systems).

Compressibility effects produce:

- shock waves
- a large increase in drag
- noise (whop whop whop)
- large pitching moments, and
- local changes in the density of the air over the blade.

The compressibility effect will be one of the determining factors in limiting the helicopter's high forward speed.

Blade design

Manufacturers may use a rotor blade designed with a swept tip to delay the onset of compressibility and allow higher speeds.

Retreating blade stall

Retreating blade stall is the tendency for the retreating blade to stall at high forward speeds; it is inherent in all present-day helicopters and is one of the factors in determining the maximum forward speed of a helicopter, more commonly known as the Vne (Velocity Never Exceed speed) which is marked on the airspeed indicator as a red line and also placarded in the flight manual and the helicopter cockpit.

Airspeed indicator   **Placard**

NEVER EXCEED SPEED - KIAS								
2200 LB TOGW & BELOW								
PRESS ALT-FT	OAT - °C							
	-30	-20	-10	0	10	20	30	40
SL								
2000	130						127	123
4000					126	122	118	114
6000			126	122	117	113	108	103
8000	126	122	117	112	107	101	96	91
10000	117	112	106	101	95	90	85	
12000	107	101	95	89	NO FLIGHT			
14000	95	89						
OVER 2200 LB TOGW, SUBTRACT 10 KIAS								
FOR AUTOROTATION, SUBTRACT 30 KIAS								

As the blades rotate while the helicopter is in forward flight, the retreating blade must produce the same amount of lift as the advancing blade. Because the velocities of the two blades are different, the blades must equal out the lift by altering the angle of attack. As speed increases:

- the advancing blade will be reducing its angle of attack to fight flapback and to keep the disc tilted forward, and
- the retreating blade will be increasing its angle of attack to keep the back of the disc high.

Eventually, a forward speed will be reached where the angle of attack over the retreating blade is so high that the airflow breaks away and the aerofoil (or a portion of it) stalls.

As a result, rotor thrust is lost, and the retreating blade will flap down further, but instead of flapping to equality, the effect will be to deepen the stall and enlarge the stalled area.

The tip of the blade will be most affected because the relative arm or length produces the largest flapping at the tip. Also, the innermost part of the blade is in reverse flow; as you move further out, the blade is only producing small amounts of lift until the rotational flow is enough to gain useful lift. The maximum angle of attack will be when the blade is ½ way around the retreating side.

Below is a diagram showing the angle of attack distribution throughout the disc while in a stalled condition.

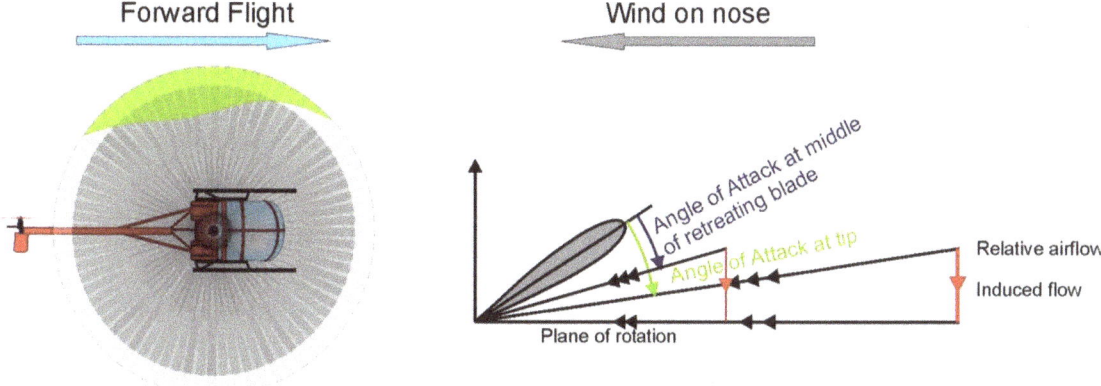

Pilot indications

Pilot indications of entering a retreating blade stall condition include:

- rotor roughness and vibration
- erratic stick forces, and
- stick shake.

If these symptoms are ignored, a tendency to pitch up will develop, followed by a roll towards the retreating blade side. There will be a partial loss of control; if the stall is severe, control may be lost altogether. The helicopter will roll left due to the stall, but the nose will want to pitch up as the other blades flap up in response.

In helicopters where the flight controls are hydraulically assisted, the pilot may encounter a jack stall. This is where the forces on the rotor blades are so strong that the hydraulics (the hydraulic pistons are referred to as 'jacks') cannot cope with the load and will be forced back, resisting what the pilot is attempting to do.

Results from

Retreating blade stall can result from:

- high forward speed (it is common to exceed the Vne when operating at high-density altitudes)
- high-density altitudes and high all-up weights resulting in high angles of attack
- manoeuvres with high 'g' loading
- excessive or abrupt control movements, or
- flying in turbulent air.

Recovery

The recovery depends upon the prevailing in-flight conditions when the stall symptoms are recognised. It will usually be made by:

- reducing forward speed
- reducing collective pitch
- reducing the severity of the manoeuvre, or
- doing all the above actions at once in a slow, smooth manner.

Secondary Controls

Secondary controls

The secondary controls are those used to control ancillary systems, devices or aids that help manage or protect the helicopters controls and power plant:

Secondary Control	Effect
Fuel Valve lever or switch	This main fuel valve allows flow from the fuel tank to the engine.
Mixture (piston only)	Controls fuel-to-air ratio to the engine and is usually only set to lean (off) or full rich (on) below 5000 feet.
Clutch switch/lever (piston only)	Either an electric switch or a mechanical lever that engages the engine with the rotor system.
Carb heat (piston only)	A lever to duct hot air into the carburettor to control carb ice.
Frictions	To hold firm the controls during ground running
Trim levers	If fitted, used to relieve forces on the cyclic in forward flight.
Force trims	If fitted, are electrical devices using magnets or electrical motors tightening or loosening bungee straps or springs used to counter any feedback forces on the cyclic and/or collective while in flight.
Rotor Brake	A lever or pull chain that operates a disc brake to help slow the rotor blades quickly after engine shutdown.
Governor Switch	In piston helicopters with a governor, there is an electrical switch to activate or override the system.
SAS	If fitted, the Stability Augmentation System (SAS) is designed to take out unwanted control movements and is designed in larger helicopters to smooth out the control inputs.

Carburettor Icing

Introduction

Every year there are several major aircraft accidents attributable to carburettor icing in a piston engine helicopter. Carburettor icing is probably the most insidious hazard confronting a pilot, and it can occur in conditions when it would appear to the uninformed that icing is not possible.

Consequently, all pilots must understand:

- what carburettor icing is
- why it occurs, and
- what preventative and remedial actions they can take.

Chapter 2 Effects of Controls

Susceptibility to carb ice

The following diagram, produced by the Civil Aviation Safety Authority (CASA) of Australia, indicates the percentage of susceptibility to carb ice based on temperature and humidity.

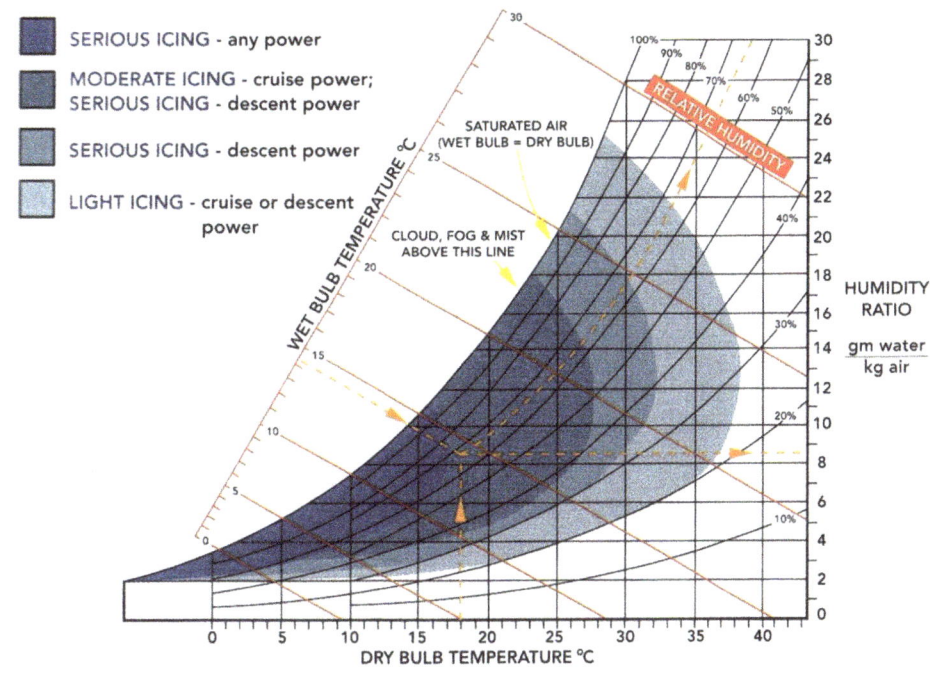

Characteristic of carburettor icing

A characteristic of carburettor icing is that it will form readily when the mixture temperature in the carburettor is between -10°C and +3°C. It can occur under almost any conditions - in cloud, precipitation or even clear air - when the outside temperature is between -10°C and +25°C. Considerable cooling can occur in the carburettor, resulting in a temperature below the dew point of the air and below the freezing point, causing ice formation on the carburettor walls and internal components.

Types of carburettor icing

There are three types of carburettor icing:

1. Impact Icing
2. Refrigeration Icing
3. Throttle Butterfly Icing

Diagram

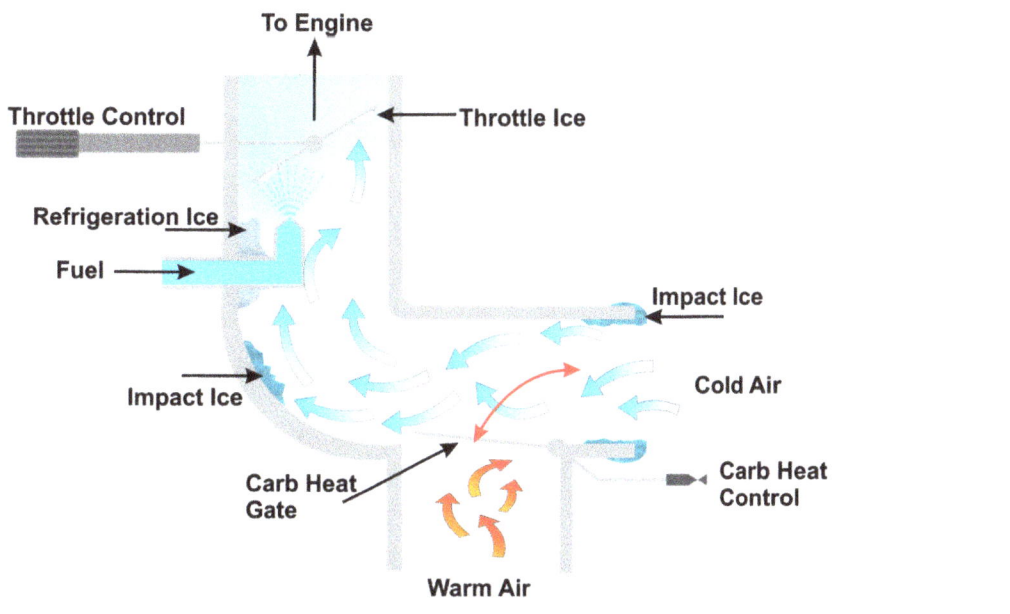

Impact icing	Impact icing results from the freezing of supercooled water droplets on impact with the induction system, progressively closing the air intake. Impact icing is normally experienced in cloud when the outside air temperature is between -15°C and 0°C. These conditions would also be conducive to airframe icing.
Refrigeration icing	Refrigeration icing results from the refrigeration effect occurring when avgas fuel evaporates from surfaces that are wet with avgas fuel. This is especially likely to occur when the water content of the air (humidity) is high, and such ice is formed upstream of the fuel diffuser (see diagram).
	Refrigeration icing can occur with relatively high outside air temperatures. Because of its design, the direct-injection type of carburettor (fuel injection system) fitted to many modern piston engines is free from refrigeration icing.
Throttle butterfly icing	Throttle butterfly icing is formed at or near the partly closed throttle butterfly and results from the condensation of water vapour in the fuel/air mixture. When the air is accelerated through the throat of the carburettor, the local reduction of pressure (venturi effect) results in adiabatic cooling, and the water vapour present consequently condenses.
	Once again, this form of icing can occur with relatively high outside air temperatures but is less likely to occur in engines where the fuel enters downstream of the throttle butterfly.
Combined effect of icing	The combined effect of refrigeration and throttle butterfly icing can be so significant as to reduce the air temperature within the carburettor by more than 25°C.
Symptoms of carburettor icing	Ice formation in the air intake and induction system gradually restricts the air supply to the engine, disturbs fuel metering, and may render moving parts inoperative. Carburettor icing will cause a drop in RPM. Rough running may accompany the decrease in power, or it may even be the only noticeable effect. Complete engine failure is likely to occur if remedial action is not taken.
Remedial action	Moving the Carb Heat control to the **WARM /HOT/ON** position will duct hot air into the induction system and prevent ice from forming in the carburettor.
	However, applying hot air will enrich the mixture and decrease the engine's volumetric efficiency; power output will decrease, and cylinder head temperatures will rise.
	Although carburettor icing occurs more readily when the engine is operating under low power conditions, different engines vary widely in their susceptibility. Instructions for the use of the Carb Heat Control in a particular type of helicopter will be found in the operator's manual or pilot's notes. All helicopters are fitted with a carb air temp gauge which is colour coded. If the temp drops to yellow, apply some carb heat.

Chapter 2 Effects of Controls

Procedure However, remedial action will generally conform to the following procedure:

Step	Action
1	Check the operation of the Carb Heat control during the run-up. The pilot should note a drop in RPM and a rise in the carb air temperature. Return the control to Cold for take-off.
2	During the flight, remain alert to the possibility of carburettor icing. Periodically exercise the Carb Heat control to ensure the carburettor is clear of any incipient ice build-up. Monitor the carb air temp gauge.
3	Before closed or low power operation, such as descent, approach for landing or practice autorotation, apply full heat for a period. The carburettor heat should be returned to Cold on the final approach and when full power is applied.
4	If carburettor icing is indicated by loss of power or rough running at any time, select full heat for a short period and then inch off to cold, observing whether the problem has been reduced or solved entirely.

Conclusion In conclusion, carburettor icing can occur with outside temperatures between -10°C and +25°C whenever there is high relative humidity or visible moisture in the air (clouds). Expect carburettor icing every time you fly, and know how to recognise and control it, should it occur.

Air Exercises: Effects of Controls

Introduction Understanding how each control works and relates to the others requires good demonstration and practice until finally obtaining understanding and control. This takes time and does not usually come naturally.

Common Faults
- tenseness on the controls
- too much attention on the instruments and not enough on flying and looking outside towards the horizon
- occasional confusion during power changes about the sequence of control inputs
- not yet understanding the relationship between magnitude and rate of control movement, and
- confusion on the radio.

Handing Over and Taking Over (HOTO) Control

Taking control over from instructor The handing over procedure, where the instructor gives control of the helicopter to the student, is a necessary procedure that eliminates confusion in the cockpit as to who is actually in command of the flying controls. The procedure may vary from school to school and can be two- or three-step process. Below is an example of the two-step process commonly used in civil flight schools.

The procedure for the instructor to give control to the student is as follows:

Step	Who	Action
1	Instructor	Asks the student to take hold of the controls by saying, "*You have control*".
2	Student	Takes hold of the controls and confirms by saying, "*I have control*".
3	Instructor	Allows the student to manipulate controls while the instructor either shadows or removes their hands and feet from the controls.

Handing control back to the instructor The procedure for the instructor to take control back from the student is as follows:

Step	Who	Action
1	Instructor	Tells the student they want control back by saying, "*I have control*".
2	Student	Once the student feels the instructor coming on the controls, replies "*You have control*".
3	Instructor	Assumes control of the helicopter, and the student takes hands and feet off the controls.

Reason for handover procedures Although the above procedure sounds long-winded, it can be vital, especially when the student is concentrating hard and does not realise they are getting in trouble.

Chapter 2 Effects of Controls

Demonstrating a manoeuvre

There are two ways to demonstrate a manoeuvre. Either the student:

- has their hands and feet lightly on the controls so they can feel what the instructor is doing, referred to as *'following through'*, or
- removes their hands and feet, simply observing the instructor.

Which one is used is up to the instructor.

No Follow-through

The procedure for the instructor to demonstrate with **no follow-through** is as follows:

Step	Who	Action
1	Instructor	Says, *"I have control, take your hands and feet off the controls"*.
2	Student	Folds their arms and takes their feet off the pedals and keeps away from the controls, and replies, *"You have control"*.

With Follow-through

The procedure for the instructor to demonstrate **with follow-through** is as follows:

Step	Who	Action
1	Instructor	Says, *"I have control, follow me through"*.
2	Student	Remains lightly on the controls to feel what the instructor is doing and replies, *"You have control, following you through"*.

The Work Cycle

What is it

The Work Cycle is the **continuous** monitoring and adjusting of the flight controls after receiving information from the:

    LOOKOUT   Looking outside. This takes most of your time.

    ATTITUDE   Confirming and correcting the helicopter's pitch and roll relative to the horizon by moving the controls.

    PERFORMANCE   Checking the instruments to confirm the Desired Performance.

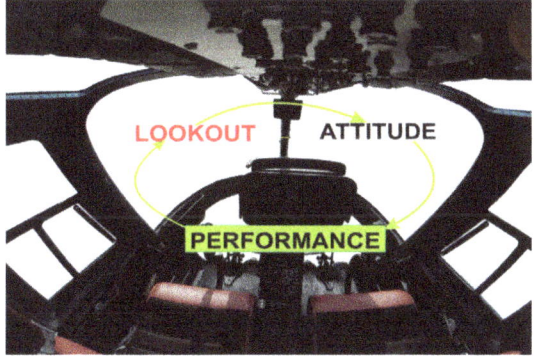

This work cycle will be the foundation of your scan in the helicopter and will, in time, become an automatic response to manage the helicopter's desired performance.

Lookout	The **lookout** involves the crew constantly moving their head and eyes to: ■ check for other aircraft or obstacles, and ■ set or confirm references. A good lookout will help the crew to: ■ make good decisions ■ manage the aircraft, and ■ identify threats and errors and manage them.
Attitude	The attitude refers to the aircraft's angle relative to the natural horizon.
Horizon	The horizon is described as that part of the earth you can see in the distance. It divides the ground from the sky. 
Horizon in the cockpit	We can now relate the horizon's position with reference points in the helicopter to give the pilot an idea of the helicopter's attitude. 

Attitudes	The attitude consists of 2 parts:
	1. The **Pitch** attitude:
	   ■ pitch **up** attitude, and
	   ■ pitch **down** attitude.
	2. The **Roll** attitude:
	   ■ **left** roll attitude, and
	   ■ **right** roll attitude.

Important terms to understand	Following on from the understanding of attitude, it is important that the student also understands the difference between the following terms:
	■ Attitude, including:
	   ■ Pitch
	   ■ Roll
	■ Angle of bank
	■ Turn
	■ Magnitude (amount of movement), and
	■ Rate (speed of movement)

Chapter 2 Effects of Controls

Pitch	The word "pitch" describes the position of the nose of the helicopter about the lateral axis.	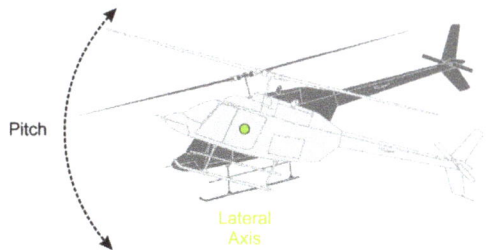
Pitch and reference points	The **pitch attitude** can be seen by looking outside to the horizon and comparing the reference points on the centre pillar.	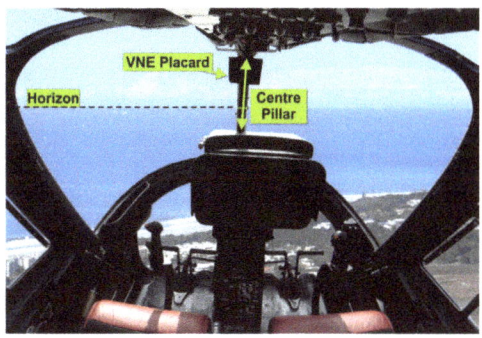

Pitch on the Attitude Indicator

A change in pitch can also be seen on the Attitude Indicator (AI):

Pitch Down   **Pitch Neutral**   **Pitch Up**

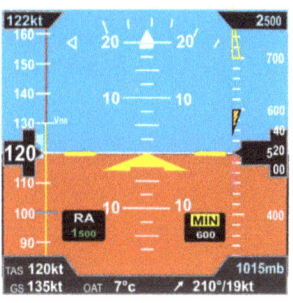 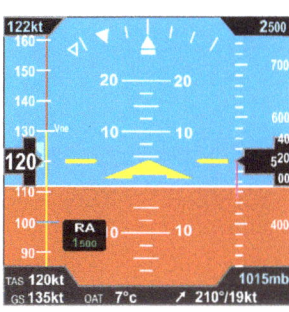

Roll	The word "roll" describes the movement of the nose of the helicopter about the longitudinal axis.	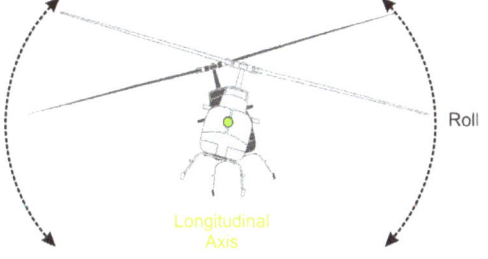
Roll and reference points	The roll attitude can be seen by looking outside to the horizon and comparing the reference points on door pillars and centre console.	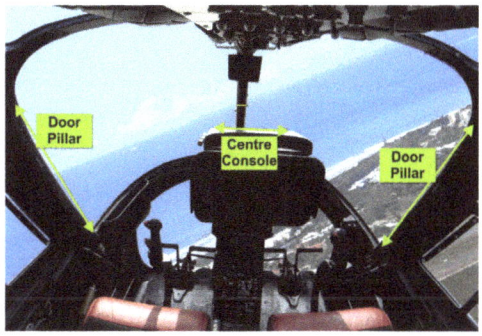

Roll on the Attitude Indicator

A change in roll can also be seen on the Attitude Indicator (AI):

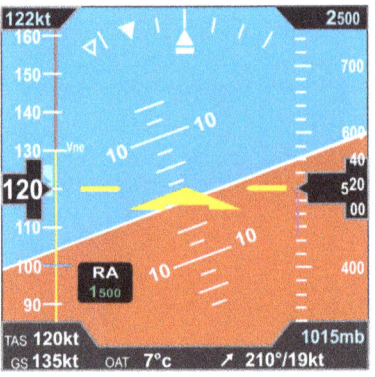

Angle of Bank

The words "angle of bank" describes the angle of the helicopter relative to the horizon.

A particular angle of bank in normal level flight (constant altitude) can be set and maintained from zero to 60 degrees.

The most common angles of bank set for training are 10, 20, 30, and 60 degrees.

For example:

A zero-angle of bank would have the helicopter in a level attitude.

A 20-degree angle of bank would have the helicopter at 20 degrees to the horizon, as shown in the image below.

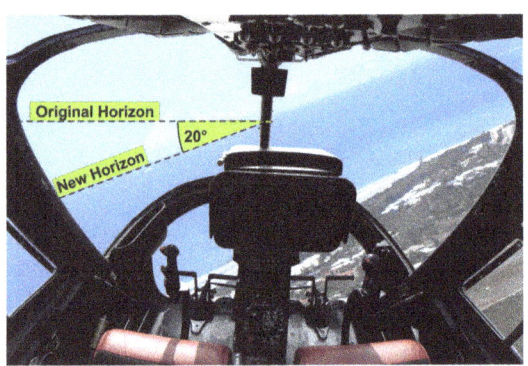

The angle of bank can be measured on the Attitude Indicator. On the top of the indicator is an arrow head, it can point to 0, 10, 20, 30, 60 and 90 degree angle of bank increment markers.

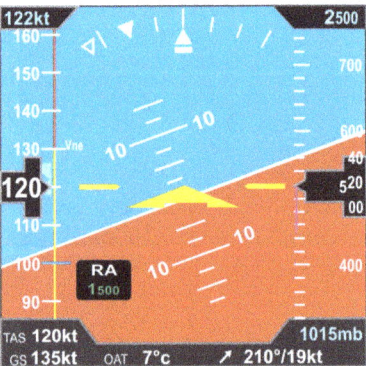

Chapter 2 Effects of Controls

Turn

The word "turn" describes a change in heading. To turn in forward flight, the pilot will roll the helicopter left or right to the desired angle of bank and then hold it there. The helicopter will then start to turn in the direction of the roll. To roll out of the turn, the pilot will return to level flight.

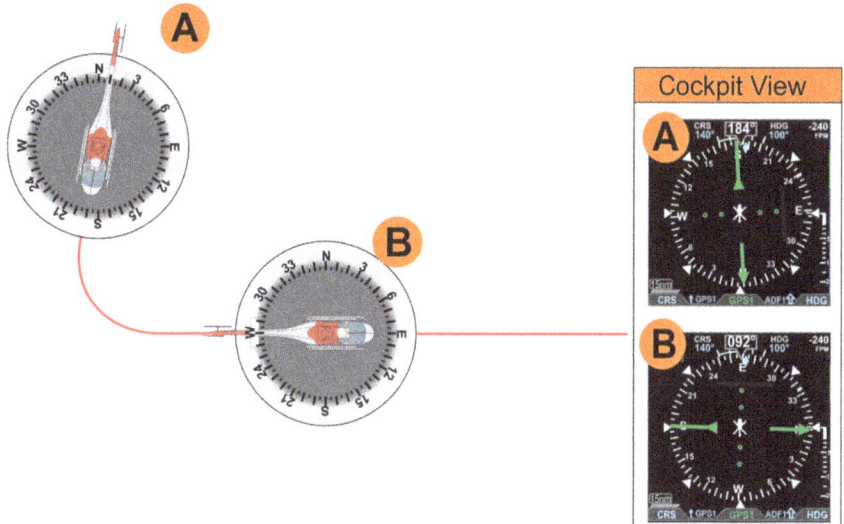

In the early stages of training, the instructor will ask for 30 degrees angle of bank for level and descending turns and 15 degrees angle of bank for climbing turns.

Air Exercise 2-1: Using the Cyclic

Demonstration Before take-off, the instructor will demonstrate the use of:

- Fuel valve
- Mixture
- Carb Heat
- Clutch
- Governor, and
- Frictions and Trim levers.

Judging using cyclic by recognising attitude

Attitude is judged by sight.

A pilot can judge the helicopter's attitude and, therefore, use of the cyclic by using their eyes to recognise attitude changes against the natural horizon.

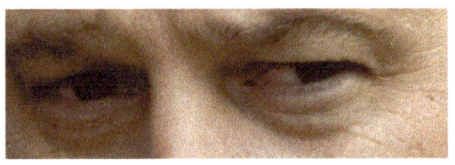

Using the Cyclic After a demonstration by the instructor, you will practice using the cyclic, following the air exercise detailed below.

Step	Action	Discussion
1	Pick a reference point inside the helicopter For example: The compass, console, a screw on the centre column, a speck of dirt on the bubble etc., and note its position relative to the horizon. The instructor will ask you to indicate where the horizon is from your perspective.	The instructor at this stage will still have control, and you will be able to see the straight and level attitude. Level and Straight Attitude / Rotor Disc / Horizon

Chapter 2 Effects of Controls

Step	Action	Discussion
2	From straight and level flight, ease cyclic forward and hold that new attitude. You will have to move cyclic forward to counter the effect of flapback as the helicopter builds up speed.	As a result, note the: - nose of the helicopter drops, and you will see the horizon move up the bubble - reference point selected in the first step will move down relative to the horizon - airspeed increases, stabilising at a higher speed - vertical speed indicator shows a rate of descent - altimeter shows that the helicopter is descending, and - RPM, which decreases initially, then increases. **Note the new attitude.** Forward Cyclic, Nose Low Horizon moves up the bubble
3	From straight and level flight, ease the cyclic back to a nose-high attitude and hold it there.	As a result, note the: - nose of the helicopter rises, and you will see the horizon move down the bubble - reference point will move up relative to the horizon - airspeed decreases and stabilises at a lower speed - vertical speed indicator shows a rate of climb - altimeter shows that the helicopter is climbing, and - RPM may tend to increase. Aft Cyclic, Nose High Horizon moves down the bubble

Step	Action	Discussion
4	From straight and level flight, ease the cyclic to the left and hold it there.	As a result, note the: - nose of the helicopter rolling to the left and holding an angle of bank while turning, depending on the amount of left cyclic used - airspeed will decrease marginally - VSI should remain at 0 - altimeter should remain constant - RPM may tend to increase, and - depending on which side of the cockpit you are sitting your new attitude will appear high or low. Left Cyclic, Left Bank In the left seat, the Horizon moves down the bubble
5	From straight and level flight, ease the cyclic to the right and hold it there.	As a result, note the: - nose of the helicopter rolling to the right and holding an angle of bank while turning depending on the amount of left cyclic used - airspeed will decrease marginally - VSI should remain at 0 - altimeter should remain constant - RPM may tend to increase, and - depending on which side of the cockpit you are sitting your new attitude will appear high or low. Right Cyclic, Right Bank In the left seat, the Horizon moves up the bubble

Chapter 2 Effects of Controls

Air Exercise 2-2: Using the Pedals

Judging pedals Balance can be felt in the seat of the pants; while you are sitting in the pilot's seat, you may feel yourself leaning left, etc. You can learn to judge whether you are out of balance without looking at the balance ball. It is usually felt as a twist and a lean in the seat of your pants.

Judge use of Pedals using the seat of your pants

Direction of rotation The following exercise is based on the main rotor blades rotating in an anti-clockwise direction. In clockwise rotating systems, the pedal movements are reversed.

Using the pedals After a demonstration by the instructor, you will practice using the tail rotor pedals following the air exercise detailed below:

Step	Action	Discussion
1	From straight and level flight, move the **left** pedal partly forward until feeling a change.	As a result, note the: ■ nose of the helicopter yaw left ■ balance ball is out to the right ■ helicopter has a right skid low lean ■ indicated airspeed will decrease, and ■ RPM may tend to decrease.

Page 2-43

Step	Action	Discussion
2	From straight and level flight, press the **right** pedal partly forward until feeling a change.	As a result, note the: ■ nose of the helicopter yaw right ■ balance ball is out to the left ■ helicopter has a right skid low lean ■ indicated airspeed will decrease, and ■ RPM may tend to decrease.

When to use pedals

When to use the pedals:

As the collective is	The throttle is	Pedal required
Raised	Increased	More left pedal required
Lowered	Decreased	More right pedal required

Otherwise, use pedals to keep the helicopter balanced. That is, keep the balance ball in the middle.

Air Exercise 2-3: Using the Collective

Judgement

You can learn to feel where power settings are without looking at the MAP gauge (for a piston) or Torque/Power gauge (for a turbine).

Initially, judge use of the collective using the position and muscle memory of your left arm

Practice

After a demonstration by the instructor, you will practice using the collective following the air exercise detailed below. This is based on a piston-engine helicopter. The principles for a turbine are the same.

Step	Action
1	From 20" MAP, lower collective to 15" MAP maintaining RPM. As a result, note the: - MAP decrease - nose of the helicopter will want to drop forward due to a decrease in flapback effect - nose of the helicopter will tend to yaw left, and - RPM will tend to increase.  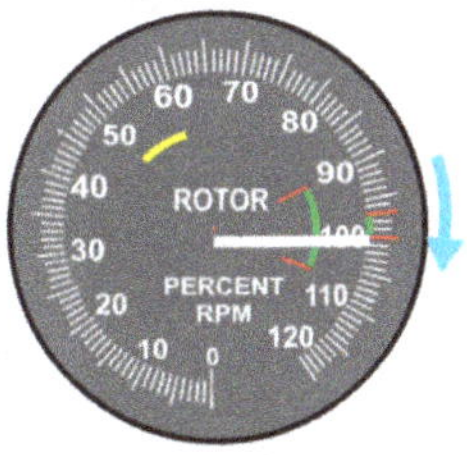 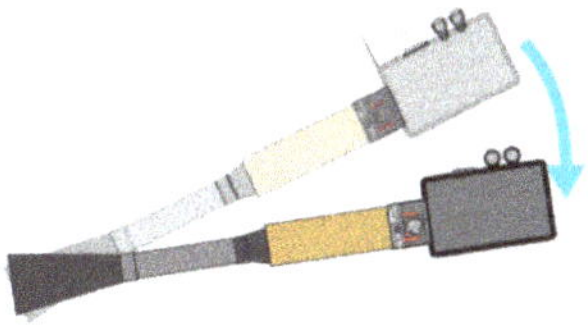 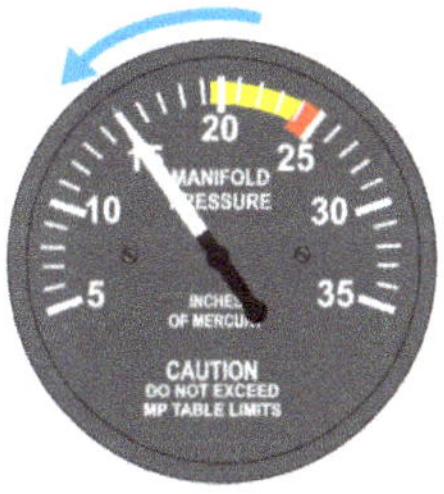
2	From 15" MAP, raise collective to 20" MAP maintaining RPM. As a result, note the: - MAP Increase - nose of the helicopter will want to come up due to an increase in flapback effect - nose of the helicopter will tend to yaw right, and - RPM will tend to decrease.   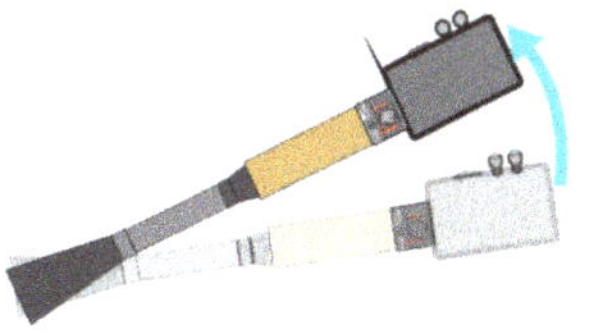 

Air Exercise 2-4: Using the Throttle

Judgement You are listening for a change in RPM. Similar to a car, where you can hear the revs going up or down.

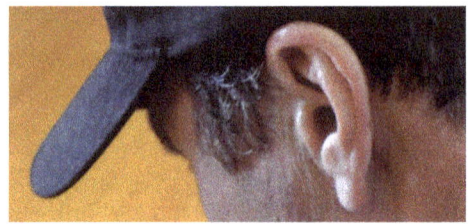

Judge use of the throttle using your ear, listening for a change in the tone of the engine

Using the throttle After a demonstration by the instructor, you will practice using the throttle following the air exercise detailed below. This is based on a piston engine. In a turbine, the pilot will not usually manipulate the throttle in flight.

Step	Action
1	With the RPM set top of the green range on the RPM Gauge, decrease RPM to the bottom of the green range by slowly and smoothly rolling off some throttle (roll inboard). Do this slowly and in very small amounts. As a result, note the: ■ RPM decreases ■ MAP decreases, and ■ helicopter yaws to the left.
2	With the RPM set bottom of the green range, increase the RPM to the top of the green range, by slowly and smoothly rolling on some throttle (roll outboard). Do this slowly and in very small amounts. As a result, note the: ■ RPM increases ■ MAP increases, and ■ helicopter yaws to the right.

Chapter 2 Effects of Controls

Demonstration 2-1: Further Effects of Controls

Demonstration
The instructor will demonstrate what happens during forward flight when one control is moved and held in position without any other controls used to compensate.

This demonstration is beneficial to show the flapback effects when using cyclic and collective.

Remember, when V^2 (Velocity or RPM) or C_L (blade pitch) is increased, you will experience flapback, which is compensated for by the pilot using cyclic to counter the effect through cyclic feathering.

If V^2 or C_L is decreased, then it seems you are experiencing flap forward. However, you are actually experiencing less flapback, and the cyclic input being applied is now too much; therefore, you are titling the disc forward.

Demonstration 2-2: Effects of Controls at the Hover

Effect of controls at hover
The instructor will demonstrate the primary effect of controls at the hover.

Cyclic

The cyclic gives horizontal movement in any direction: backward, sidewards, forwards, or a combination. It controls the helicopter's position over the ground.

Action by instructor	Result	Diagram
From the still air hover ease cyclic slightly: ■ Forward	The nose of the helicopter will drop, there will be a small delay, and the helicopter will then start to move forward.	Forward Cyclic, Nose Low Rotor Disc Horizon Horizon moves up the bubble
■ Aft	The nose of the helicopter will rise, there will be a small delay, and the helicopter will start to move backward.	Aft Cyclic, Nose High Rotor Disc Horizon Horizon moves down the bubble

Action by instructor	Result	Diagram
■ Left	The fuselage will tilt left, there will be a small delay, and then the helicopter will start to move left.	Left Cyclic, Left Bank / Rotor Disc / Horizon — In the left seat, the Horizon moves down the bubble
■ Right	The fuselage will tilt right, there will be a small delay, and then the helicopter will start to move right.	Right Cyclic, Right Bank / Rotor Disc / Horizon — In the left seat, the Horizon moves up the bubble

Collective

Action by instructor	Result	Diagram
From the still air hover: ■ raise collective	When the collective is raised, the helicopter will rise vertically.	Raise collective
■ lower collective	When the collective is lowered, the helicopter will descend vertically.	Lower collective

Anti-torque Pedals

Action by instructor	Result	Diagram
From the still air hover move some: ■ left pedal	When the left pedal is pushed, the helicopter will turn about its axis over a spot to the left.	
■ right pedal	When the right pedal is pushed, the helicopter will turn about its axis over a spot to the right.	

Throttle

Action by instructor	Result	Diagram
From the still air hover: ■ Gently increase throttle to top of the green (roll outboard)	When the throttle is increased, RPM and MAP will increase causing the helicopter to yaw right and slowly gain height.	
■ Gently decrease throttle to bottom of the green (roll inboard).	When the throttle is decreased, RPM and MAP will decrease causing the helicopter to yaw left and begin sinking.	

3

Elementary Handling

Aim	To coordinate moving the flight controls to achieve a desired performance.
Objectives	On completion of this lesson, the student will be able to: ■ recite the Performance Equation ■ make power changes, and ■ conduct turns.
Motivation	Making the helicopter maintain a straight and level attitude, hold height, climb, descend, turn, and adjust speed accurately and as you want or need, allows you to have mastery over this flying machine. Gaining the skills, knowledge and attitudes to do this is what elementary handling is all about.

Preparation: Elementary Handling

Introduction

Understanding some of the theories and terms behind control movement allows the pilot to control the helicopter better. Using the Performance Equation is useful for visual flying and is fundamental when flying in reduced visibility when instrument flying. This section covers the theory behind:

- the Performance Equation, and
- power changes allowing speed changes, climbing, descending, levelling off and turns.

Performance Equation

The Performance Equation describes what a pilot must do to get a desired response (performance) from the helicopter. This equation sets the basis for how the controls interact with each other.

The Performance Equation is described below as:

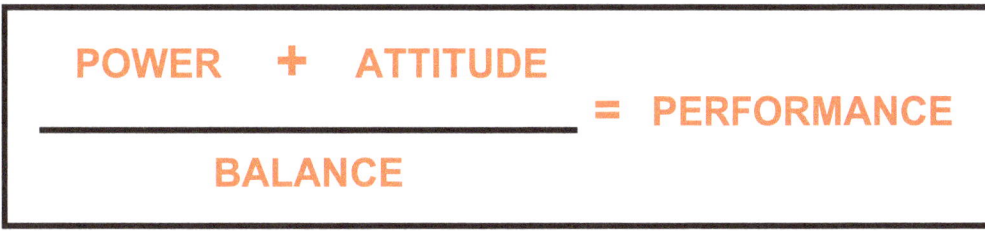

Power

Power is adjusted by using the collective and is generally measured as manifold pressure (MAP) in a piston engine and torque (Tq) in a turbine engine.

Adjusted using

Collective

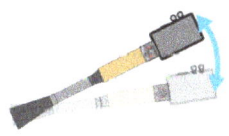

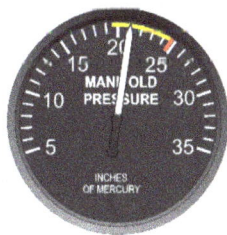

Indicated by

Piston Engine:

Manifold Pressure (MAP)

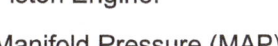

Turbine Engine:
Torque (Tq)

Chapter 3 Elementary Handling

Attitude

Attitude is adjusted by the cyclic and is observed by the pilot looking out towards the natural horizon and displayed in the cockpit by the attitude indicator (pitch and roll).

Adjusted using

Cyclic

Determined by

Looking outside at the natural horizon

Or inside by looking at the Artificial Horizon (AH)

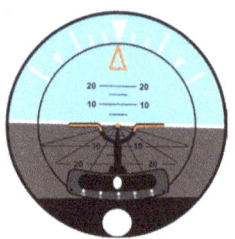

Balance

Balance is adjusted by the pedals and is felt in the seat of the pants by the pilot and displayed in the cockpit when in forward flight by the balance ball.

Adjusted using

Pedals

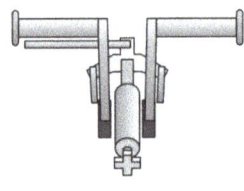
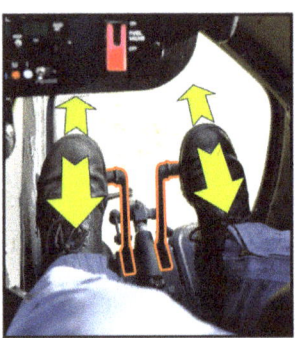

Indicated by the balance ball

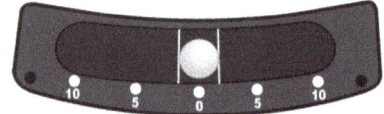

Mike Becker's Helicopter Handbook

Performance
: **Performance** for a VFR pilot is primarily determined by **visual cues** looking outside the cockpit with glances inside (referred to as the 'work cycle') at the following instruments:

- heading indicator (straight or turning)
- altimeter (level, climbing, or descending), and
- airspeed indicator (cruise, climb, or descent).

Compass   **Altimeter**   **Airspeed Indicator**

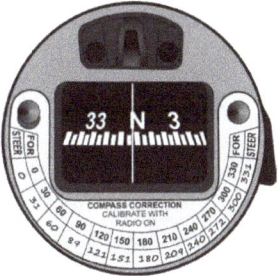  

Secondary instruments
: Secondary instruments, if available, give additional information, including the vertical speed indicator (VSI), turn indicator, and, if fitted, another more accurate form of heading indication, such as a Directional Gyro (DG) or the Horizontal Situation Indicator (HSI).

Vertical Speed Indicator (VSI)   **Turn Indicator**   **HSI**

 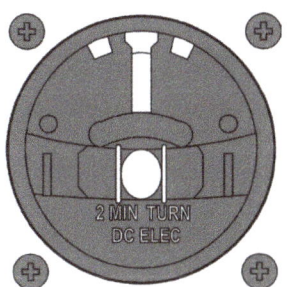 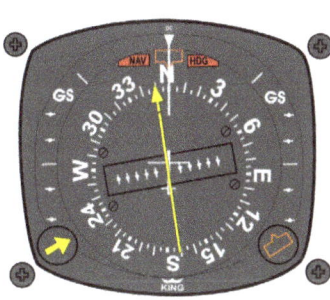

Summary
: *Power + Attitude while in balance will give a Desired Performance.*

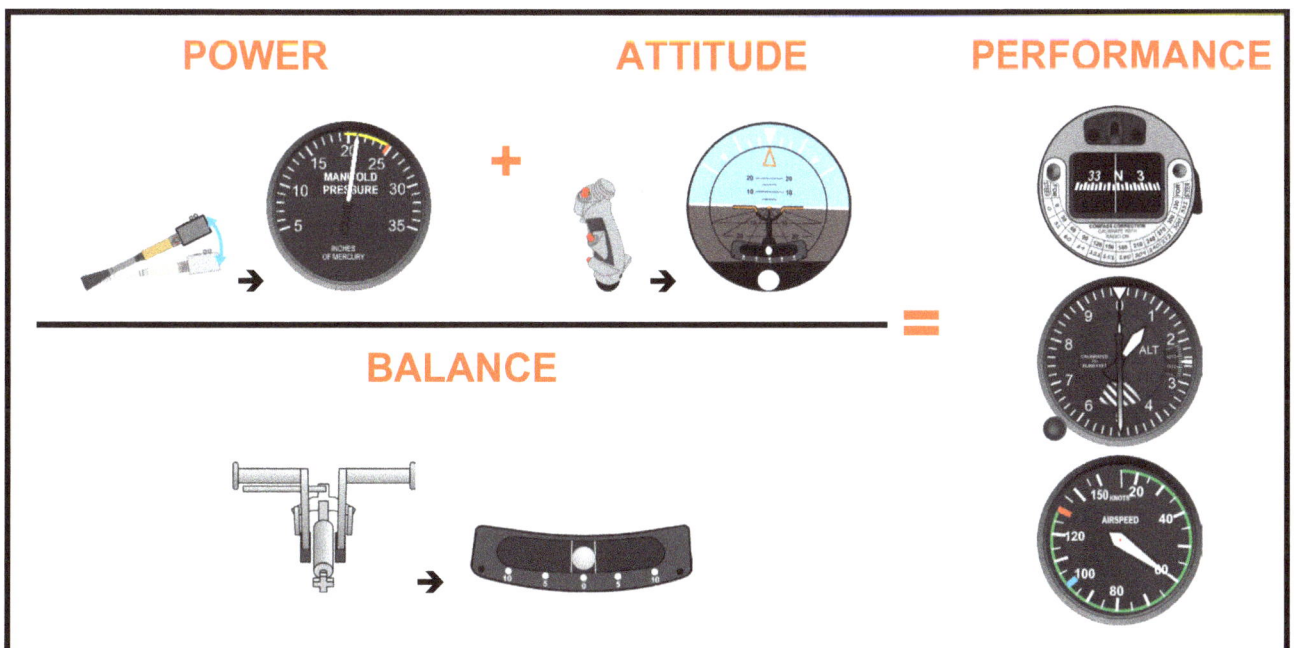

Chapter 3 Elementary Handling

Forces in Forward Flight

Forces in forward flight

When the helicopter travels forward at uniform speed, the horizontal component of rotor thrust will be balanced by the parasite drag of the fuselage.

As parasite drag increases with V^2, the faster the helicopter moves forward, the greater must become the tilt of the disc to provide the necessary horizontal thrust.

At the same time, for level flight, the vertical component of rotor thrust must remain equal to the weight. This can only be achieved by increasing Total Rotor Thrust using the collective/throttle.

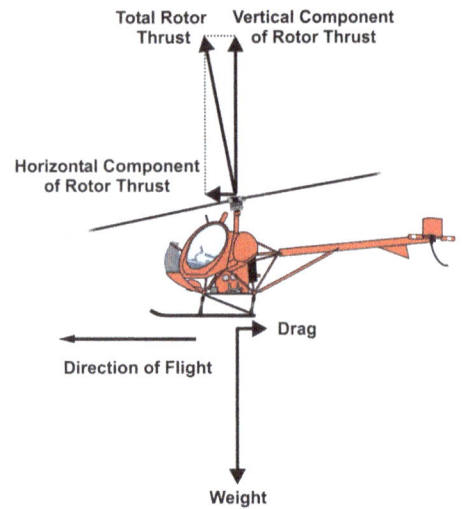

Forces in the Climb

Forces in the climb

For the helicopter to climb, the vertical component of rotor thrust must be greater than the weight. This is achieved by reducing the disc tilt or by increasing Total Rotor Thrust. In practice, the climb is normally initiated by a combination of these actions.

For the helicopter to climb from forward flight:

- apply some aft cyclic to place the helicopter in a climbing attitude
- increase power (collective/throttle) by raising collective and maintaining RRPM using the throttle, and
- use the anti-torque pedals (tail rotor pedals) to balance the change in the torque reaction.

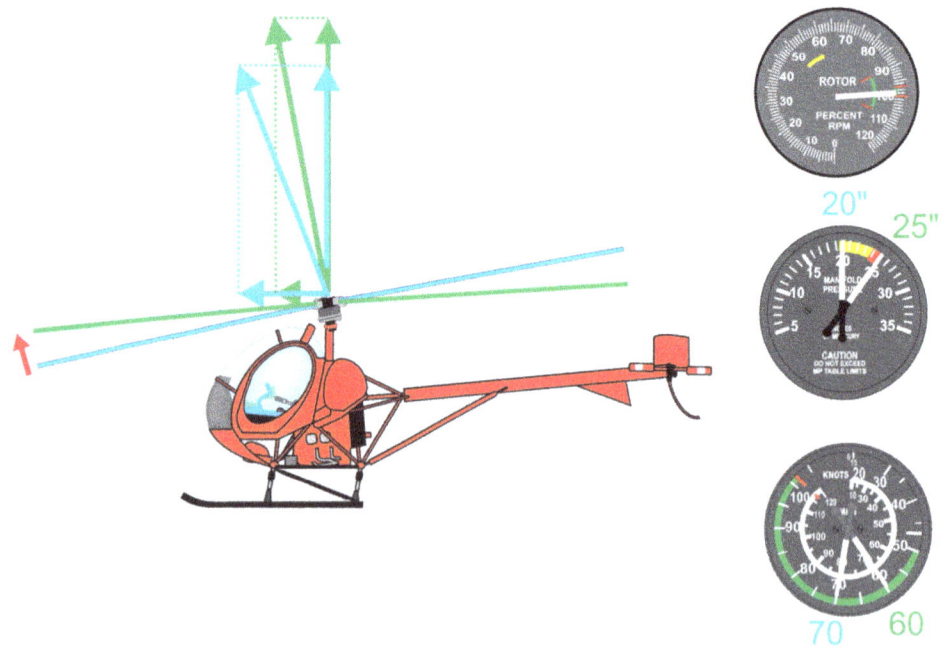

Forces in a Descent

Forces in a descent

For the helicopter to descend, the vertical component of rotor thrust must be less than the weight. This is achieved by increasing the disc tilt or decreasing the Total Rotor Thrust. Since a descent also implies that you wish to slow down, tilting the disc is not desired; therefore, a descent is usually controlled by collective.

For the helicopter to descend from forward flight:

- reduce power by lowering the collective (maintain RRPM using the throttle)
- apply aft cyclic to control the effects of flap forward and maintain constant attitude for descent, normally around 40-60 kts, depending on the helicopter type, and
- use the anti-toque pedals to balance the change in torque.

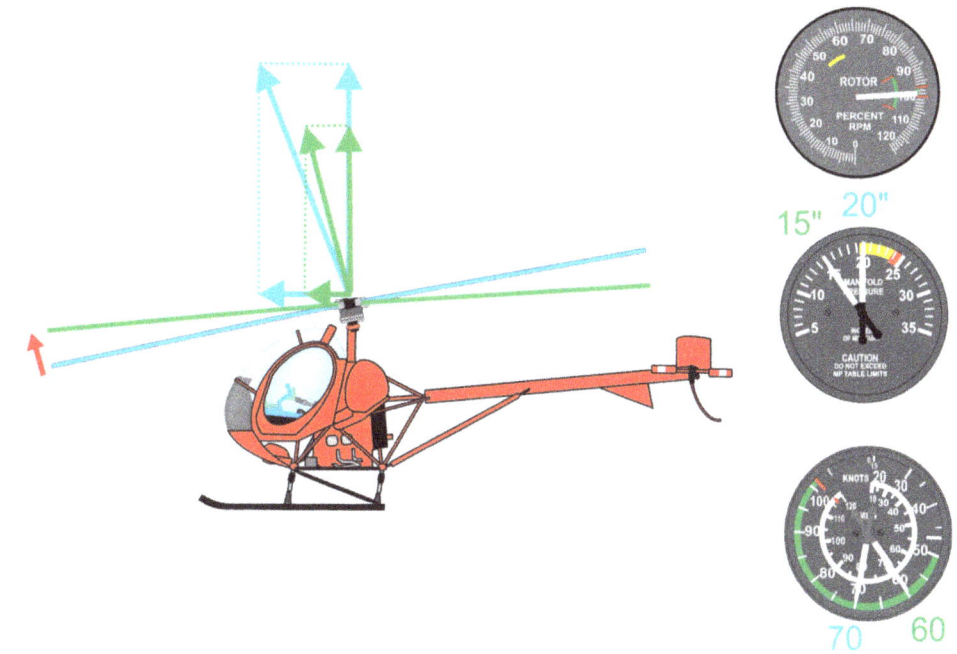

Forces in a Turn

Introduction

Like a fixed-wing aircraft, turns in a helicopter are made by banking.

In forward flight, the rotor disc is tilted forward, which also tilts Total Rotor Thrust. When the helicopter is in a bank, the rotor disc is tilted sideways.

Use of cyclic

Tilting the rotor disc sideways (using the cyclic) causes the Total Rotor Thrust to be tilted sideways, which in turn, is divided into two components:

- the vertical component of rotor thrust opposing weight, and
- the horizontal component of rotor thrust opposing centrifugal force.

The horizontal component of rotor thrust pulls the helicopter in the desired direction of turn. As the angle of bank increases, the Total Rotor Thrust is tilted more toward the horizontal, thus causing the rate of turn to increase because more thrust is acting horizontally.

Chapter 3 Elementary Handling

Use of collective Since tilting the rotor disc sideways causes the Total Rotor Thrust to act more horizontally, the vertical component of rotor thrust will decrease.

The collective pitch must be increased to balance weight and maintain altitude to compensate for the decreased vertical component.

The steeper the angle of bank, the greater the amount of collective pitch required to maintain altitude. Thus, with an increase in bank and a greater collective pitch, the Total Rotor Thrust will be increased, and the rate of turn will be faster.

Use of pedals In many helicopters, turn and bank indicators are not fitted, and balance must be maintained by 'seat of the pants flying'. To keep balanced, that is, to keep the tail behind the nose, you need to apply a small amount of right pedal during a right turn and a small amount of left pedal during a left turn.

Angles of bank can be judged by the tip path plane's angle or the console's attitude on the horizon.

Note: The tip path plane is the tip of the rotor blades, as seen by the pilot.

Summary A turn is produced by banking the helicopter, thus allowing the horizontal component of rotor thrust to pull the helicopter from its straight course.

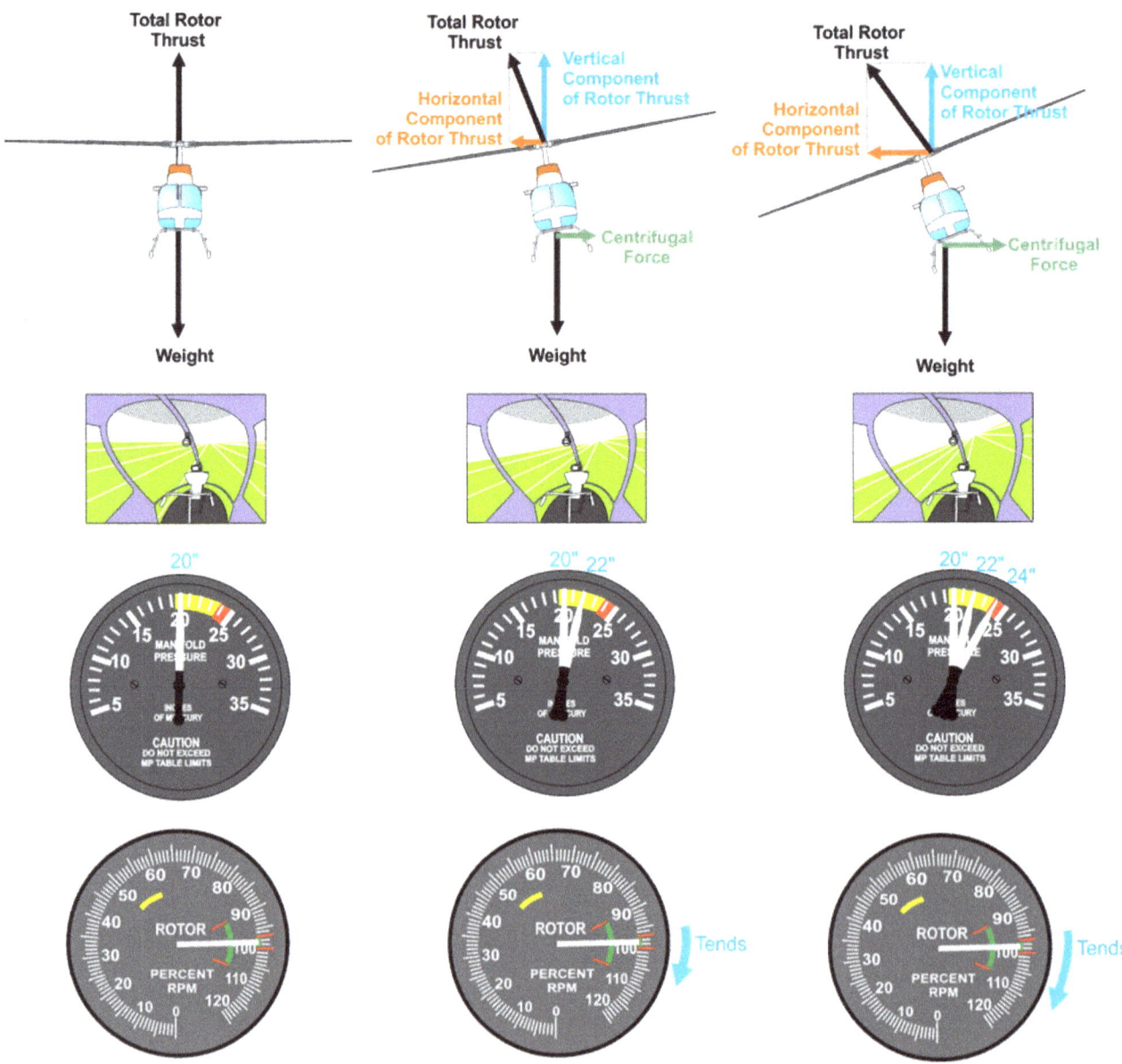

Power Changes

Introduction As previously mentioned, as pitch is increased, rotor thrust and rotor drag are increased; therefore, an increase in power is required to maintain rotor RRPM.

What is power In a piston engine helicopter, power is a combination of manifold pressure (MAP) and Engine RPM.

A turbine helicopter is similar, where power is a combination of Torque (Tq) and Engine RPM.

The Engine RPM and Rotor RPM are in a fixed ratio to each other set by the manufacturer and must remain in a narrow operating band. Rotor and Engine RPM are presented on the dual tachometer, MAP and Tq are displayed on their respective gauges, as illustrated below.

For the purposes of this chapter, we will focus mainly on the piston engine.

Maintaining rotor RPM To increase power, the collective is raised. This increases the pitch of the rotor blades, therefore, increasing Total Rotor Thrust. To maintain Rotor RPM in a piston engine, a cam system automatically opens the throttle butterfly as the collective is raised, which increases engine power to overcome the added rotor drag caused by raising the collective. Unfortunately, the cam system in some helicopters is not very accurate (such as the B47), and the pilot must override the system by using the throttle. Other helicopters (such as the R22) have very accurate throttle/collective correlation systems and require minimal throttle manipulation by the pilot, especially if they also have a governor system.

Chapter 3 Elementary Handling

Increasing power

So, to increase power from 18" to 22" MAP, the pilot raises the collective lever. If the RPM rises or falls, the pilot checks this using the throttle. On reaching 22", the collective action is stopped. If at 22" the RPM is not at the correct figure, an RPM change will have to be made while maintaining constant MAP.

To change RPM at constant MAP, other factors need consideration. In this example, if the throttle was closed to reduce RPM, the MAP would also decrease from the required 22". Therefore, to decrease RPM, the throttle is closed, and at the same time, the collective is raised slightly to maintain MAP, which also increases RPM.

The power and RPM changes at this stage of instruction require small control movements and, commonly, students' over-control. The MAP and RPM gauges will demand much attention.

Fully closed throttle

With the throttle fully closed, the throttle cam is inoperative. This allows the collective lever to be raised without opening the throttle. The effectiveness of the cam control varies with power settings at different stages of the collective's movement.

Torque reaction

MAP and RPM changes also produce a change of torque which must be compensated for by tail rotor thrust to maintain balanced flight. An increase in torque will require left pedal, and a decrease will require right pedal. (All helicopters with an anti-clockwise rotating main rotor act like this).

Summary

The primary control for:

- manifold air pressure (MAP) is the collective pitch lever, and
- RPM is the throttle.

Since the collective pitch control also influences RPM and the throttle also influences manifold pressure, each is considered to be a secondary control of the other.

Therefore, the pilot must analyse both the tachometer (RPM indicator) and manifold pressure gauge to determine which control to use and how much.

Questions

Below are some simple problems which best illustrate this relationship. Assume the correct settings should be RPM at mid-cruise and MAP at 22".

What would you do to correct for the following?

(The answers are on the following page.)

A: Low RPM,
low manifold pressure

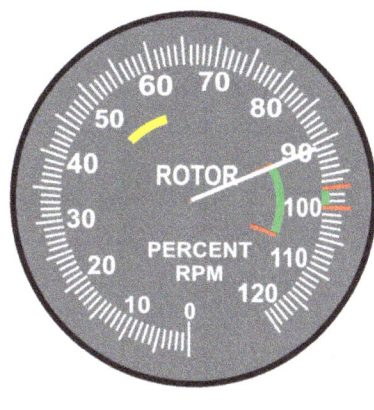 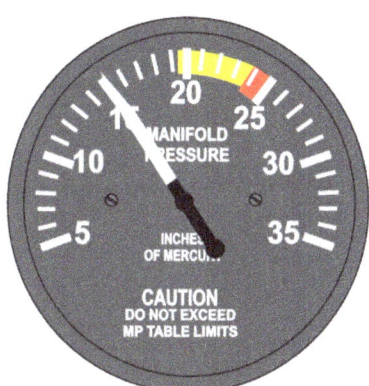

B: Low RPM,
high manifold pressure

 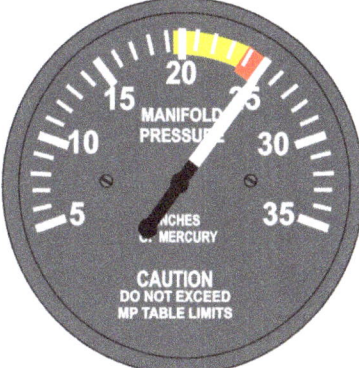

C: High RPM,
low manifold pressure

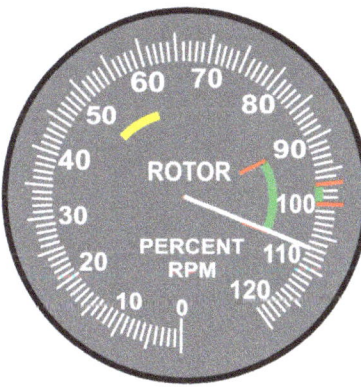 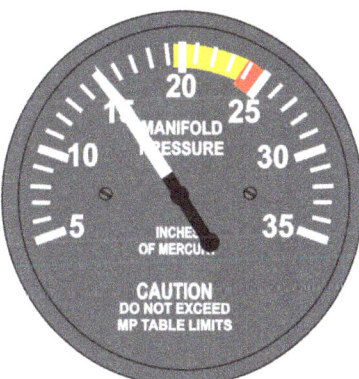

D: High RPM,
high manifold pressure

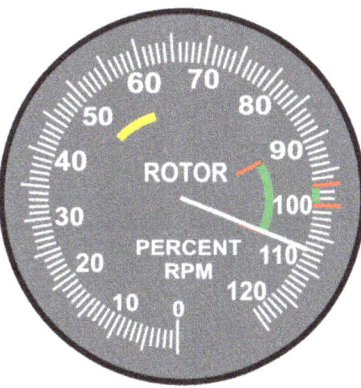 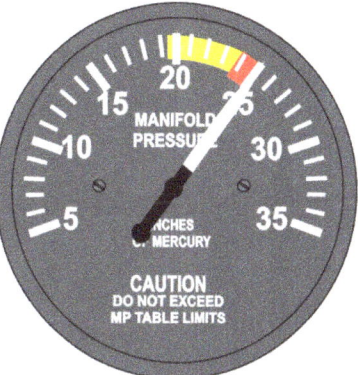

Chapter 3 Elementary Handling

Answers

A: Increasing the throttle will increase the RRPM and the manifold pressure.

B: Lowering the collective pitch will reduce the manifold pressure, decrease the drag on the rotor, and increase the RRPM.

C: Raising the collective pitch will increase manifold pressure, increase drag on the rotor, and decrease the RRPM.

D: Decreasing the throttle reduces the RRPM and the manifold pressure.

Note: The R22 is now sold standard with an automatic governor, designed to maintain the Engine RPM in a narrow operating band, without using the throttle. This is great for safety, as pilots should never now encounter low or high RRPM situations. The negative side is if the R22 loses the throttle for general helicopter flight training, and pilot training in this type only does not gain valuable experience using the throttle.

Air Exercises: Elementary Handling

Introduction

The air exercises will be split into two parts:

1. Power changes, including:
   - maintaining straight and level while adjusting speed
   - climbing
   - descending
   - levelling out, and
2. Turns.

Airmanship

- Observe the handover-takeover (HOTO) procedure
- Before commencing each new exercise, have a good lookout **above**, **below** and at the **same level**
- Monitor temperatures and pressures of the engine and transmission
- Manage the RRPM and Power
- Start learning the radio calls.

Common faults

- No lookout
- Over-controlling, especially with throttle and pedals
- Concentrating on the instruments instead of looking outside
- Incorrect sequencing of the control inputs, particularly when levelling out from a climb where the student often lowers collective before gaining the required airspeed
- Failure to hold a control input and wait for it to happen
- Lack of yaw control (balance).

Air Exercise 3-1: Power Changes

Aircraft control

Aircraft control is the term used to describe the **sequence, magnitude,** and **rate** at which a pilot moves the primary flight controls to achieve a desired result, more commonly referred to as a 'Desired Performance'. The control movement will be based on how a pilot interprets the visual information by looking outside towards the natural horizon.

An instrument pilot must do a similar thing but with sole reference to the attitude instruments because they cannot see the natural horizon (which can be taught later).

The amount of control input required to manipulate the helicopter will depend on the pilot's ability to accurately interpret what is being seen and translate this into a control input. Misinterpretation can result in improper control inputs.

Performance (or Desired Performance) refers to the state you want the helicopter to be in. That is, hovering, straight and level, climbing, descending, or turning.

It is up to the pilot to set the desired **power** and **attitude** while in **balance** to achieve the **desired performance**.

To do this in the correct order, we follow a common sequence of:

Power - Attitude - Balance (PAB)

Chapter 3 Elementary Handling

Increase speed and maintain altitude

Starting at 50 kts and 1000 ft, increase speed to 70 kts.

Power – Attitude – Balance (PAB)

Collective	Raise some collective in anticipation of the requirement to maintain 1000 ft.
Cyclic	As collective is raised, ease the cyclic slightly forward. The nose will drop, and speed will begin to increase.
Throttle	Adjust if necessary to maintain RRPM.
Pedals	Use left pedal to maintain balance throughout.

Decrease speed – maintain altitude

Starting at 70 kts at 1000 ft, decrease speed to 50 kts.

Power – Attitude – Balance (PAB)

Collective	Lower some collective in anticipation of the requirement to maintain 1000 ft.
Cyclic	As collective is lowered, ease the cyclic slightly aft. The nose will rise, and speed will begin to decrease.
Throttle	Adjust throttle, if required, to maintain RRPM.
Pedals	Use right pedal to maintain balance throughout.

Climb and maintain speed

Starting from a cruise speed at 1000 ft, initiate a climb at 60 kts to 2000 ft using climb power settings as specified by the instructor. (Usually around 25" MAP and 100% rotor RPM)

Power – Attitude – Balance (PAB)

Collective	Raise collective to climb power.
Cyclic	As collective is raised, ease the cyclic aft until the nose attitude is set for 60 kts.
Throttle	Adjust throttle, if required, to maintain RRPM.
Pedals	Use left pedal to maintain balance throughout.

Levelling out from climb

Level out from the climb at 2000 ft and increase speed to cruise speed.

This exercise requires some anticipation. If you wait until you get to 2000 ft before you start thinking about levelling out, you will overshoot the desired height and end up levelling out at, say, 2300 ft - not good. The secret, therefore, is to start levelling out about 50 ft before the target height; in this case, start levelling at 1950 ft.

Anticipate - Power – Attitude – Balance (APAB)

Anticipate	Anticipate by maintaining climb power but easing the cyclic forward to accelerate by converting the thrust used for climbing to accelerating.
Collective	As speed accelerates to cruise and the target altitude is achieved, lower collective to cruise power.
Cyclic	Adjust to set the cruise attitude.
Throttle	Adjust throttle, if required, to maintain RRPM.
Pedals	Use pedals to maintain balance throughout.

Descend and maintain speed	Starting from cruise speed and 2000 ft, initiate a descent at 60 kts to 1000 ft. Use descent power settings specified by the instructor (usually 15" MAP and 100% rotor RPM.)

Power – Attitude – Balance (PAB)

Carb Heat	ON during lower power settings.
Collective	Lower to descent power setting (15").
Throttle	Adjust throttle, if required, to maintain RRPM.
Cyclic	As collective is lowered, ease the cyclic aft until the nose attitude is set for 60 kts.
Pedals	Use right pedal to maintain balance.

Levelling out from a descent	Level out from the descent at 1000 ft and increase speed to cruise speed. As with levelling out from the climb, this exercise requires some anticipation. Start bringing in some power (collective) and increasing speed (cyclic) about 50 ft before your target altitude of 1000 ft.

Anticipate - Power – Attitude – Balance (APAB)

Carb Heat	Off
Collective	Raise to cruise power setting (22").
Throttle	Adjust throttle, if required, to maintain RRPM.
Cyclic	As collective is raised, ease the cyclic forward to increase speed.
Pedals	Use left pedal to maintain balance.

Air Exercise 3-2: Turns

Introduction	Level turns at 60 kts and 1000 ft and 30° angle of bank. The secret to flying smoothly is to look outside the helicopter at the natural horizon and not get fixated inside on the instruments.
Left turn	Initiate a left turn at 60 kts, 1000 ft and 30° angle of bank.

Cyclic	Ease left to the desired angle of bank and hold. Ease forward or aft to control airspeed.
Collective	Normally, raise slightly to maintain altitude. (Note: Use your own judgement here - remember collective is used to maintain altitude.)
Throttle	If raising collective, you will normally have to increase throttle; however, in a turn, you are increasing the "G" loading on the disc, which increases rotor RRPM and may require some throttle to be reduced. In a turn, use throttle to maintain RRPM as required.
Pedals	Use pedal as required to maintain balance throughout.

Chapter 3 Elementary Handling

Right turn Initiate a right turn at 60 kts, 1000 ft and 30° angle of bank.

Cyclic	Ease right to desired angle of bank and hold. Ease forward or aft to control airspeed.
Collective	Normally, raise slightly to maintain altitude.
Throttle	Adjust throttle, if required, to maintain RRPM.
Pedals	Use pedal as required to maintain balance throughout.

Levelling out of turn Level out of turn.

Cyclic	Ease back to centre.
Collective	Lower to original position.
Throttle	Adjust throttle, if required, to maintain RRPM.
Pedals	Use pedal as required to maintain balance throughout.

Air Exercise 3-3: Climbing and Descending Turns

Climbing/ descending turns You will use skills learned in the previous exercises during the climbing and descending turns. The instructor will expect you to be thinking for yourself and, by now, coming to grips with coordination at the controls.

Climbing turns Climbing turn at 60 kts.

Power – Attitude – Balance (PAB)

Collective	Raise to climb power.
Cyclic	Ease cyclic slightly aft to set a 60 kts attitude then roll right or left (as required) to the desired angle of bank and hold.
Throttle	Maintain RRPM.
Pedals	Maintain balance.

Descending turns Descending turn at 60 kts.

Power – Attitude – Balance (PAB)

Collective	Lower to descent power.
Cyclic	Ease cyclic slightly aft to set a 60 kts attitude then roll right or left (as required) to the desired angle of bank and hold.
Throttle	Maintain RRPM.
Pedals	Maintain balance.
Carb Heat	As required.

4

Hover

Aim To accurately control the helicopter's position and height over a surface.

Objectives On completion of this lesson, the student will be able to:

- state where to look while at the hover
- explain hovering In-Ground Effect (IGE) and Out of Ground Effect (OGE)
- hover by maintaining a constant height, heading and position over a selected spot
- move the helicopter forwards, sideways and backwards across the ground, maintaining a constant height and heading at speeds below translational lift
- turn the helicopter through 360° over a spot maintaining a constant height and rate of turn, and be able to start and stop the turn when required
- complete hover patterns, and
- manage an engine failure while at the hover.

Motivation Being able to control the helicopter accurately at the hover is what separates the helicopter from all other flying aircraft. It is what makes the helicopter unique and fun to fly. It also means the helicopter can do jobs that other aircraft cannot do, such as precision sling loads, picking up and dropping people and cargo from very tight spaces or sloping ground, and conducting rescue operations. Hovering is a delicate skill that must be learned through hands-on experience. No amount of theory knowledge will prepare a student for the mental and physical effort it takes to hover; however, being well prepared by reading through this material will undoubtedly help give perspective on the art of hovering and perhaps explain why the helicopter is, initially, so tricky to control.

Preparation: Hover

The Work Cycle

Introduction

Although the primary flight controls move in the same manner as in forward flight, the sensitivity, magnitude, and rate they need to be used in the hover will differ and can vary on every flight, given the differences in the wind, weight, power, surface, and manoeuvre.

Balancing

Hovering has been compared to trying to stand on a ball. It takes many small movements at just the right time to get it right. It takes practice, and even when it can be done, you can never relax as something always tries to throw you off balance.

Control movements

Sometimes you need small movements; sometimes, you need big movements.

The control movements required to hover are broken down as follows:

- **Cyclic** controls the position of the helicopter over the ground
- **Collective** controls the height of the helicopter above the ground, and
- **Pedals** control the direction or the way the nose points.

To achieve this, the student will use the *Work Cycle* so that the **primary focus** is:

- looking outside and recognising the changes, and then
- responding to them with minimal time spent looking inside at the instruments.

The Work Cycle

The work cycle is used to manage the sequences required to hover and manoeuvre at the hover. We know that 90% of getting the helicopter under control relies on looking outside and setting the correct attitude, and only 10% requires confirmation by looking inside at the performance instruments.

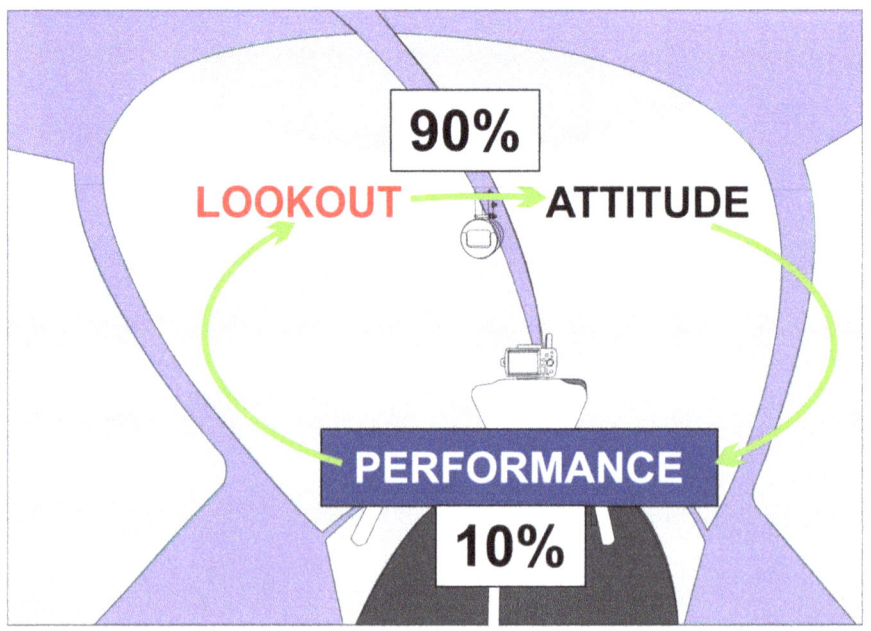

Chapter 4 Hover

What is important to monitor?

When very close to the ground, the attitude instruments (airspeed indicator, altimeter and VSI, artificial horizon, balance ball and heading indicator) become less important as they are not accurate enough for the fine control movements needed to hover. Often their readings are irrelevant compared to what you can see outside.

Who cares what the airspeed or the altimeter is reading when you are 5 ft from the ground in a hover with a 20 kts wind?

What is important is that you are controlling the helicopter with reference to what you are seeing and achieving the desired performance from the helicopter relative to the ground.

Summary of monitoring

What is important to you, as a pilot, changes close to the ground compared to forward flight, as summarised in the following table:

What is important to monitor	
In forward flight	**At the hover**
Airspeed (Attitude)	Ground speed (movement across the ground)
Rate of climb or descent	Rate of closure with the ground up or down
Altitude	Height above the ground
Heading	Direction (which way the nose points)
Balance	Not applicable
Power (MAP or Tq)	Power (MAP or Tq)

Lookout

The pilot must adjust the lookout to achieve the desired performance at the hover. Instead of referencing only the horizon for attitude, it becomes essential to develop a lookout that takes in the *near* distance (ground), the *middle* distance (close obstacles and reference points) and the *far* distance (horizon).

A constant and vigilant lookout is done by constantly moving the head and eyes. Looking at each position will give the pilot different bits of information and help them understand the complete picture of what is happening to the helicopter, making the correct control movement to achieve the desired performance.

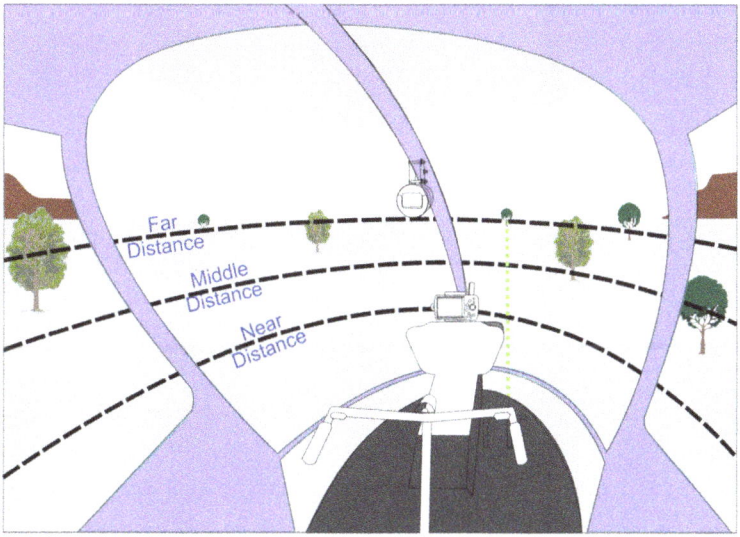

Far distance

Select features in the far distance out on the horizon line and relate them to some reference features in the helicopter. Any change in the helicopter's *attitude* can then be instantly recognised. The pilot can then make the appropriate cyclic inputs to prevent the change in attitude, avoiding unwanted movement. You need to catch it early to control the ground position.

70% of the lookout is focused on the **far distance** to achieve a stable hover.

The far distance is used to determine:

- nose high and low, left or right attitudes which left unchecked can progress into moving in that direction, and
- the heading alignment point so that left and right movements due to yaw can be recognised.

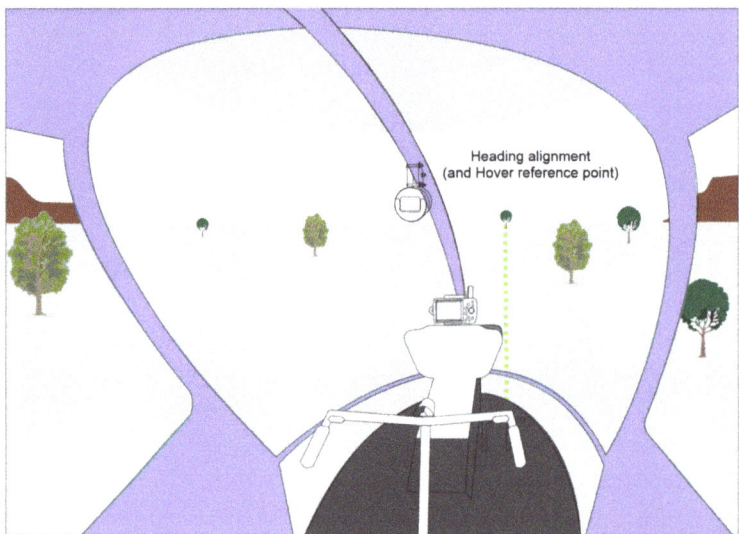

For example:

Consider driving a car down the highway; where do you look?

You will drive off the road if you look at your feet (near distance).

If you look at the car just in front of you (middle distance), you cannot see what is coming up and may not be able to react fast enough.

If you look down the road where you are travelling (far distance), you can better anticipate what is coming up and what you will need to do next.

In a helicopter, to hover like an expert, the pilot needs to focus most of the attention on the far distance to recognise changes in the *attitude* before the associated movement.

Chapter 4 Hover

Middle Distance

Selecting features in the middle distance, approximately **25-50 metres away**, allows the pilot to gain more detailed information on the surroundings and how the helicopter is moving.

Approximately **15%** of the lookout can be focused on the middle distance.

The middle distance is used to determine:

- any sideways, forwards or backwards movement, and
- what is coming up, and how they may have to respond.

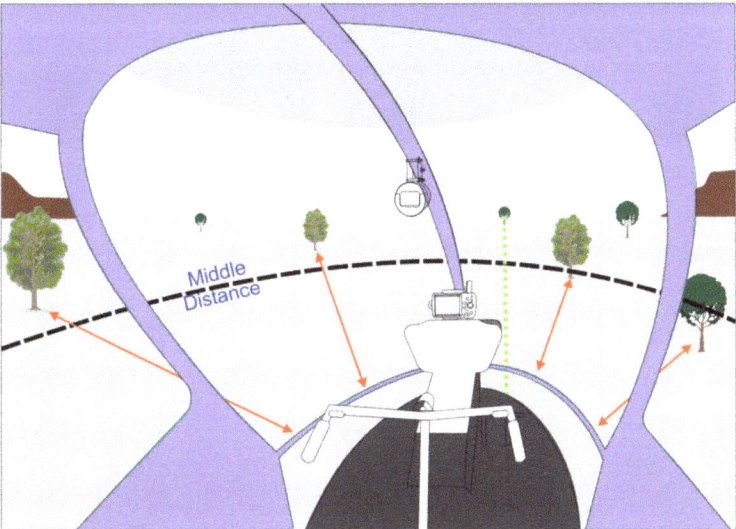

Much of the middle-distance information can be gained by looking into the far distance as the pilot's peripheral vision will include the middle-distance detail.

Near Distance

Near distance detail of **fewer than 25 metres** allows the pilot to gain accurate information on:

- the height of the helicopter's skids Above Ground Level (AGL)
- any movement across the ground
- the relative position and proximity of the helicopter to any obstacles, and
- the way the nose is pointing and where the tail is relative to these obstacles.

As the helicopter gets closer to the ground, the sight picture appears to get smaller. As the helicopter moves away from the ground, the sight picture appears to open up and get bigger. Understanding this perspective helps the pilot manage the height AGL and the relative position over the hover point.

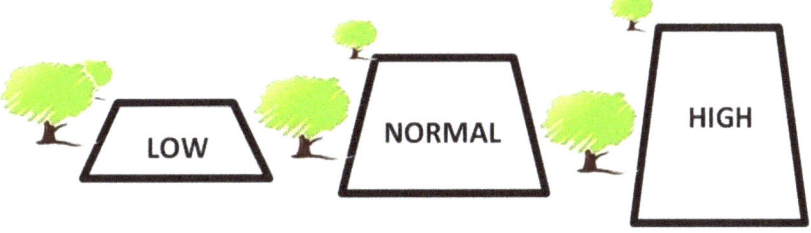

Just like driving in the car and staring at your feet, you do not want to focus on the near distance for too long before looking up at the far and middle distances again.

Approximately **10%** of the lookout can be focused on the near distance.

Scanning technique

To manage the helicopter proficiently close to the ground, the pilot will have to develop a lookout technique that allows them to focus on what is important at that moment while maintaining situational awareness.

This takes time and practice to develop.

The pilot needs to be constantly moving the head and eyes to take in relevant information, notice changes or trends, then make control inputs and repeat the whole process over and over again.

For example:

To understand the concept of a scanning technique, let us consider driving a car and about to pass a slower car on the highway.

Before passing, you look into the far distance to see if anything is coming and mentally judge what you have to do next.

You then check the middle distance to see if there are any immediate obstacles or vehicles that are going to affect you. You then glance back to the far distance to ensure nothing has changed and you are still driving on the correct side of the road.

You then look at the near distance at the car you want to pass as well as look in the rear-view mirror to make sure no one else is passing you and that it is all clear to make a move. Glance inside at the speed and RPM (performance) to make sure you can do it then you make a move.

As you go, you look into the far distance to make sure you are travelling straight down the road and can see if anything is coming in the opposite direction. Small quick glances at the middle and near distances to check you are not too close and nothing has changed. Ignore the speed and RPM gauges as you will do whatever it takes to pass, so they are irrelevant at that moment. Once passed, you look in the near distance to make sure you can slide back into the correct lane without hitting anyone, then back to the middle distance to see what other obstacles there are before moving back into the lane and focusing again on the far distance.

Hovering and manoeuvring at the hover are exactly the same. The pilot will need to be constantly looking in different places depending on what is happening.

Chapter 4 Hover

Performance information

The only performance information that is going to be important to monitor and manage is the RPM and power (MAP or Tq). The attitude instruments are almost irrelevant at this time.

Small quick glances at the power instruments are expected at the hover, but never look inside for too long as the helicopter has no stability and your constant attention and focus are required outside.

Approximately **5-10%** of the work cycle is dedicated to cross-checking the power gauges when at the hover.

Summary

Cyclic controls ground position

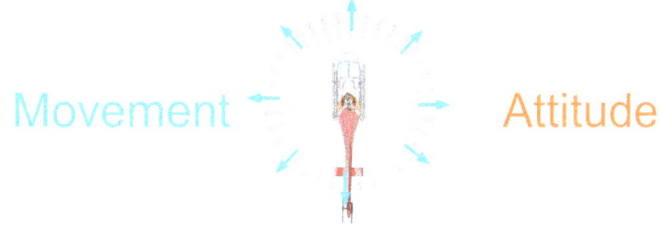

Collective controls height AGL

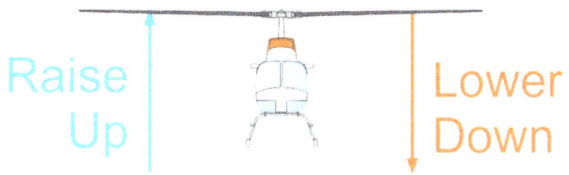

Pedals control direction

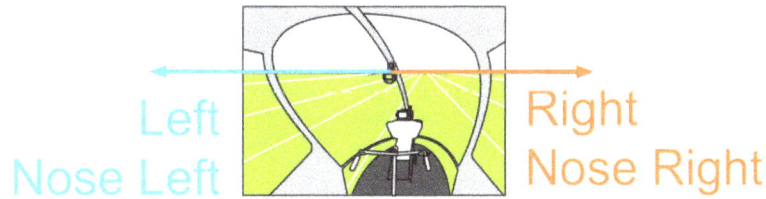

Throttle controls Engine RPM

Ground Effect at the Hover

Ground effect When a helicopter hovers near the ground at heights less than approximately ½ - ⅔ of the rotor diameter in winds less than 10 kts, it can maintain flight using less power than if it is in a high hover at heights greater than approximately ½ - ⅔ of the rotor diameter. The amount of "lift" or Total Rotor Thrust required is exactly the same because the helicopter's weight has not changed, but the power required by the engine to drive the rotor system when hovering In Ground Effect (IGE) is less.

Why When hovering In Ground Effect (IGE), the engine requires less power because the induced flow from the top of the rotor no longer moves straight down and quickly away, but changes its path as it meets the ground. This change of direction causes a decrease in the velocity of the induced flow, increasing the angle of attack and producing a greater rotor thrust for a smaller collective pitch setting.

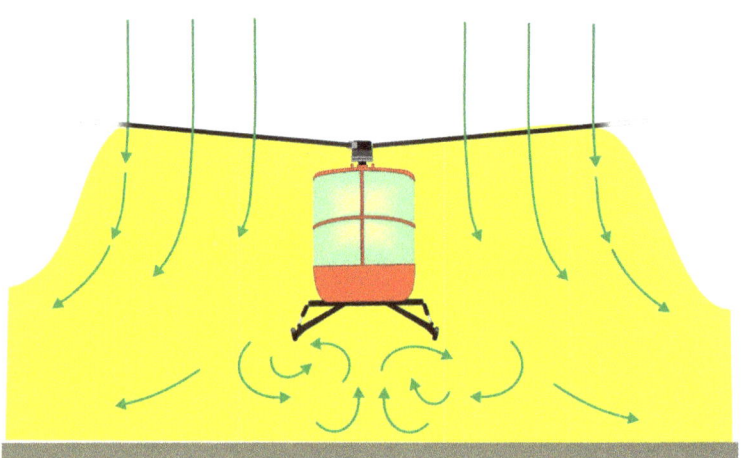

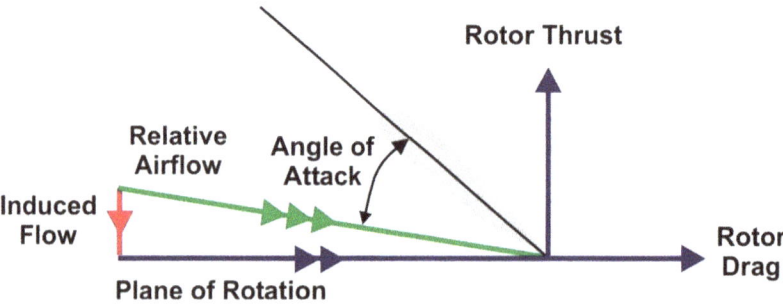

Maintain hover height To maintain hover height, the pilot needs to decrease the pitch angle by lowering some collective. With less rotor drag, less power is required.

The helicopter is now considered to be hovering **In Ground Effect (IGE)**.

Ground cushion The helicopter is sitting on a cushion of air called the **ground cushion** caused by the back pressure of the induced flow as it meets the surface.

Chapter 4 Hover

Out of Ground Effect (IGE)

When the helicopter is hovering out of the ground cushion, **'Out of Ground Effect' (OGE)**, the induced flow will increase again, decreasing the angle of attack and thereby producing less rotor thrust for the same pitch setting. To maintain rotor thrust, the pitch angle will need to be increased by raising collective, increasing rotor drag, requiring more engine power.

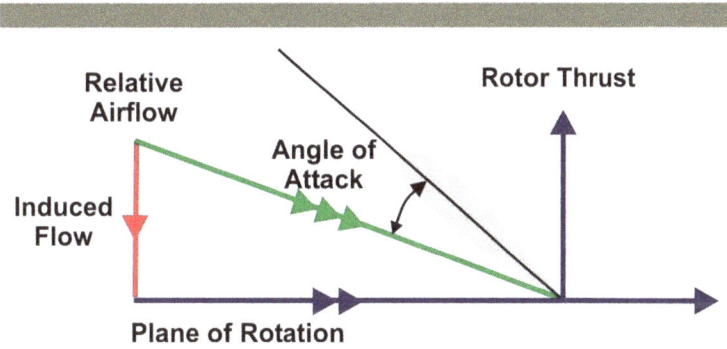

Varying Ground Effect

The height at which the helicopter is hovering above the ground will determine the usefulness and strength of the ground cushion.

The closer the helicopter is to the ground or surface, the more intense the ground cushion and the greater the effect. The further away the helicopter is from the ground or surface, the cushioning effect starts to decrease.

Ground cushion disappears at height

Ground cushion disappears completely at a height equal to approximately ½ - ⅔ of the total rotor disc diameter.

Ground effect influenced by

Ground effect is also influenced by:

- **Nature of the ground** the helicopter is hovering over. Hard, smooth and flat is best. Water, long grass or bush are worst as they absorb the downwash and reduce the cushioning effects.
- **Slope of the ground.** This will produce an uneven ground cushion.
- **Wind.** The ground cushion is displaced or blown downwind as wind strength increases. The ground cushion is most obvious at 0 kts wind and starts to lose its effectiveness at speeds above translational lift (12-15 kts).

The Effect of Wind at the Hover

No wind   In nil wind conditions, Total Rotor Thrust must balance weight to maintain constant height.

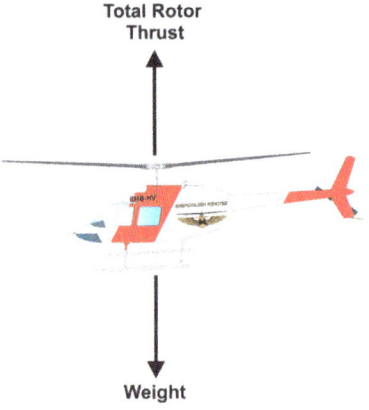

With wind   With wind conditions, the cyclic is displaced to windward to maintain the hover over a spot.

This tilts the Total Rotor Thrust, so some collective must also be used to maintain a constant height.

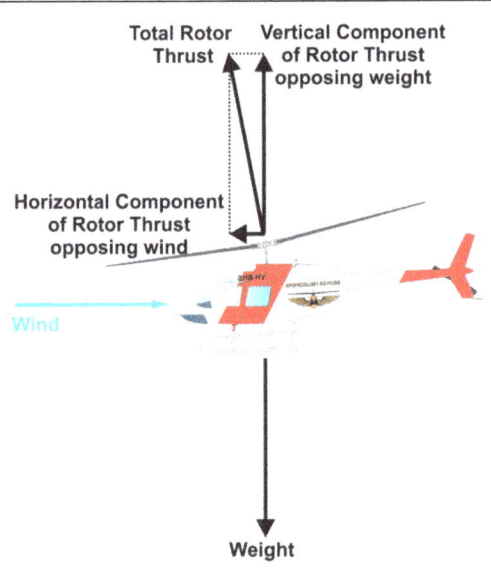

Airspeed   The helicopter considers wind the same as gaining airspeed.

With an increase in airflow over the rotor disc, there is a decrease in induced flow. This will result in an increase in the angle of attack for the same pitch angle. This will allow the pilot to lower the collective and decrease the rotor drag while at the same time maintaining the same amount of lift.

Winds greater than 10 kts but less than 15 kts will give an advantage at the hover by reducing the power required.

Winds greater than 15 kts will have the rotor disc passing through translational lift while still at the hover, reducing the required power.

Wind is always considered an advantage giving an increased power margin and better directional control due to streamlining (or weathercocking of the fuselage) when at the hover.

Chapter 4 Hover

Ground cushion and nil wind In nil wind conditions, the ground cushion will be uniform around the entire helicopter.

Ground cushion and wind The ground cushion will always be blown downwind. This can often be seen by looking at the ground and noting how the downwash affects it.

As the wind increases, the ground cushion will steadily be moved backwards until the rotor disc passes through translational lift; at this point, the helicopter will no longer benefit from the ground cushion, but the wind compensates for this.

Knowing this will help the pilot make good decisions for take-offs and landings.

Nil wind When there is nil wind, the downwash effect is equal.

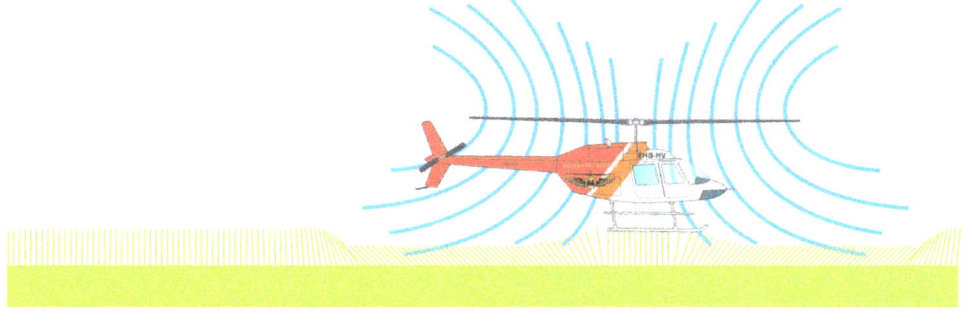

Wind 5 kts At 5 kts, the wind starts to blow the downwash downwind, and the airflow passing through the rotor disc starts to be affected by the oncoming wind.

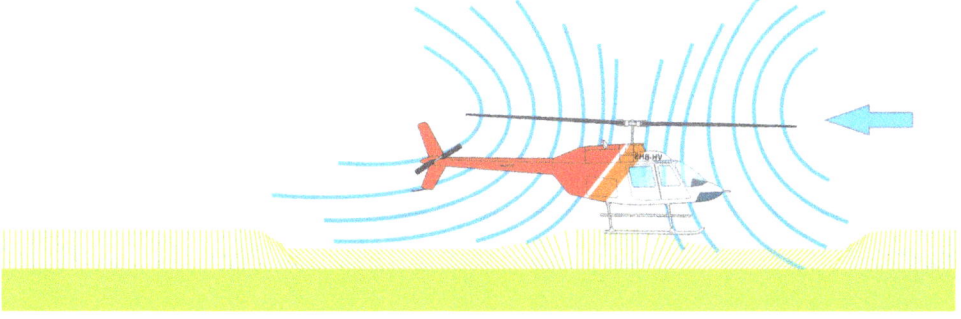

Wind 10 kts — At 10 kts, the helicopter is not yet experiencing translational lift, but the downwash has been affected: it is blown downwind, and any forward movement will see the helicopter enter clean air and pass through translational lift.

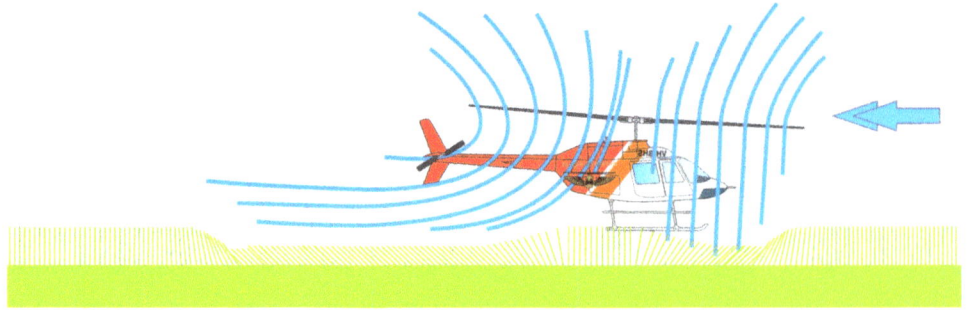

Wind 15 kts — At 15 kts, the helicopter has passed through translational lift. All of the downwash is blown behind the helicopter. Even though the helicopter is hovering over the ground, aerodynamically, it is flying through the air.

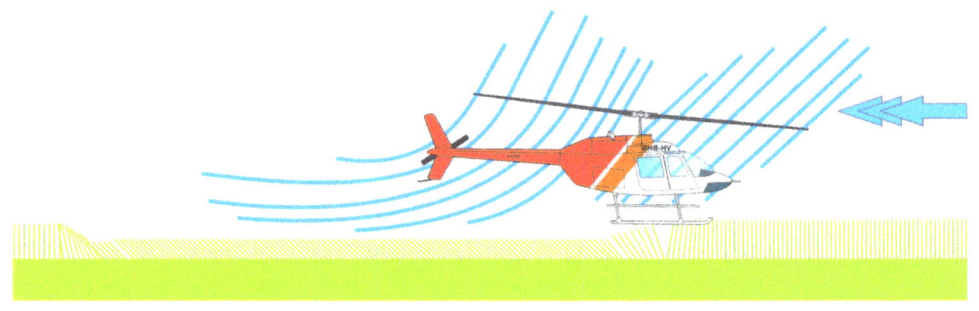

Weathercock — As the wind blows past the helicopter, it will always try to straighten the helicopter. This is called weathercocking effect and works in exactly the same manner as the windsock or a weathervane.

Weathervane **Windsock** **Weathercocking**

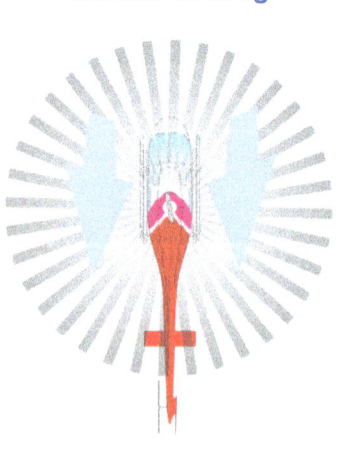

Chapter 4 Hover

Wind on the nose

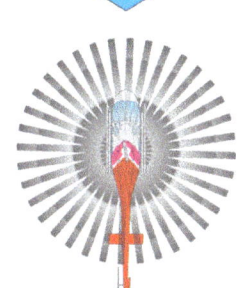

With the wind on the nose, the cyclic will be slightly forward into the wind.

The helicopter will experience weather cocking of the fuselage, and the pilot will not have to work as hard on the pedals.

The vertical fin attached to the tail will also start to produce some lift to assist the directional control.

The helicopter is most stable when hovering into wind.

Wind from the right

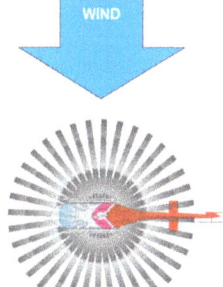

As the helicopter is turned side onto the wind, the cyclic will be slightly held to the right into the wind.

More of the fuselage surface will be exposed to the wind, producing more drag. This drag will try to make the helicopter weathercock back into the wind.

The pilot will need more left pedal with the wind 90 degrees on the right to hold the hover position or maintain the turn to the left.

In very strong winds, the pilot may reach the pedal limit and not be able to complete the turn.

Wind on the tail

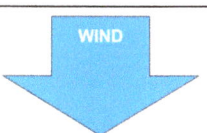
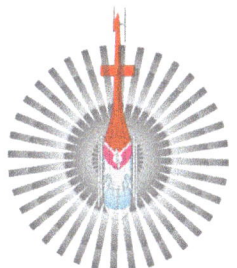

As the tail reaches the 180-degree point, the cyclic will be moved aft into the wind.

The direction of the wind on the fuselage will change from one side of the fuselage to the other. This will require the pilot to use opposite pedal to control the rate of the turn.

If maintaining the hover downwind, the tail can be very twitchy, and the pilot will have to work very hard on the pedals as the wind moves from one side of the fuselage to the other.

Wind from the left

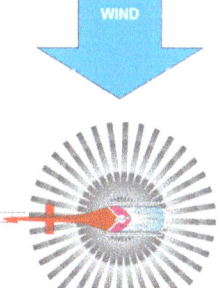

As the helicopter is again turned side onto the wind, the cyclic will be slightly held to the left into the wind.

More of the fuselage surface will be exposed to the wind, producing more drag. This drag will try to make the helicopter weathercock back into the wind.

The pilot will need less left pedal with the wind 90 degrees on the left as the wind will oppose the torque reaction.

With the wind on the left (in anticlockwise rotating systems), the tail rotor blades' aerodynamic airflow may be disrupted, so it can be very twitchy and harder to hover with wind on the left.

Engine Failure at the Hover and Hover Taxi

Introduction When at the hover, there is always the possibility that the engine may fail. This will require immediate action by the pilot to allow the helicopter to land.

Indications Indications of an Engine Failure while at the Hover (EFAH) or Engine Failure while at the Hover Taxi (EFAT) are similar to an engine failure in forward flight and includes:

Symptom	When it is obvious
Change in engine noise (sound)	Both real and practice engine failure at the hover
Yaw left	Both real and practice engine failure at the hover
LOW RPM warning light and horn	Both real and practice engine failure at the hover
ENGINE OUT warning light	Real engine failure only
Generator Fail warning light	Real engine failure only
Change in engine instruments	Both real and practice engine failure at the hover
Additionally, with an engine failure at the hover the helicopter will want to drift left.	Both real and practice engine failure at the hover

Tail rotor drift In powered flight, tail rotor drift will cause the helicopter to drift right. This is countered by the pilot applying left cyclic.

When the engine fails, tail rotor drift goes away immediately, but the helicopter will start to drift left because the pilot still has left cyclic applied.

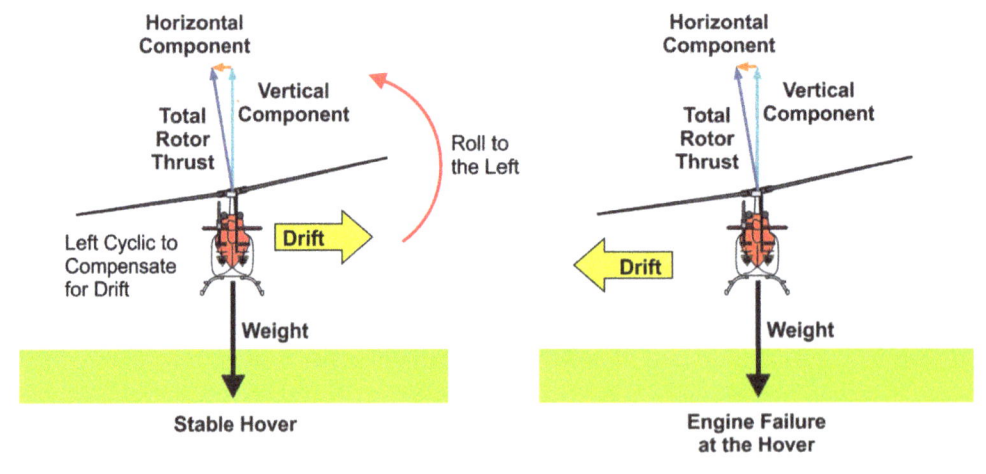

Chapter 4 Hover

NOT an autorotation

With an engine failure at the hover or hover taxi, the helicopter will **NOT** enter autorotation as this takes several hundred feet of altitude to establish a rate of descent airflow. Instead, the pilot must use the remaining energy stored as Rotor RPM to cushion the helicopter onto the ground.

The **collective is not immediately lowered**. Instead, the **collective is held in place** until pulled up to cushion the helicopter onto the ground.

Immediate actions

If the engine fails at the hover or hover taxi, the helicopter will react immediately, and the pilot will have to respond automatically by memory. There is no time for checklists and no time for thinking about it.

The pilot shall conduct the following actions on experiencing an engine failure at the hover or hover taxi:

Action	Description
PEDAL	Apply pedal to keep the helicopter straight and prevent it from turning.
SETTLE	Let the helicopter settle towards the ground. **Hold the collective in place** **Do NOT move the collective up or down** Hold the collective in its current position and let the helicopter descend towards the ground as the Rotor RPM decays. The rate of Rotor RPM decay will vary depending on the inertia built into the rotor system and the current amount of blade pitch. This Rotor RPM decay can be slow for high-inertia rotor systems (e.g. Bell206) and fast for low-inertia rotor systems (e.g. R22). Maintain ground position by anticipating using some forward and right cyclic to prevent the helicopter from drifting left.
PULL	When within 1 ft of the ground, start to pull the collective up at a rate that equals the descent. - Descending slowly, pull the collective up slowly to cushion. - Descending quickly, pull the collective up quickly to cushion. This will allow the remaining Rotor RPM to provide additional rotor thrust to cushion the helicopter onto the ground. *Remember, there is no engine, so there is no MAP or Tq limit. The collective can be pulled up as required.*

At stable hover

An engine failure at a stable hover will result in a vertical landing with no forward movement. Once the skids touch the ground, the collective may be lowered slowly and smoothly to the bottom stop.

In Hover Taxi

An engine failure while in a hover taxi will result in vertical and horizontal movement on landing, so the helicopter will move forward and slide along the ground.

This will require discipline by the pilot in maintaining the cyclic slightly forward of centre as the helicopter slides forward. Do not use aft cyclic to slow the helicopter, as this will only result in the blades striking the tail boom.

The faster the taxi, the longer the slide, the longer the pilot needs to hold some forward cyclic, and the longer the delay before lowering the collective.

If the collective is lowered too much, too early, while the helicopter is still sliding along the ground, the weight of the helicopter may cause the skids to dig into the ground, and the helicopter may roll over, so caution and care must be exercised when conducting an engine failure at the hover taxi.

It is also important that the pilot keeps the skids aligned with the direction of travel so that on ground contact, the helicopter does not roll over.

Air Exercises: Hover

Introduction	The air exercises will be divided into the following parts:

- The hover
- Horizontal movement at the hover (sideways, backwards and forwards)
- Hover turns through 360 degrees
- Hover Patterns
- Engine Failure at the hover and hover taxi

Airmanship

- Be sure of and observe the handover/takeover (HOTO) procedure
- Maintain a good lookout and situational awareness
- Protect the tail by maintaining a minimum 5 ft AGL skid height hover
- Keep a safe distance from obstacles on the ground (minimum 10 meters)
- Monitor temperature and pressures of the engine and transmission
- Manage RPM and Power
- Memorise some of the aircraft's power and wind limits
- Start talking to the instructor and announcing what you are doing
- Be aware of the effect of the rotor wash on the ground.

Common faults

- Poor lookout and no situational awareness of the surroundings
- Over-controlling of the cyclic and collective
- Undercontrolling the pedals
- Very slow to recognise attitude changes and the required control inputs within a time frame
- Very tense on the pedals and lack of directional control
- Concentrating too much on the instruments instead of looking outside
- Not anticipating what is coming up next
- Failure to hold in a control input and waiting for a response.

Air Exercise 4-1: Effect of controls at the Hover

Introduction   The instructor will demonstrate each of the controls at the hover. After the demonstration, the instructor will initially give the student one control at a time to get a "feel" for what the control does and start understanding where to "look". This will then be built on until the student can manage all three controls at the same time.

Work cycle

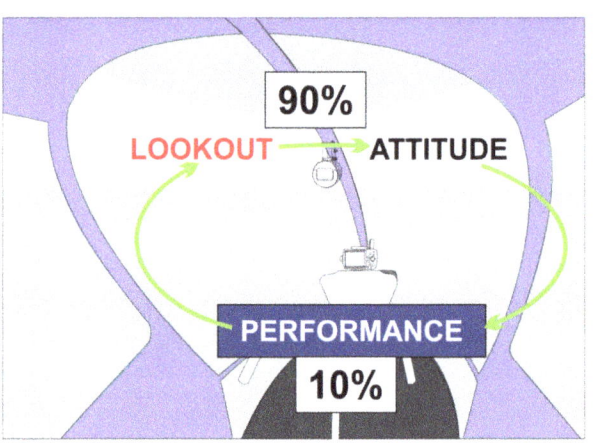

Process   Move the head and eyes to achieve the following.

Look	At the...	And...
Outside	**far** distance (horizon)	note the attitude
Outside	**far** and **middle** distances	pick some hover reference points: ■ Use the cyclic to maintain or adjust the attitude as required and manage the ground position. Hold it there. ■ Use pedals to control the yaw and keep the nose of the helicopter pointing at the nominated alignment point in the distance.
Outside	**middle** distance	continually assess the nominated reference points so you can see if there is movement: ■ Fine-tune the ground position with cyclic. ■ Fine-tune the direction with pedals. ■ Use the collective to control the height AGL.
Outside	**near** distance	to judge height AGL: ■ Use collective to fine-tune height AGL with reference to the ground.
Outside		repeat the whole process making adjustments as you go
Inside	Instrument panel	glance in at the RRPM, Power, Ts and Ps.

Chapter 4 Hover

Summary   The table below summarises the sequence of controls when hovering.

Step	Action	Discussion
1	LOOKOUT	Select hover reference points and relate them to reference points in the helicopter.
2	PEDALS	Pedals control YAW. Input left pedal, and the helicopter will turn about the mast in a circle to the left. Input right pedal, and the helicopter will turn about the mast in a circle to the right. Use the pedals to adjust YAW left or right to keep the nose of the helicopter pointing at the hover reference point.
3	COLLECTIVE	The collective controls height AGL. Raising the collective will make the helicopter go up. Lowering the collective will make the helicopter go down. Use the collective to adjust the power up or down to maintain a constant height AGL.
4	THROTTLE	As required to manage RRPM.
5	CYCLIC	The cyclic controls the position of the helicopter over the ground. - Forward cyclic - the nose drops, there is a small lag, then the helicopter accelerates forward. - Aft cyclic - the nose lifts, there is a small lag, then the the helicopter accelerates backwards. - Left cyclic - the fuselage tilts left, there is a smaller lag, then the helicopter moves left. - Right cyclic - the fuselage tilts right, there is a smaller lag, then the helicopter moves right. Use the cyclic to move the helicopter across the ground in any direction at speeds less than a fast-walking pace.

Air Exercise 4-2: Manoeuvring at the Hover

Introduction   From the hover, the instructor will direct the student through a series of turns to help increase hover skills.

Turns about the mast

Turns about the mast   Also referred to as pedal turns or spot turns, the helicopter will turn about the mast in either a left or right direction while at the hover.

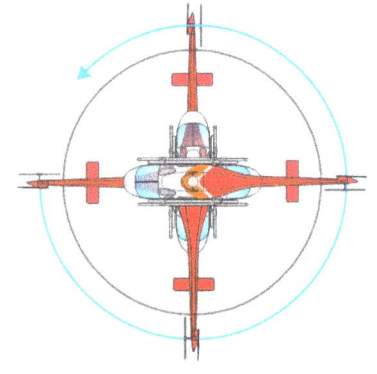

Standard turns   The standard turns are:

90 degrees (Left or Right)	180 degrees (Left or Right)	360 degrees (Left or Right)
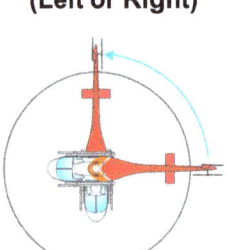	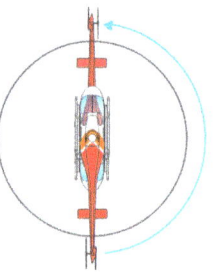	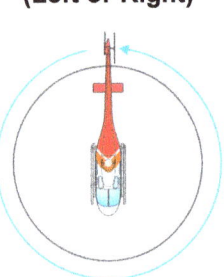

Step	Action	Discussion
1	LOOKOUT	Area clear.
2	ANNOUNCE	*"Pedal turn to the Left (or Right)"*
3	PEDALS	Use some left pedal to start turning the nose of the helicopter to the left. This will mean the tail is sweeping to the right, so that is the area that the lookout should have cleared. Use the pedals as required to control the magnitude and rate of turn and counter any wind effect.
4	CYCLIC	Use the cyclic to control the position of the helicopter over the ground. Remember that any wind will try to blow the helicopter downwind, so the cyclic shall be positioned constantly into the wind as the helicopter turns.
5	COLLECTIVE	Use the collective to manage the height as the helicopter turns.
6	PEDALS	As the reference point approaches to stop the turn, use some opposite pedal to stop the turn and return to a stable hover.

Chapter 4 Hover

Hover Taxi

Hover taxi

Hover taxi is where the helicopter is taxied (moved) forward at speeds less than translational lift (15 kts) and within the ground cushion. The pedals shall be used to keep the skids aligned with the direction of travel (ignore balance) and the cyclic used to control the forward movement and speed. Collective controls the height AGL.

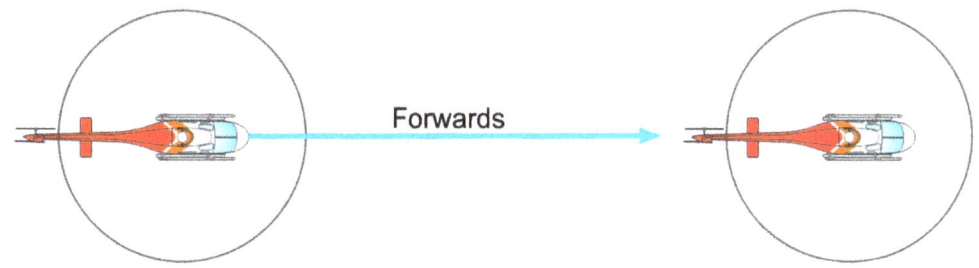

Step	Action	Discussion
1	LOOKOUT	Area clear.
2	ANNOUNCE	*"Hover taxi"*.
3	CYCLIC	Ease the cyclic forward to allow the helicopter to start moving forward. ■ More forward cyclic and it will go faster. ■ Less forward cyclic and it will go slower.
4	COLLECTIVE	Use the collective to control the height AGL.
5	THROTTLE	As required to manage RRPM.
6	PEDALS	Use the pedals to maintain the skids aligned with the direction of travel.
7	CYCLIC	To stop the taxi, use a small amount of aft cyclic to stop the forward movement and return to the hover.

Sideways flight

Sideways flight Sideways flight is where the helicopter slides sideways left or right while the nose maintains its original heading. The sideways movement is typically at speeds less than translational lift (15 kts) and within the ground cushion. The pedals shall be used to keep the nose pointing in the original heading and the cyclic used to control the sideways movement. Collective controls the height AGL.

Sideways to the left

Step	Action	Discussion
1	LOOKOUT	Area clear.
2	ANNOUNCE	*"Sliding left"*
3	CYCLIC	Ease the cyclic left to allow the helicopter to start moving left. ■ More left cyclic and it will go faster. ■ Less left cyclic and it will go slower.
4	COLLECTIVE	Use the collective to control the height AGL.
5	THROTTLE	As required to manage RRPM.
6	PEDALS	Use the pedals to maintain the original direction (heading).
7	CYCLIC	To stop the sideways movement, use a small amount of right cyclic to stop the left movement and return to the hover.

Sideways to the right

Step	Action	Discussion
1	LOOKOUT	Area clear.
2	ANNOUNCE	*"Sliding right"*
3	CYCLIC	Ease the cyclic right to allow the helicopter to start moving right. ■ More right cyclic and it will go faster. ■ Less right cyclic and it will go slower.
4	COLLECTIVE	Use the collective to control the height AGL.
5	THROTTLE	As required to manage RRPM.
6	PEDALS	Use the pedals to maintain the original direction (heading).
7	CYCLIC	To stop the sideways movement, use a small amount of left cyclic to stop the left movement and return to the hover.

Backwards flight

Backwards flight

Backwards flight is where the helicopter is taxied backwards. The backwards movement is typically at speeds less than translational lift (15 kts) and within the ground cushion.

Because it is very difficult to see behind, the student should first complete a 90-degree clearing turn to check the area behind, nominate a reference point to go back to, and then turn back into wind. Before moving, the **hover height** will be **increased by an extra 5 ft** to protect the tail.

Use some aft cyclic to start the helicopter moving backwards, use the pedals to control the weathercocking and keep the helicopter straight. Collective controls the height AGL.

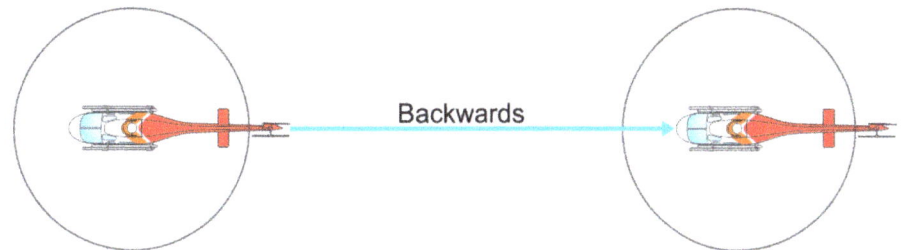

Backwards taxi

Step	Action	Discussion
1	LOOKOUT	Clearing turn to see what is behind. Area clear. Pick a reference point so the pilot will stop the backwards movement when arriving over it. Turn back into the wind. Increase height by 5 ft AGL (total of 10 ft AGL).
2	ANNOUNCE	*"Taxing back"*
3	CYCLIC	Ease the cyclic aft to allow the helicopter to start moving aft. ■ More aft cyclic and it will go faster. ■ Less aft cyclic and it will go slower.
4	COLLECTIVE	Use the collective to control the height AGL.
5	THROTTLE	As required to manage RRPM.
6	PEDALS	Use the pedals to maintain the original direction (heading).
7	CYCLIC	To stop the backwards movement, use a small amount of forward cyclic to stop the aft movement and return to the hover.

Turns about the Nose

Turns about the nose

Turns about the nose is where the helicopter will turn about an imaginary point in front of the nose of the helicopter. In effect, the helicopter will travel around the outside of a big imaginary circle.

The cyclic initiates a sideways movement in the circle's direction; the pedals keep pointing towards the centre of the imaginary circle, and the collective controls the height AGL.

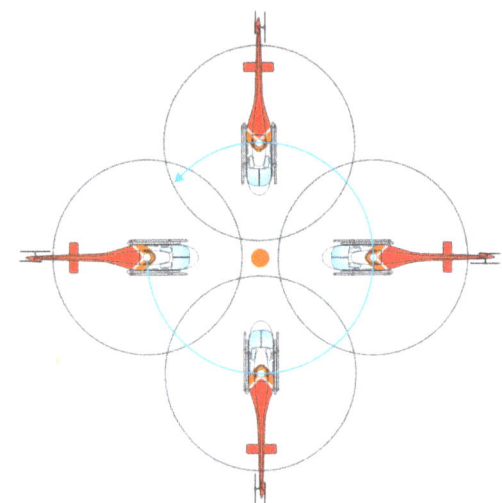

Turn about the nose: Left or Right

Step	Action	Discussion
1	**LOOKOUT**	Area clear.
2	**ANNOUNCE**	*"Turn about the nose to the left/right"*
3	**CYCLIC**	Use some cyclic to start the helicopter moving. Turning left – left cyclic Turning right – right cyclic
4	**PEDALS**	At the same time, use some pedal to orientate the fuselage so that the nose continues to point towards the centre of the imaginary circle. Turning left – right pedal Turning right – left pedal
5	**CYCLIC AND PEDALS**	Coordinate the use of both these controls to achieve the desired performance.
6	**COLLECTIVE**	Use the collective to manage the height as the helicopter turns.
7	**THROTTLE**	As required to manage RRPM.
8	**CYCLIC**	As the reference point approaches to stop the turn, use some opposite cyclic to stop the turn and return to a stable hover.

Turns about the tail

Turn about the tail This is where the helicopter will turn left or right with the axis of the turn being the tail.

In effect, the helicopter will travel around the inside of a big imaginary circle.

The cyclic is used to initiate a sideways movement in the direction of the circle, the pedals are used to keep the tail over the centre of the imaginary circle and the collective is used to control the height AGL.

This exercise is an essential tool when manoeuvring inside a confined area where the pilot needs to protect the tail from obstacles.

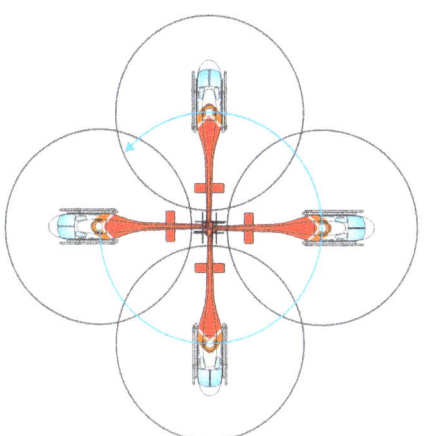

Turns about the tail: Left or Right

Step	Action	Discussion
1	LOOKOUT	Area clear.
2	ANNOUNCE	*"Turn about the tail to the left/right"*
3	CYCLIC	Use some cyclic to start the helicopter moving. Turning left – left cyclic Turning right – right cyclic
4	PEDALS	At the same time, use some pedal (left or right as required) to orientate the fuselage so that the tail stays over the centre of the imaginary circle.
5	CYCLIC AND PEDALS	Coordinate the use of both these controls to achieve the desired performance.
6	COLLECTIVE	Use the collective to manage the height as the helicopter turns.
7	THROTTLE	As required to manage RRPM.
8	CYCLIC	As the reference point approaches to stop the turn, use some opposite cyclic to stop the turn and return to a stable hover.

Pattern Hover

Pattern hover A pattern hover is where the previous exercises are put together to form a hover pattern at the instructor's discretion. Using a designated helipad or some other markers, the following patterns may be used.

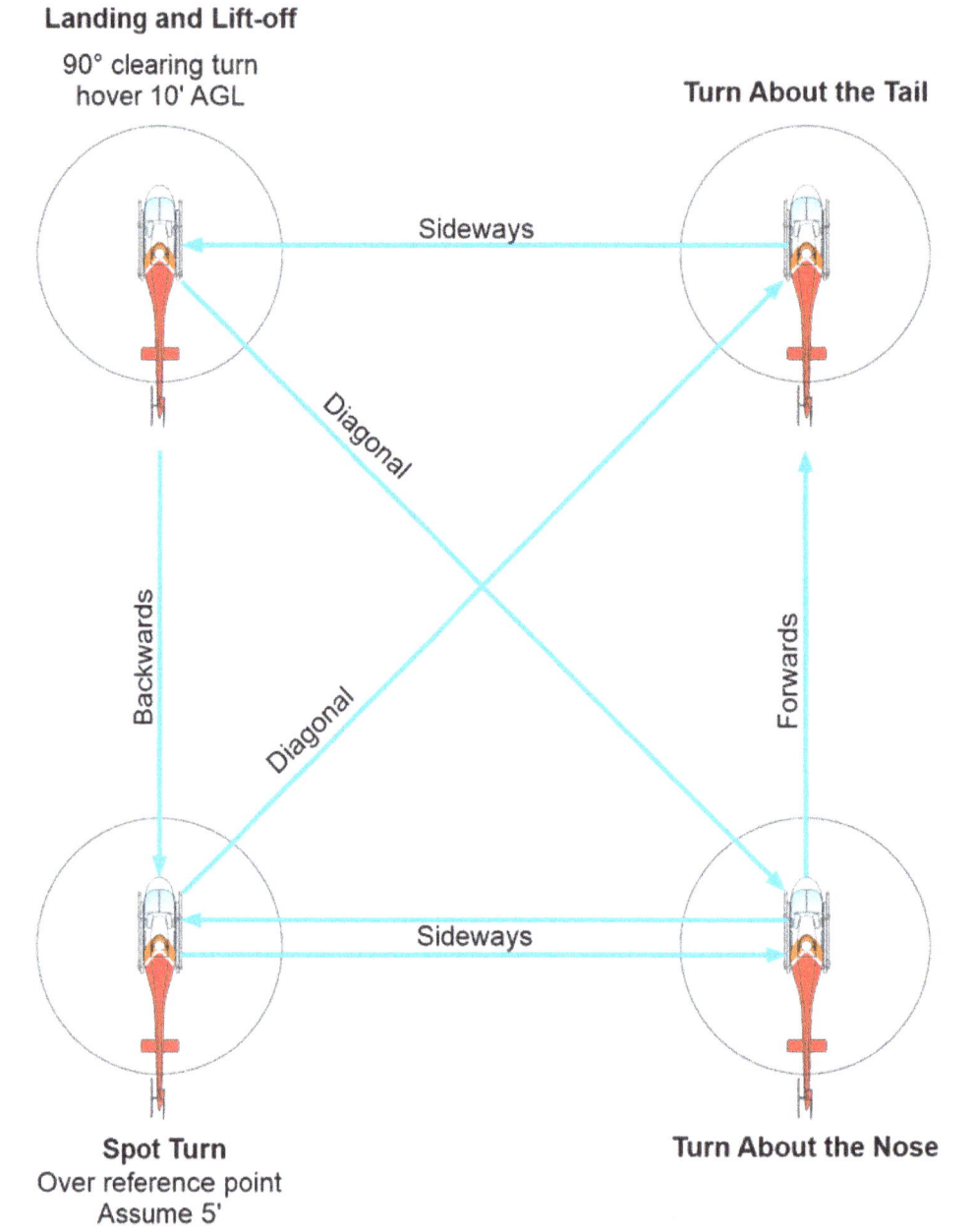

Summary

Because the hover sessions can be very intense, they are usually broken up with some transitions and circuit training to allow the student to relax, get some fresh air in the cabin and have a small break.

Chapter 4 Hover

Air Exercise 4-3: Engine Failure at the Hover and Hover Taxi

Demonstration

The instructor will demonstrate an:

- engine failure while at a stable hover, and
- engine failure while taxiing forward below translational lift speeds (less than 10 kts).

After the demonstration, the instructor will position the helicopter in a suitable area and give the student the controls to practice.

From the Hover

Conduct the following steps to complete an Engine Failure at the Hover (EFAH) practice.

Step	Action	Discussion
1	PREPARATION	Conduct a clearing turn and ensure the area is clear. Position the helicopter over a flat, even firm surface (preferably a runway-type environment). Configure the helicopter in a steady 5 ft skid height hover into wind.
2	Handover/Takeover	Confirm who is at the controls and which pilot is responsible for the throttle.
3	INSTRUCTOR WILL ANNOUNCE	"3, 2, 1, Practice engine failure"
4	PEDAL	Maintain heading with pedals to control the YAW.
5	SETTLE	Freeze the collective in its current position and allow the helicopter to "SETTLE" towards the ground. **DO NOT PUSH THE COLLECTIVE DOWN** Maintain ground position with the cyclic. This will typically require some forward and right cyclic to counter the left drift the helicopter may experience with an engine failure at the hover.
6	PULL	Once the helicopter is approximately 1 ft skid height above the ground, "PULL" the collective up as required to cushion the landing.
7	ON LANDING	Lower the collective and secure the helicopter.

Diagram

From the Hover Taxi	Conduct the following steps to complete an Engine Failure at the Hover Taxi (EFAHT) practice.

Step	Action	Discussion
1	PREPARATION	Ensure the area is clear and the surface is firm and smooth.
		Position the helicopter over a flat, even firm surface (preferably a runway-type environment). Configure the helicopter at 5 ft AGL or less, into wind, with a ground speed of 10 kts or less.
2	Handover/Takeover	Confirm who is at the controls and which pilot is responsible for the throttle.
3	INSTRUCTOR WILL ANNOUNCE	*"3, 2, 1, Practice engine failure"*
4	PEDAL	Maintain heading with pedals to control the YAW.
		Use the pedals as required, to keep the skids aligned with the direction of travel as the skids slide across the ground.
5	SETTLE	Freeze the collective in its current position and allow the helicopter to "SETTLE" towards the ground.
		DO NOT PUSH THE COLLECTIVE DOWN
		Allow the helicopter to continue moving forward and prevent any sideways drift with cyclic.
		DO NOT USE AFT CYCLIC
		Do not try to use aft cyclic, as this can cause the heels of the skids to hit the ground first, which will then bounce the helicopter onto its nose. The pilot will automatically use more aft cyclic, which can result in the blades striking the tail.
6	PULL	Once the helicopter is approximately 1 ft skid height above the ground, "PULL" the collective up as required to cushion the landing.
7	ON LANDING	As the helicopter comes to a complete stop, then lower the collective slowly and smoothly and secure the helicopter.

Diagram

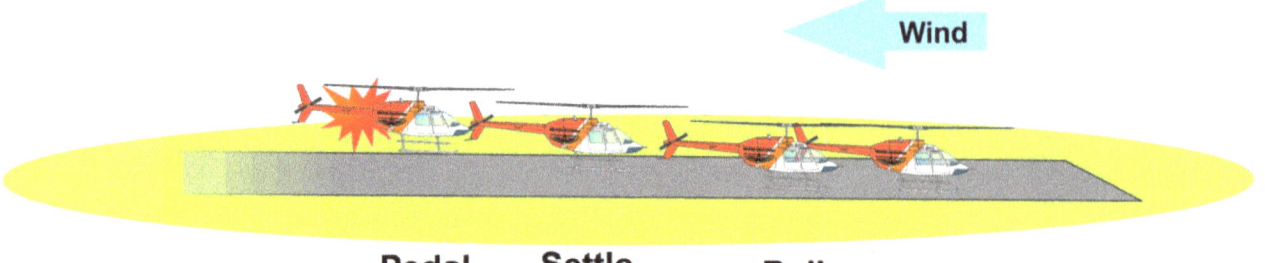

5

Lift-off, Transition and Landing

Aim To lift-off, land and transition to and from forward flight to the hover.

Objectives On completion of this lesson, the student will be able to:

- recite the Pre-lift-off checks
- recite the Hover checks
- recite the short finals checks
- state when to use each of the checks listed above
- lift the helicopter off the ground to a stable hover
- transition into forward flight up to 50 kts
- from 50 kts, transition back to the hover, and
- land the helicopter back onto the ground.

Motivation A helicopter has no inherent stability; it does not want to fly all by itself. It relies on small, subtle inputs from the pilot to accurately adjust the flight controls to achieve a successful lift-off and landing every single time.

Although with experience, the pilot gets better, even the most experienced should not allow complacency to creep in because every lift-off and landing is new and unique.

The process of moving from a hover into forward flight and from forward flight back to the hover means there will be dramatic changes in the aerodynamic forces affecting the helicopter. The pilot must manage these changes, referred to as a transition as we 'transition' from one flying configuration to another.

Preparation: Lift-off and Landing

Introduction

Because lifting off and landing require the skids to come into contact with a fixed surface, any slope, sideways or backwards movement of the helicopter over the ground can progress into Dynamic Rollover. Dynamic Rollover can start for many different reasons. Some are predictable due to known aerodynamic principles and countered by design or pilot input; others can be caused by the pilot not recognising a trend or overcontrolling.

Lifting off and landing are two of the most critical phases of flight, requiring the pilot's full attention and accurate flying.

Translating Tendency (Tail Rotor Drift)

Translating tendency (Tail rotor drift)

Unless compensated for by design, all conventional helicopters will suffer from translating tendency (also called tail rotor drift).

As the collective is raised to increase power, the torque reaction between the main rotor blades and the engine will increase. To counter the torque reaction (shown below as a couple YY), the pilot will need to apply more left pedal and, thereby, increase tail rotor thrust (shown below as a moment ZZ). This will cause a pull on the fuselage to the right, as both the torque and the tail rotor thrust combine to form an imbalance of forces trying to "drift" the helicopter to the right.

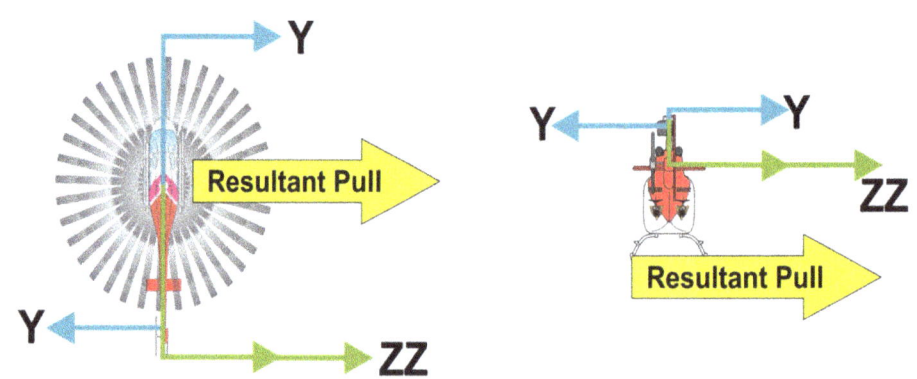

Countering translating tendency

To counter tail rotor drift, the pilot must apply some left cyclic to counter the resultant pull.

If the pilot does not apply some left cyclic, the helicopter could experience a roll to the right as the helicopter starts to pivot about the right skid caused initially by the right drift.

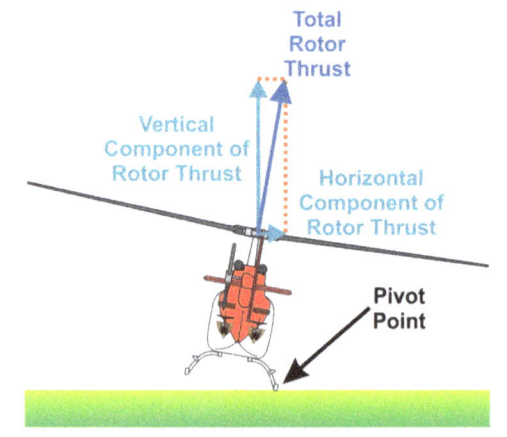

Chapter 5 Lift-off, Transition and Landing

Tail rotor roll

If the pilot uses some left cyclic, the helicopter should lift evenly off the ground, but because the rotor disc is tilted to the left, the fuselage may also sit in the hover slightly left skid low. This is referred to as tail rotor roll.

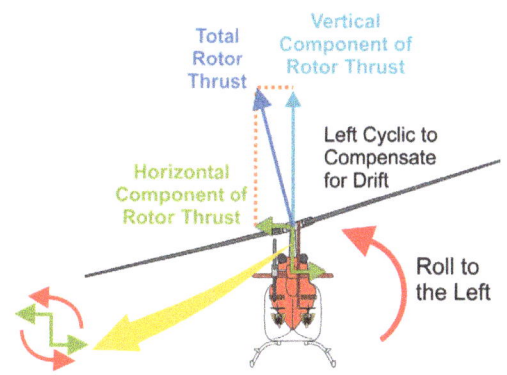

Balance of opposing forces

The downside is that if the pilot uses too much left cyclic, the helicopter may want to **pivot** and **roll** about the left skid.

Both lifting and landing the helicopter to and from the hover can be a delicate balance between these opposing forces and requires the delicate management of the rotor disc with the cyclic.

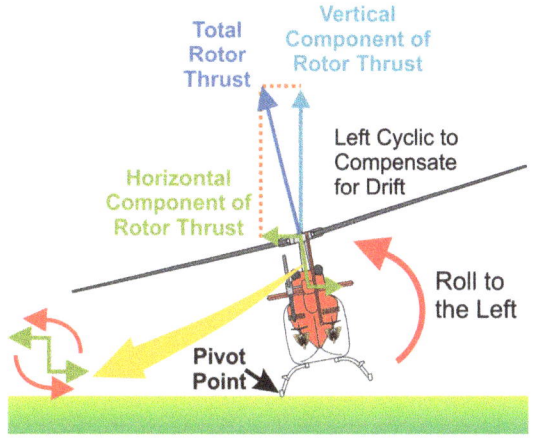

Dynamic Rollover

Not correctly managing the lift-off or landing can result in an uncontrollable rollover of the fuselage as it pivots about one of the skids.

This is referred to as Dynamic Rollover. Small two-bladed helicopters are particularly susceptible to Dynamic Rollover. This is partly due to the rotor head design, their skids being reasonably close together, and the rotor disc being relatively high, giving a high and narrow Centre of Gravity.

Any rolling sensation experienced on the lift-off or landing requires the pilot to **lower the collective immediately** and use some **opposite cyclic**.

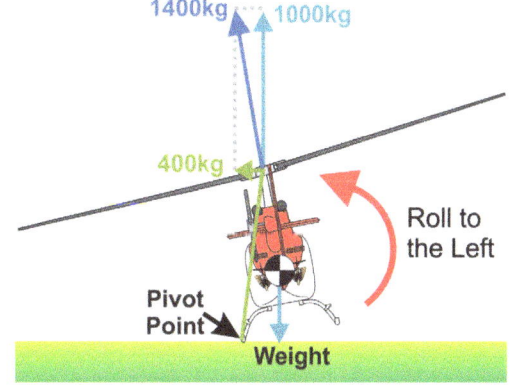

Dynamic Rollover

When a helicopter is lifting off the ground (or a surface), landing, or hovering and one skid or wheel touches a solid surface or obstacle, the helicopter may begin a rolling motion about the point of contact, which under certain circumstances cannot be controlled.

This is called Dynamic Rollover and can occur in any direction (left, right, forwards or backwards) and usually results in the helicopter's destruction.

Types of Rollover

To help understand Dynamic Rollover, let's consider the two types of Rollover:

- Static Rollover, where *Static* means lacking in movement, and
- Dynamic Rollover, where *Dynamic* means moving or a force that generates movement.

The difference is important to understand as what makes a static helicopter rollover is completely different to what may cause a helicopter to roll over when moving (Dynamic Rollover).

Static Rollover

To understand *Static Rollover*, consider a helicopter on the ground with the blades not turning; therefore, no rotor thrust is being produced. There is no movement.

For all intents and purposes, the helicopter is a lump of metal, plastic and fibreglass with the weight of all its parts bearing down towards the earth about its Centre of Gravity.

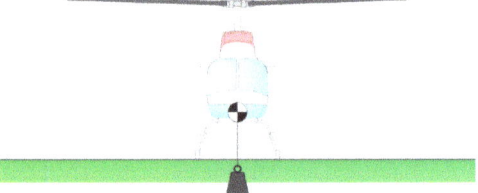

Pivot point

If a bottle jack (used to jack up a car when you have a flat tyre) was placed under one of the skids, the helicopter could be moved. One skid would be raised; the other would remain on the ground and act as a pivot point. The helicopter would roll around the pivot point.

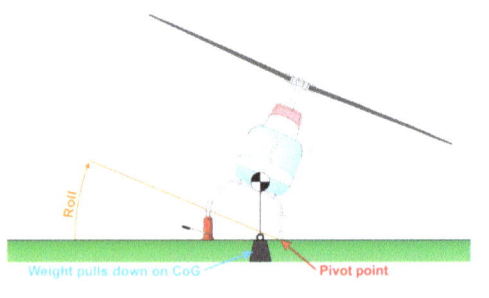

---

[1] http://s118.photobucket.com/user/ichris7/media/914632_m0w600h392e0t1v24248_II0A7085_zps85aabdff.jpg.html

Chapter 5 Lift-off, Transition and Landing

Pivot balance point

The interesting point to note here is that the helicopter will not roll over onto its side until the Centre of Gravity of the helicopter passes through an imaginary line (pivot balance datum) running vertically upwards from the pivot point.

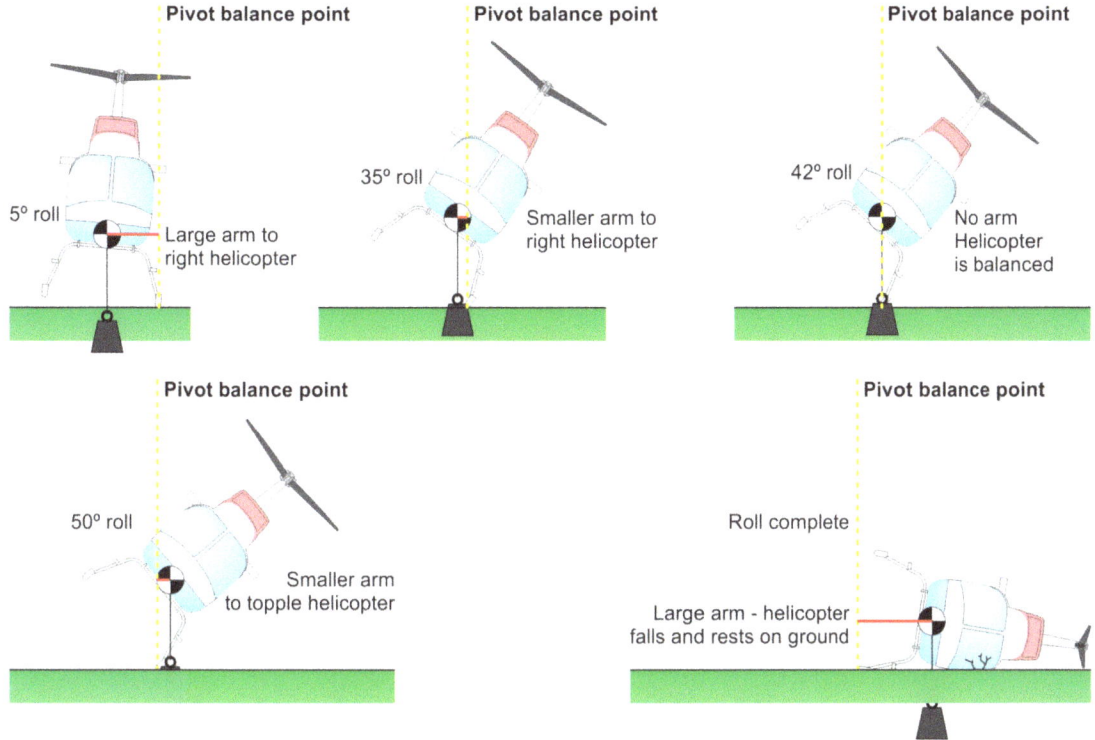

Restoring force

At any time before reaching the balance point, the bottle jack could be removed, and gravity, acting on the helicopter's weight, will cause the helicopter to fall back onto its skids.

The restoring force created by the moment of *weight x the distance from the pivot point,* will reduce the closer the *weight* is to the imaginary vertical line.

Once the helicopter's Centre of Gravity reaches the vertical line, it balances on one skid. Once the Centre of Gravity goes slightly over the vertical line, the helicopter will roll uncontrollably onto its side as gravity now acts on the weight and pulls it over, so the restoring force accelerates the roll.

Example

This whole process can be demonstrated by leaning back on a chair.

The closer the person on the chair gets to the balance point, the slower any recovery. Once past the balance point, the chair tips over.

Conclusion

The R22 and Bell 206 helicopters (as examples) can reach a static critical angle of approximately 42° before they reach the point where they will roll over. Any time before reaching that critical point, the helicopter will return back to level once the bottle jack is removed. Most other helicopters will have a similar angle.

What has been described above is Static Rollover, where the helicopter is not under power, and the rotor blades are not turning or producing any thrust.

So the questions to ask are:

- Why a helicopter experiencing Dynamic Rollover will reach a critical angle as low as 5° before control is lost and the helicopter rolls uncontrollably over onto its side?
- Why does it happen so fast?

Dynamic Rollover

Consider a helicopter lifting off the ground, and one skid becomes stuck.

The pilot continues to increase rotor thrust by raising the collective, and the helicopter tries to lift off but now commences a roll towards the stuck skid (the pivot point). The weight on the skids is less, so the restoring moment is getting less.

Assuming the pilot does not move the cyclic, the rotor disc will now be tilted slightly towards the stuck skid, giving a horizontal component of rotor thrust.

Increasing Rotor Thrust

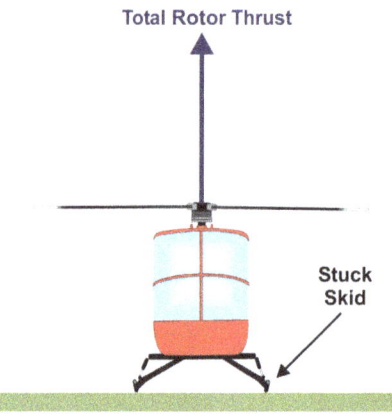

Roll Begins

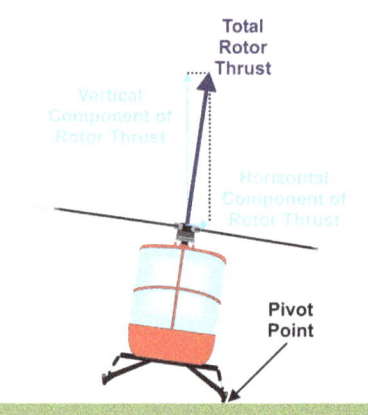

This horizontal component acts from the rotor head along the length of the arm to the pivot point. You have effectively created a moment or a lever trying to roll the helicopter over.

As it starts to roll, the helicopter experiences an increase in momentum and will want to continue to roll unless another force is applied to stop it.

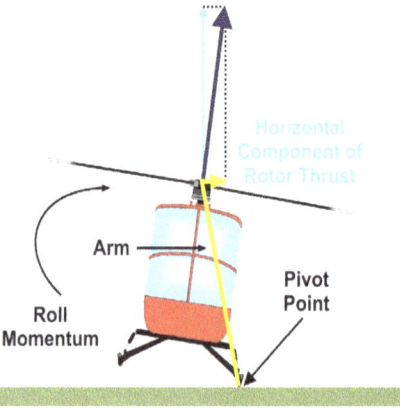

Chapter 5 Lift-off, Transition and Landing

Looking at the numbers

To look at the numbers, let's assume for our hypothetical helicopter:

- the cyclic is designed to handle no more than a roll rate of 10° per second or 10000 units of roll momentum, whichever comes first
- weighs 1000 kg all up weight, and
- the distance from the pivot point (skid) to the rotor head is 3 meters.

As we lift off, the rotor thrust would increase to almost 1000 kg of thrust (it wouldn't get off the ground otherwise).

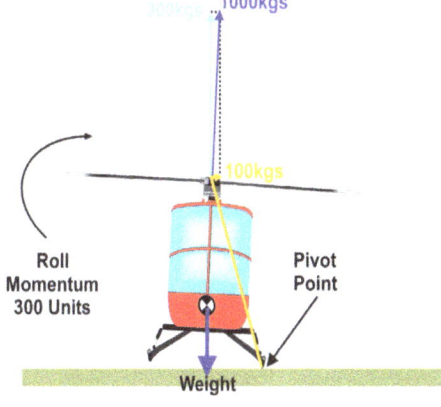

Mathematically it could be described as follows:

    Horizontal component of rotor thrust
x   **the arm to the pivot point**
x   **the roll rate squared**
=   **units of roll momentum being experienced by the helicopter**

Therefore:

    100 kg
x   **3 m**
x   **$1°^2$ of roll per second**
=   **300 units of roll momentum (less than 10,000 units)**

At this point, the roll is controllable with cyclic, and the pilot is not in danger.

Let's say the pilot ignores the early warning signs and continues to pull the collective without arresting the roll.

The helicopter now rolls over further, its rate of roll increases, and everything starts to happen very quickly.

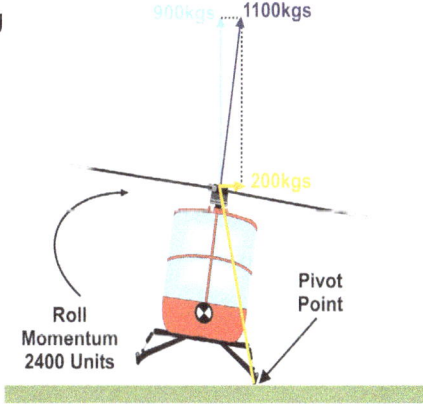

Mathematically it would look something like this:

    200 kg
x   **3 m**
x   **$2°^2$ of roll per second**
=   **2400 units of roll momentum**

The roll momentum has not just doubled; it has **increased by a factor of eight (8)**.

At this point, the helicopter is well on its way; it may still be controllable if the helicopter's weight is sufficient to overcome the roll momentum, but it is definitely getting scary.

The pilot needs to lower the collective and use some opposite cyclic and hope that the reorientation and eradication of some rotor thrust and the weight of the helicopter will allow the helicopter to fall back level to the ground.

In most cases, though, the pilot starts to panic a little and pulls the collective up more, hoping to get the helicopter off the ground and free the stuck skid.

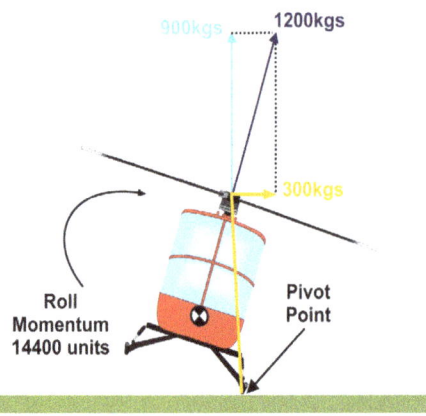

Mathematically it would look something like this:

        300 kg
x   3 m
x $4°^2$ of roll per second
=   14,400 units of roll momentum

Conclusion

The helicopter is definitely a goner now, and there is no recovery. The helicopter is rolling over due to the excessive roll momentum even though it has only rolled a total say 7° from level.

The pilot has experienced Dynamic Rollover. As illustrated, the units of roll momentum were increasing at a rapid exponential rate. Just wait till you experience it.

Cyclic Limits and Dynamic Rollover

In free flight

In normal free flight (both skids off the ground), the pilot can use the cyclic to counter any roll. Using the cyclic produces a couple between the horizontal component of rotor thrust and the helicopter's Centre of Gravity. The effectiveness of the cyclic in controlling the rate of the roll is limited by the:

- amount of total disk flapping, and
- roll rate of the helicopter.

With any roll outside of the design limits, using the cyclic to control that roll will have limited or no effect.

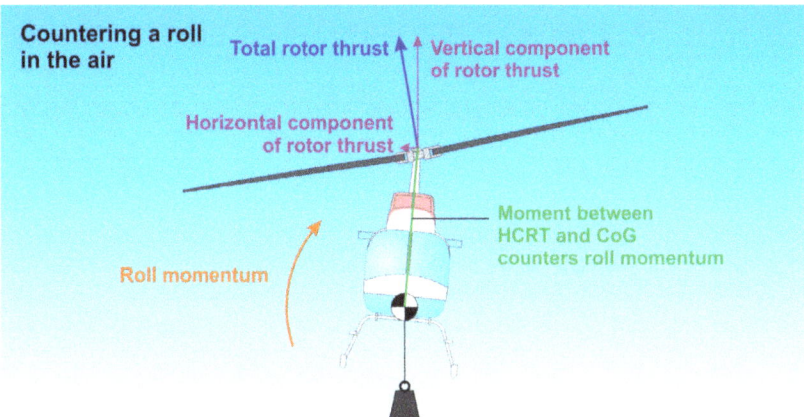

With a stuck skid

Additionally, the designer has created the cyclic authority based on the required roll rate being about the aircraft's Centre of Gravity (CofG).

With a skid stuck on the ground, the pivot point is now the centre of the roll (not the helicopter's CofG), and the cyclic's authority to counter this roll will be much reduced in terms of its ability to apply a rotational force (torque) around this axis to stop any undesired roll.

For this reason, the cyclic will have a limited effect on countering a Dynamic Rollover condition.

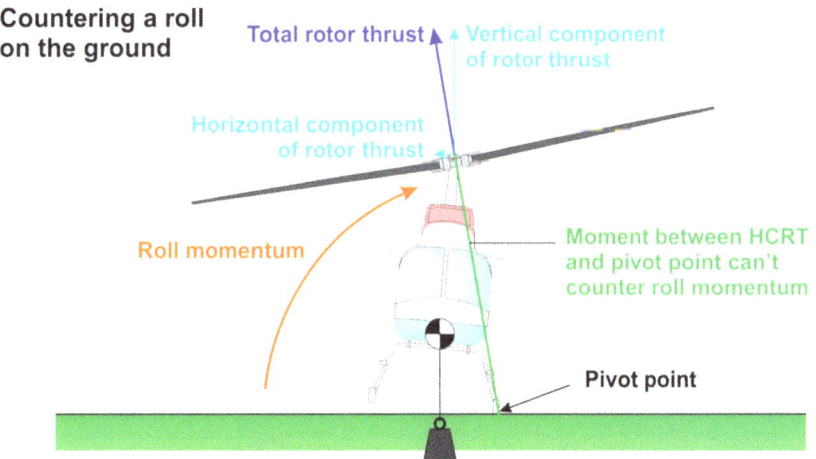

Rotor head designs

Different rotor head designs give varying degrees of response to the pilot's cyclic inputs and how the fuselage responds to these changes, so pilots will need to know how their particular helicopter responds to inputs and fly it accordingly. This also means some rotor head designs are better suited for certain tasks.

Mike Becker's Helicopter Handbook

Scenarios that May Contribute to Dynamic Rollover

Scenarios   Several common scenarios that may contribute to Dynamic Rollover include:

- too much cyclic when lifting off or landing on sloping ground
- experiencing tail rotor drift or roll on lift-off or landing
- lifting off close to an obstacle, and
- when emplaning and deplaning.

Each of these scenarios is illustrated below.

Too much cyclic   Overuse of the cyclic when lifting off or landing on sloping ground. The roll can be amplified either way, depending on the pilot's use of cyclic and collective.

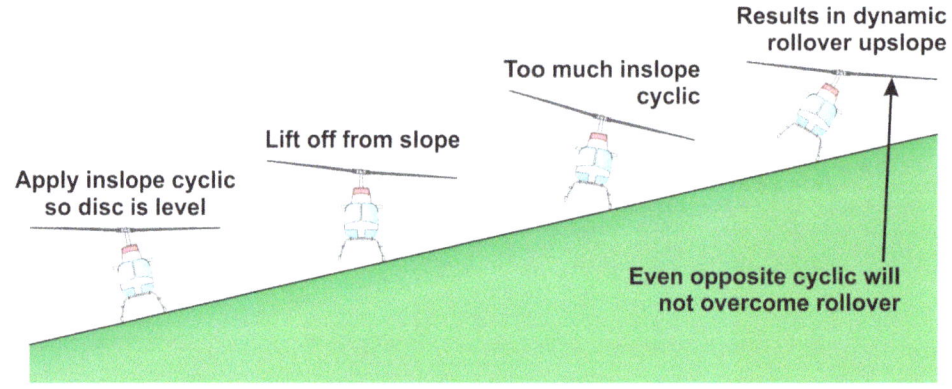

Tail rotor drift or roll   Experiencing tail rotor drift or roll on lift-off or landing and not compensating with the cyclic allowing the helicopter to drift inadvertently.

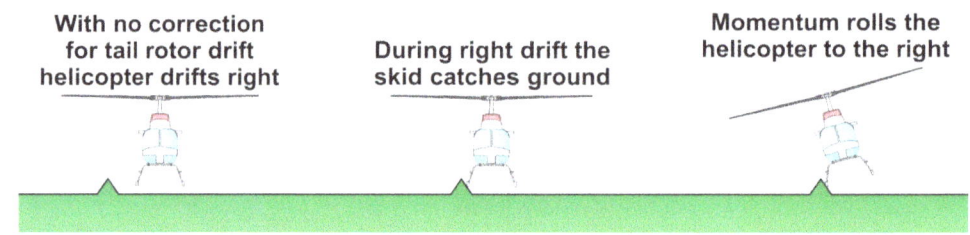

Near obstacles   Lifting off close to an obstacle, such as a refuelling drum, and not correcting for any sideways drift causing the helicopter to bump into the obstacle and produce a roll.

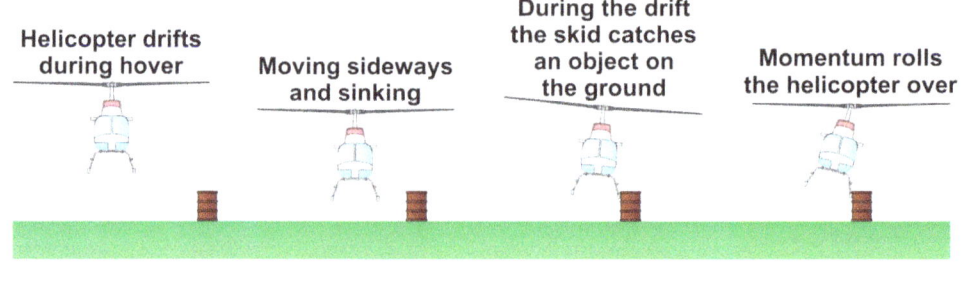

Chapter 5 Lift-off, Transition and Landing

Emplaning and deplaning	When emplaning and deplaning, the pilot can experience a Dynamic Rollover if the weight on the skid is greater than the cyclic can manage.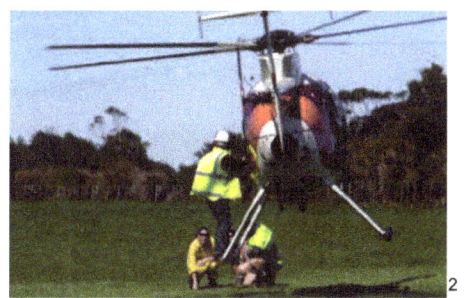

This requires training, and the pilot and crew emplaning and deplaning having knowledge and understanding of the helicopter's limits, as some rotor head designs are better than others for these tasks.

For example: When emplaning and deplaning using a fully articulated rotor head, such as in an MD500, the pilot will have much more control compared to using a teetering head, such as in a Bell 206. |

Preventing Dynamic Rollover

Recognise the risk	Understand that every lift-off and landing and every time a piece of the helicopter comes into contact with a solid object, Dynamic Rollover is a possibility, and careful management of the helicopter is required.
Precautions	Whenever lifting off and landing, do so slowly to feel the helicopter lift-off and land onto the surface.
Situational awareness	When taxiing, it is imperative the pilot and crew maintain situational awareness, as making contact with a solid object while moving means momentum will speed up any rollover process and make any recovery difficult, if not impossible. 
Slopes	If on a slope, operate within your capability limits (and design limits of the helicopter) and know how wind affects your helicopter. Be selective in slope selection, and do not overuse the cyclic.
Hovering and taxiing	If hovering or taxiing, do so at 5 ft rather than 3 ft AGL to reduce the possibility of inadvertently touching a solid object. If limited on power, minimise the hovering manoeuvres and keep the skids aligned with the direction of travel at a low hover.

Low hover 3-foot hover   **Higher 5-foot hover**

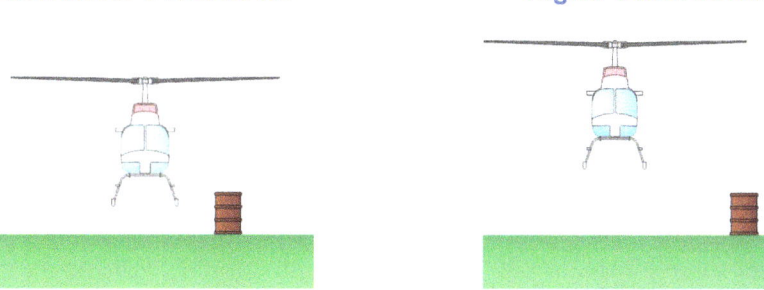

---

[2] Image courtesy of Precision Helicopters Limited / http://www.precisionhelicopters.com/personnel-training.php
[3] http://www.aerosieger.de/images/news_picupload/pic_sid5288-0-norm.jpg

Mike Becker's Helicopter Handbook

Recover from Dynamic Rollover

Recovery

If at any time a rolling sensation or a skid or wheel becomes stuck and the pilot feels the helicopter wanting to pivot about a point:

Lower the collective

to reduce the rotor thrust and allow the helicopter's weight to bring the skids back onto the ground.

The use of some opposite cyclic is appropriate, and the pilot also needs to keep straight with the pedals; however, these may have limited to no effect on countering the roll. As a general rule:

the faster the roll rate, the faster the collective needs to be lowered

---

Ground Resonance

What is it?

Ground resonance is *"a vibration of large amplitude resulting from a forced or self-induced vibration to a mass in contact with or resting upon the ground"* (surface).

Ground resonance usually only applies to 3 or more bladed rotor systems that are free to lead and lag about the dragging hinge, a Starflex coupling, or similar.

What causes ground resonance

Ground resonance is caused by the Centre of Gravity of the disc being out of alignment with the mast, which can be caused by:

- blades of unequal weight or balance
- faulty drag dampers
- faulty tracking
- incorrect oleo pressures and/or tyre pressures
- landings on hard flat surfaces
- smooth landings without lowering the collective enough to get weight on the oleos
- hard uneven landings, and
- pilot over controlling and stick stirring of the cyclic.

Effect of ground resonance

As a result, the blades tend to become unaligned with each other, shifting the Centre of Gravity, as shown in the diagram below.

When touching down, the pilot will recognise ground resonance from an increasing rocking motion or oscillation of the fuselage. If corrective action is not taken, the amplitude can increase rapidly to the point where it will be uncontrollable, and the helicopter will shake to pieces or rollover. This resonance is not so noticeable while in the air; however, when in ground contact, it can start slowly or incredibly quickly and can shake the helicopter to pieces.

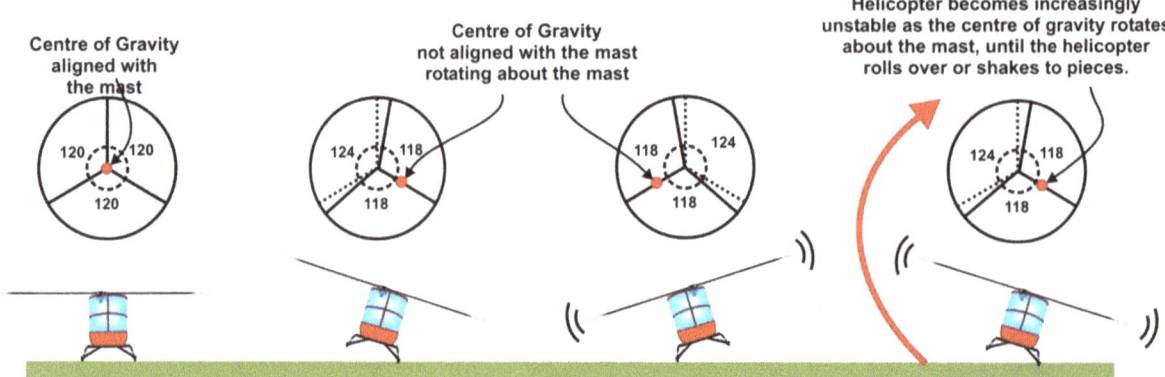

Chapter 5 Lift-off, Transition and Landing

Recovery

Usually, you will experience ground resonance **on initial touchdown**.

1. If experiencing ground resonance on initial touchdown, **immediately lift off** to the hover and try landing again.
2. If ground resonance is still encountered, go for a fly and load the disc by doing some steep turns to try and get the blades back in proper alignment.
3. If this does not help, you may want to land on a surface that will absorb some of the vibrations, such as soft grass, sand or mud. Avoid hard surfaces, such as concrete, bitumen or hard ground, that assist the vibrations.

If ground resonance is still experienced and you must land, conduct a simulated engine failure at the hover. This will quickly cause the Rotor RPM to decay. Once the helicopter has touched the ground, lower the collective fully and hang on. If there is a rotor brake installed, apply it.

Positive landing

Experienced pilots of 3-bladed systems or more prefer a positive landing; once the skids touch the ground, they continue to lower the collective another 2-3 inches and then pause; holding the helicopter light on the skids can cause ground resonance.

Lowering all the collective too quickly can cause RRPM to decay (the correlator will reduce the RRPM) and make a lift-off if resonance is encountered more difficult.

Wind

Wind

Wind will affect the helicopter on the ground in the same manner as in the hover. The difference is the pilot cannot feel the effects of the wind while the helicopter's weight is holding it securely on the ground.

Wind on lift-off

Before lift-off, the pilot must know where the wind is coming from and position the rotor disc into the wind with the cyclic. This will be an educated guess until the helicopter gets light on the skids, and the pilot starts to feel the effects of the wind and compensates accordingly.

When on the ground, the disc should always be positioned level with the ground as this is the safest position when allowing crew and passengers to embark and disembark when the rotor blades are turning.

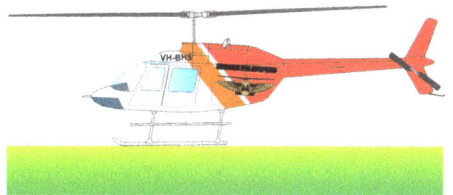

Adjusting for wind

If the wind is on the nose, the pilot can position the cyclic forward into the wind in anticipation of the lift-off and estimate what will help the helicopter maintain a constant ground position as it lifts vertically.

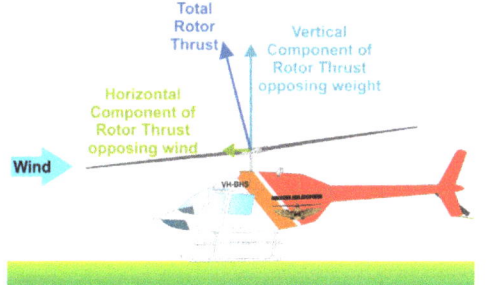

The same technique must be used for wind blowing in any other direction (from the side or rear).

Centre of Gravity Effects on the Cyclic

A helicopter's Centre of Gravity (CofG) determines the helicopter's attitude at the hover longitudinally and laterally. Depending on the rotor head design, the rotor disc may move independently of the fuselage (teetering and articulated heads) or be directly related to the fuselage (rigid rotor heads). Loading the helicopter correctly and determining the All-Up Weight (AUW) and Centre of Gravity (CofG) before lift-off is essential. Each helicopter will have a specific CofG range and AUW limit. Operating outside of these limits can result in a loss of control. If the CofG seems wrong during Lift-off and the cyclic is being moved to an unusual position, then the safest course of action is to abort the lift-off immediately by lowering the collective and re-evaluating before the next lift-off.

The effects of weight distribution on the cyclic when lifting off and coming to a hover are described below. For ease of demonstration, we will only describe a two-bladed teetering system where the fuselage and rotor head can work independently.

To calculate the actual CofG for a particular flight, refer to the Weight and Balance chapter in this book.

Longitudinal Centre of Gravity

Longitudinal CofG relates to the fore and aft movement about the lateral axis. Think of the length of the fuselage as a giant see-saw with the balance point in the centre.

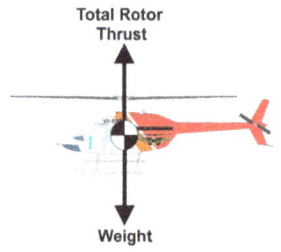

Central Centre of Gravity

If the helicopter has a CofG within the design limits of the helicopter, then as the pilot lifts off the ground to come to hover, there is little requirement to adjust the cyclic, and the helicopter should be easy to hover.

**Aft Centre of Gravity
not corrected with cyclic**

If the helicopter has an aft CofG outside the design limits, then as the pilot lifts off the ground to come to the hover, and without correcting with cyclic, the helicopter will want to hover nose high, move backwards, and sink towards the ground.

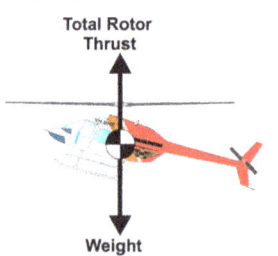

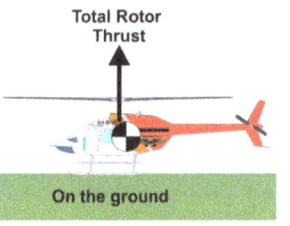

Chapter 5 Lift-off, Transition and Landing

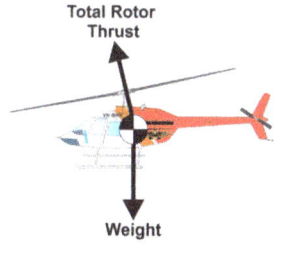

Aft Centre of Gravity corrected with cyclic

If the helicopter has an aft CofG outside the design limits, then as the pilot lifts off, the pilot uses forward cyclic to correct for the movement illustrated above. However, in an attempt to hold ground position, if the CofG is outside of the design limits, the pilot may reach the forward longitudinal limit of the cyclic and not be able to hold the ground position.

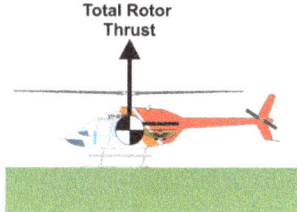

Additionally, the rotor head could also then bump the mast with the extremely tilted rotor head. This can lead to a loss of control.

Lateral Centre of Gravity

Lateral CofG relates to the left and right movement about the longitudinal axis. Think of the pilot sitting in the helicopter, and the helicopter wants to roll left or right around an imaginary line running the length of the helicopter.

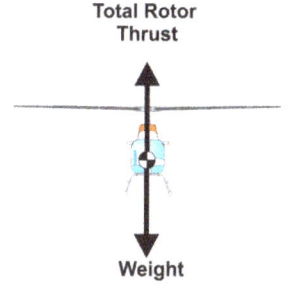

Central Centre of Gravity

If the helicopter has a CofG within the design limits, then as the pilot lifts off the ground to come to the hover, there is little requirement to adjust the cyclic, and the helicopter should be easy to hover.

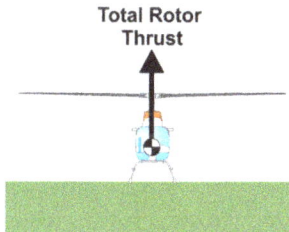

Right Centre of Gravity not corrected with cyclic

As the pilot lifts off the ground to come to the hover, without correcting with the cyclic, the helicopter will want to roll about the skid where the weight is concentrated (roll left in this case).

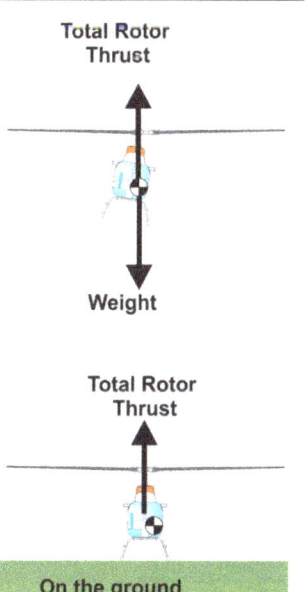

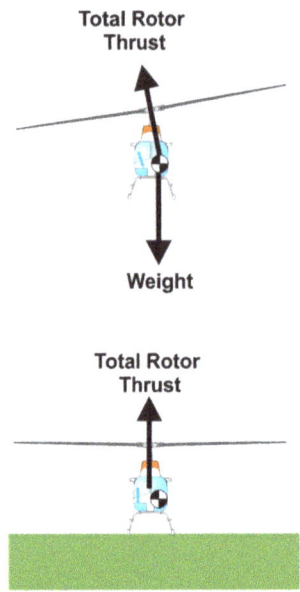

Right Centre of Gravity corrected with cyclic

If the CofG is within limits and the pilot corrects with opposite cyclic, the helicopter will hang at the hover left skid low.

If the CofG is outside of the limit, then the pilot could reach the cyclic lateral limit and not be able to correct the roll. This can result in the helicopter rolling over.

Slope

Every lift-off and landing should be considered a slope lift-off and landing. This will put the pilot in the right frame of mind to control the helicopter. Sometimes a combination of cross slope and up or down slope must be accepted. The same knowledge, skills and techniques used to lift-off and land on flat ground are used for slope operations. The most significant difference when operating on slopes is there is a greater risk of rolling over or reaching a cyclic limit due to the pilot mishandling the flight controls.

For this reason, a high level of skill and attention is required to lift-off and land from a sloping surface safely and successfully. On short finals and when approaching the hover, the pilot should assess the landing area and consider the slope.

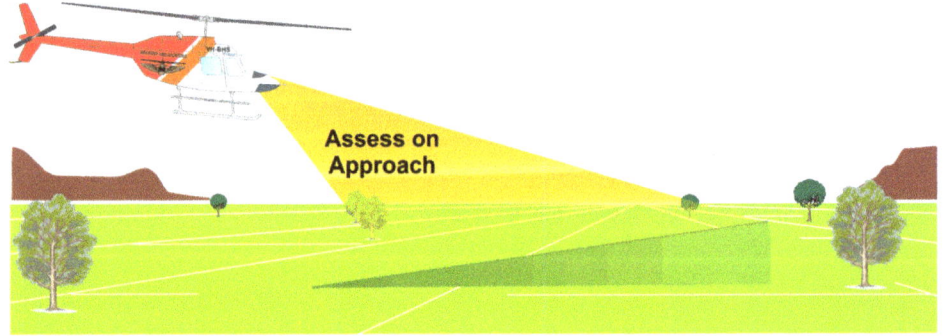

Chapter 5 Lift-off, Transition and Landing

Effect on Rotor thrust

When operating on a slope, additional forces are acting on the helicopter that the pilot needs to control. The weight of gravity acting on the helicopter is split into the:

- vertical component pulling it down towards the centre of the earth, and
- horizontal component trying to make it slide down the slope.

To counter, the pilot will use opposite cyclic, which will split the Rotor Thrust vector into the:

- vertical component opposing the weight, and
- horizontal component opposing the slide.

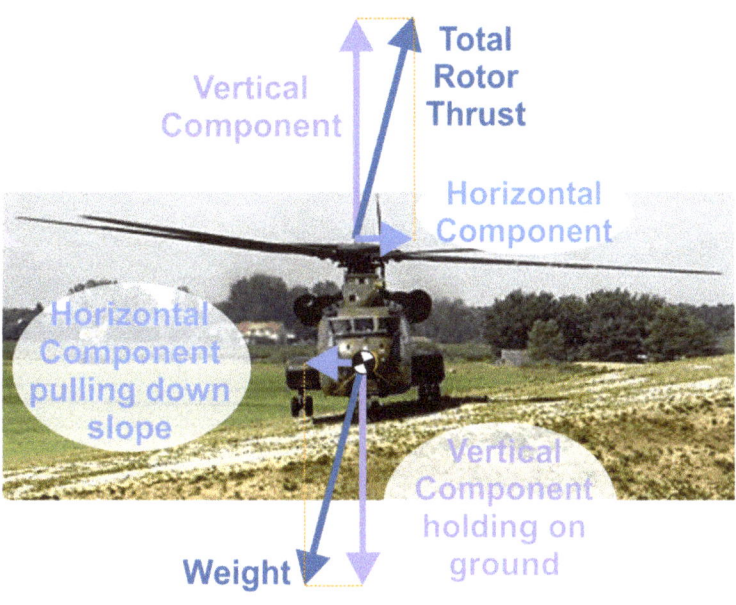

Dynamic Rollover

Just like a normal lift-off or landing from a flat, even surface, the risk of not managing the lift-off or landing correctly can be an uncontrollable rollover of the fuselage as it pivots about one of the skids.

When lifting off or landing on a slope and tilting the rotor disc, the pilot is **intentionally inducing the first stage of Dynamic Rollover,** which is to create a pivot point.

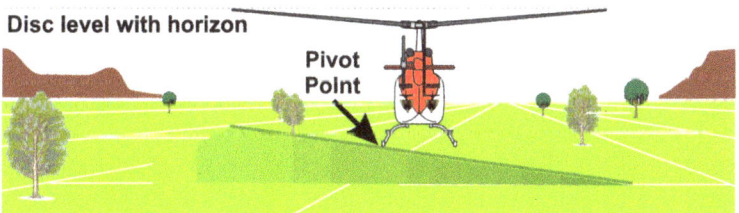

The pilot must be precise on the flight controls to maintain control.

Cyclic limits

The cyclic can only move the disc within a small range. Additionally, other design limitations may restrict the amount of disc tilt (for example, a rigid rotor system will actually be bending the mast when the fuselage is on the ground and the disc is being tilted.) These design limits will limit the degree of slope that can be landed and lifted from.

The left and right, up and down limits may all vary and are a limit the pilot needs to commit to memory even though gauging any slope is an estimation only by the pilot at the time.

These limits are found in the Limitations Section of the RFM.

Up and down slope

Up and downslope limits are designed to protect the tail.

Upslope   **Downslope**

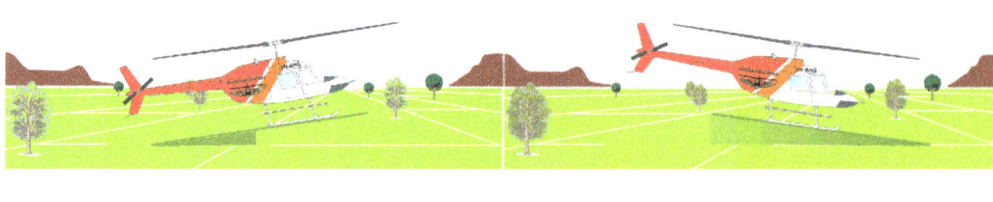

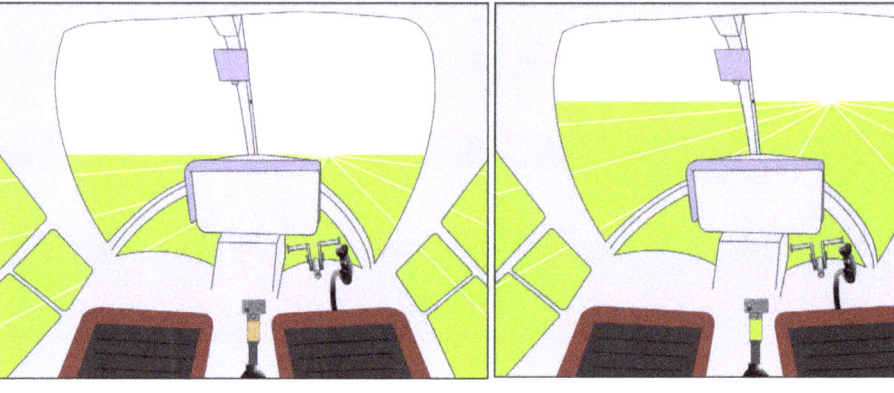

Cross slope

Cross-slope lift-offs and landings are preferred and usually have a larger limit (compared to up or downslope), although the limits left and right can be different.

This is due to tail rotor roll. In anti-clockwise rotor systems viewed from behind, the helicopter will already want to hang left skid low. This allows for a greater slope descending from right to left as the cyclic will have more travel. The opposite would be true of helicopters with clockwise rotating systems.

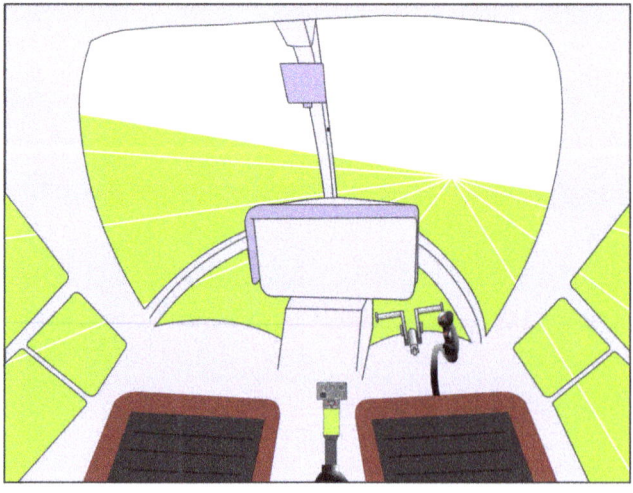

Chapter 5 Lift-off, Transition and Landing

Main and tail rotor limits	When lifting off or landing from a slope, the ground will always rise in one direction and fall in the other. How much slope there is can lead to additional hazards. The main and tail rotor blades can get closer to touching the ground, which will cause damage before the pilot recognises the danger.	
Tail upslope	When the tail is up-slope, the pilot cannot see the tail rotor, so when lifting off or landing with the tail up the slope, there is a greater risk that the tail rotor can come into contact with the ground.	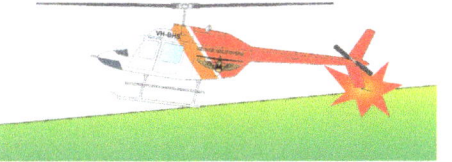
	The pilot may also falsely perceive that the helicopter is much further away from the ground, as the pilot is looking out the front where the ground is falling away.	
Tail downslope	When lifting off or landing with the nose up the slope, the CoG can cause the helicopter to tip backwards onto its tail, causing the tail rotor to come into contact with the ground. Further, overuse of cyclic could cause the rotor head to bump the mast.	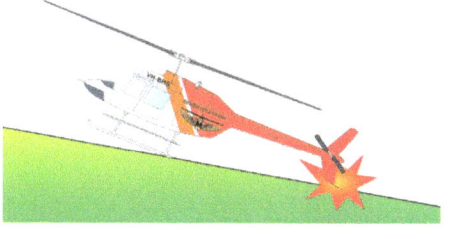
Flight Control Obstruction	When landing on a slope, the flight controls, and in particular the cyclic, may be getting close to one of its limits (forwards, backwards, sideways).	
	If the pilot is wearing a kneeboard, particularly on the left leg, it can get caught between the cyclic and the collective. Pilots who are big with long legs may also have some difficulty as the leg itself may come between the cyclic and the collective.	
	Before lifting off and landing on a slope, the crew are to conduct a brief considering the cockpit ergonomics and should try to organise themselves so that a slope can be managed.	
Manoeuvring near slopes	If having to manoeuvre while at the hover to position the helicopter near a slope, then **always** protect the tail and make any hovering turns so that the tail moves downslope (away from the slope) and not into the slope.	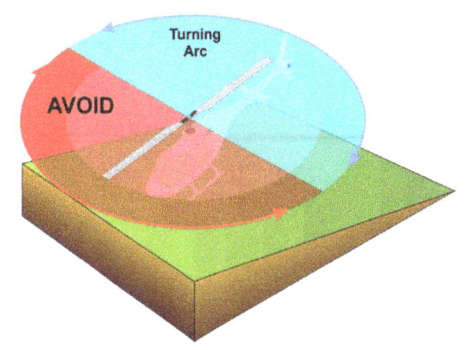

Recognising the slope limit

Regardless of any published slope limits in the RFM, the actual limit on the day may vary and oftentimes be **less** than that published. This could be because of the pilot's capability level, wind, CofG loading or other factors. The pilot needs to recognise the limit on the day when conducting a slope operation.

The first indication of a slope is the pilot's assessment. If it appears too steep before landing, then find another landing area. Other indications may include:

- The cyclic reaches its physical limit and cannot be moved any further.
- Another sign that the cyclic is getting close to its physical limit is the rotor coming into contact with the rotor mast and 'bumping' it. This mast bumping can be heard and felt. In normal circumstances, this will not be damaging at this point but is a warning that a limit has been reached, even if the pilot feels there is more cyclic movement to be had.
- The helicopter starts to slide down the hill regardless of the cyclic input. This is most common when landing nose or tail into the slope.
- With both skids on the ground, the helicopter starts to roll in the downhill direction.
- The tail stinger touches the ground. The pilot will feel this as a slight bump from the rear. This is common to a tail up-slope landing but should be avoided.
- If the power required starts to increase more than expected, this indicates that the slope is too steep, wind is having an adverse effect, and the ground effect is lost down the slope.

If a limit is recognised, the landing shall be aborted and another site selected.

HEFFR Check: Pre-Liftoff and Pre-Landing

Pre-liftoff checks Before lifting off, the crew shall run through the *"Pre-liftoff checks"* as follows:

	ITEM	ACTION
1	**H** *Hatches*	Secured.
2	*Harness*	On and Secured.
3	*Homework*	- Administration completed, Documentation in hand - SAR noted (Search and Rescue Time) - Fuel Time On Ground (FOG) calculated
4	*Helipad*	- Clear and taxi brief given. - Slope

5	**E** *Engine and instruments*	- Warning lights are out
		- Ts and Ps are in the green (normal)
		- Throttle FULL and RPM set
		- Instruments are set. This may include (if installed):
		  - AIs erect and no OFF flags
		  - Altimeters set and cross-checked ($\pm$100 ft for VFR. $\pm$60 ft for IFR.)
		  - Assigned altitude indicators and RAD ALT SET
		  - HSI and Compass aligned within 10 degrees

6	**F** *Fuel*	Amount of fuel stated. Fuel Time on the Ground (FOG) noted.
7	**F** *Frictions and locks*	OFF.
8	*Force trims*	SET as required.
9	**R** *Radios and Radiation*	Radios, navigation aids and transponder SET. ATIS obtained and Radio call made
10	*Risks*	Threat Error Management. Threats and errors are discussed among the crew, and mitigating strategies are in place while still on the ground. This is to assist and speed up the TEM discussion between the crew when at the hover before departure and conducting the Pre Departure Brief (PWPTEM).

Hover Checks

Hover checks   Once at the hover, the crew shall run through the *"Hover checks"* as follows:

	ITEM	ACTION
1	*Controls*	**Feel normal** *(if not, then land)*
2	*C of G*	**Feels normal** *(if not, then land)*
3	*Warning lights*	**OUT** *(if not, then land)*
4	*Ts and Ps*	**In the green or normal** *(if not, then land)*
5	*RRPM*	**Top of the Green** *(if not, then fix it)*
6	*Power required*	**Power required to hover is _____ giving a margin of _____** (A)
7	*Pedal Position*	**Noted** *(as a reference for later on in an emergency)*

(A) **NOTE: Power Margin**

The power margin is the difference between the power required to hover and the full power available. This is discussed in detail in the chapter on Limited Power operations.

For example, in a piston helicopter hovering at 22" MAP but a maximum available of 25" MAP, the margin would be 3" MAP.

In a turbine helicopter hovering at 75% Tq but a maximum available of 100% Tq, the margin would be 25% Tq.

Chapter 5 Lift-off, Transition and Landing

PWPTEM Check: Pre Departure Brief

Pre-take-off checks

The Pre-Departure Brief is a general departure check conducted before take-off, using the acronym "**PWPTEM**", pronounced "*Pow-tem*" or abbreviated to simply **PWP**

Because a transition from the hover into forward flight is part of the take-off procedure, it is recommended to complete the Pre-Departure Brief to help develop a positive habit.

	ITEM	ACTION
1	**P** Power	We have a power margin of _____.
2	**W** Wind	The wind is coming from xxx direction at xx knots, giving us an into wind departure.
3	**P** Plan	My plan is to conduct a_____(type of departure) and climb straight ahead to XXX feet before conducting a left/right hand turn and setting heading for XXX on a climb to XXX. (Pilot will need to state what they are doing here). Obstacles include _____ Last safe point of hover is _____ Errors include_____ Any questions?
4	**T** Threats	Any wires in the area. Any obstacles in the area. Any identified threats relevant to the take-off and flight.
5	**E** Errors	Brief any crew, air traffic, operations, internal, or external errors that may be relevant to the flight.
6	**M** Management	Questions: Ask the crew for their input and cover any additional TEM points.

Short Finals ARP Check: Pre-Landing

Short final check   The Short Finals Check is done before committing to the landing, using the acronym *"ARP"*.

The transition from forward flight back to the hover is the same procedure used on short finals when making an approach to land. Therefore, the crew should complete the Short Finals check to help develop a positive habit.

	ITEM	ACTION
1	**A** *Airspeed over ground speed*	Check that the airspeed is greater than the groundspeed, confirming that the wind is on the nose.
		If it is not, then go around.
		This can be done by either:
		■ Referencing the ground speed displayed on the GPS when compared to the IAS on the ASI, or
		■ Referencing the perceived speed of the ground as it passes underneath you compared to the IAS on the ASI. This is based on your previous experience.
2	**R** *Rate of Descent (ROD)*(A)	Confirm that the rate of descent (ROD) is under control.
		If the ROD is not in control, then go around.
3	**P** *Power in hand*(B)	Power in hand. Based on the ROD and the power margin, confirm there is sufficient power available to come to the hover.
		If not, then go around.
4	**P** *Pad*	The helipad is clear, and you have identified the termination point.

(A)   **NOTE: Rate of Descent Under Control**

Rate of Descent (ROD) under control can be indicated on the Vertical Speed Indicator (VSI). However, it is really a knowledge and feeling by the pilot that the power is increasing, and you feel *you are taking* the helicopter to the landing area rather than the *helicopter is taking you*.

If you feel the helicopter is too fast and you are still lowering the collective, you are not yet in control of the ROD. If you feel the helicopter is relatively slow and you are raising the collective and using forward cyclic to advance to the landing area, then you are more in control of the ROD.

For a more detailed description, refer to the Circuits Chapter in this book.

Chapter 5 Lift-off, Transition and Landing

(B) **NOTE: Power in Hand**

Piston example:

With hover power of 23" MAP and power available of 25" MAP with a ROD of 500 ft per minute at 30 kts and you are pulling:

- 16" MAP, there is sufficient power in hand to continue the approach and terminate at the hover.
- 23" MAP, then there is not sufficient power in hand to continue the approach and terminate at the hover and you need to conduct a go-around early.

Turbine example:

With hover power of 90% Tq and power available of 100% Tq with a ROD of 500 ft per minute at 30 kts and you are pulling:

- 40% Tq, there is sufficient power in hand to continue the approach and terminate at the hover.
- 90% Tq then there is not sufficient power in hand to continue the approach and terminate at the hover and you need to conduct a go-around early.

For a more detailed description, refer to the Circuits Chapter in this book.

Transitions

Transitions

A transition is that phase of flight where the helicopter moves (or transitions) from:

- the hover to forward flight, and
- forward flight back to the hover.

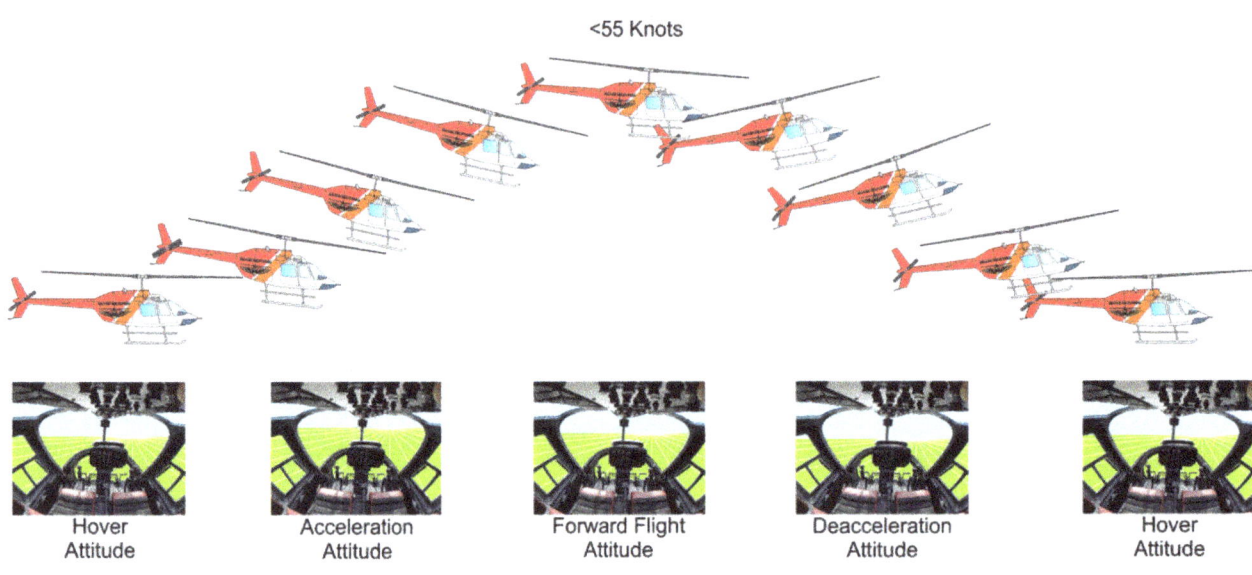

The exercise in transitions is used to help the pilot develop the control skills required to recognise the changes in power and attitude caused by the changing aerodynamics acting on the rotor head and fuselage. It allows the pilot to learn the skills to manage the helicopter while making the take-off and approach to land as smooth as possible.

Translational lift

As the helicopter starts to move forward from the hover into forward flight, two distinct changes occur:

- the helicopter will start to lose the ground cushion, and
- the wind will increase over the disc, due to the forward movement.

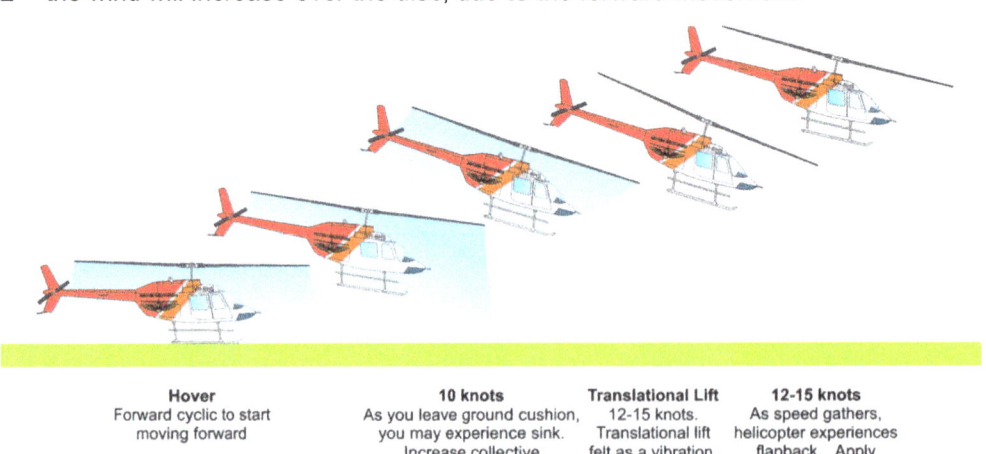

These two changes culminate at one point when the helicopter reaches an airspeed of 12-15 kts. At this point, the rotor disc will pass through Effective Translational Lift (ETL). The helicopter will shudder, and the angle of attack on the rotor blades will increase, giving the pilot some additional rotor thrust for the same power setting. At this point, the helicopter 'feels' like it wants to fly.

Chapter 5 Lift-off, Transition and Landing

Forward cyclic to counter flapback	If the pilot at that point does not respond with more forward cyclic, the disc will flapback relative to the oncoming wind, and the helicopter will lose translational lift and return to the hover.
Power	Also, prior to reaching 12-15 kts IAS, the power requirements will increase as the helicopter loses ground cushion but does not yet have the beneficial effects of the wind.
Difficult phase	This is the most challenging phase of the take-off to master, as there are no instruments to tell the pilot what to do. Instead, it must be experienced, and the pilot will have to gain a *feel* about compensating for the commencement of translational lift and the subsequent effects on the rotor disk due to the increasing wind effect.
From forward flight to hover	The same process happens in reverse when a helicopter moves from forward flight back to the hover. As the helicopter slows down, it loses translational lift and starts to enter the ground cushion. At the point of losing translational lift, the power requirements increase, causing the disk to again flapback relative to the oncoming wind requiring forward cyclic. The pilot has to be able to manage the transition so the helicopter keeps moving forwards and down towards the nominated hover point.

Transverse Flow Effect or Inflow Roll

What is it?

Transverse Flow Effect, also known as Inflow Roll, is the change to the induced flow which occurs when translational lift is not uniform over the whole disc area.

Air that is moving horizontally towards the disc will cause the:

- greatest reduction in induced flow at the front of the disc, and
- smallest reduction at the rear of the disc

This is because the action of the rotors is continuously pulling down air passing across the top of the disc and will be affected more than the clean air entering the front of the disc, and the rear blade is higher than the front blade.

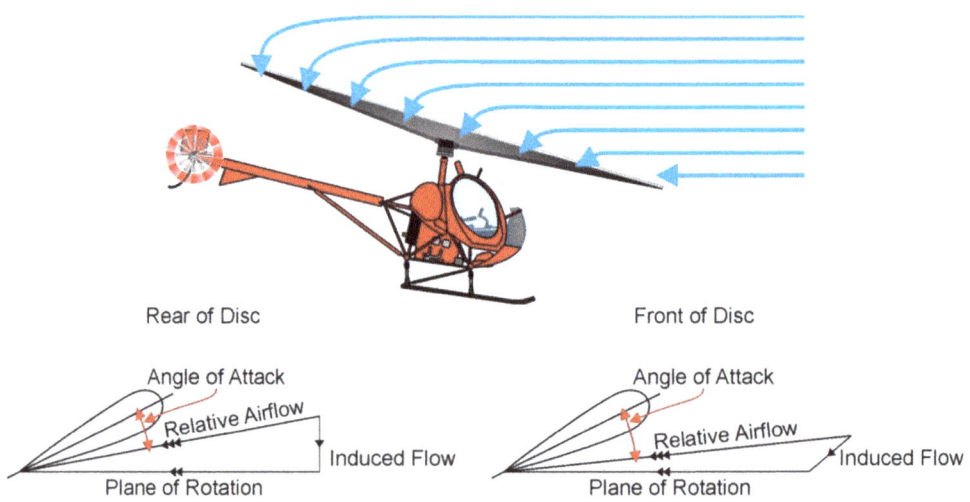

At these different angles of attack, more rotor thrust is produced toward the front of the disc than toward the rear. Due to gyroscopic precession, this change in angle of attack and, therefore, lift, will be felt 90 degrees later in the direction of rotation with the final effect being a roll of the helicopter to the advancing blade side (to the right in anti-clockwise rotating systems).

At lower airspeeds

The Transverse Flow Effect (inflow roll) is more noticeable at lower airspeeds around 15-30 kts and needs to be corrected for using some cyclic. The effect is barely noticeable, and the pilot will tend to automatically compensate for the change in attitude if looking outside towards the horizon.

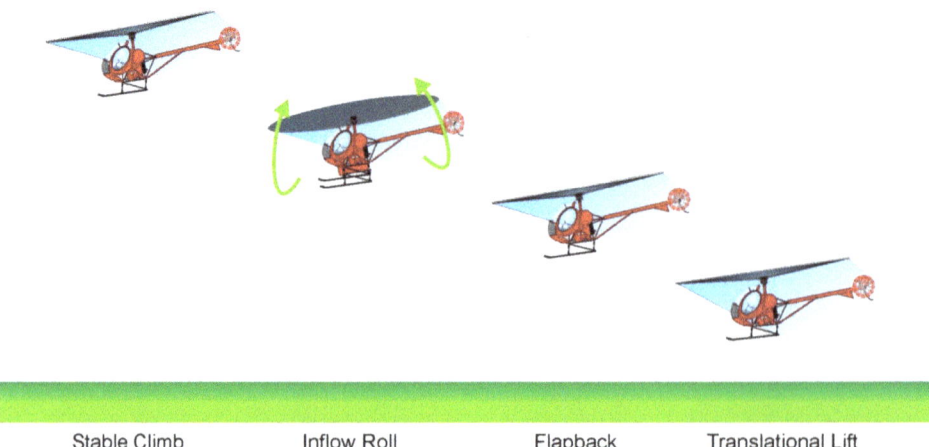

Stable Climb	Inflow Roll	Flapback	Translational Lift
Climb speed	15 - 20 knots Inflow roll to the advancing blade side	As speed increases helicopter experiences flapback. Apply more forward cyclic.	10 - 15 knots Translational lift felt as a vibration.

Air Exercises: Lift-off, Transition and Landing

Introduction

The air exercises will be divided into the following segments:

- Lift-off and Landing from a flat, level surface
- Transitions
- Lift-off and landing from a sloping, uneven surface

Airmanship

- Memorise the pre-liftoff, hover, pre-departure and short finals checks
- Observe handover procedure
- Maintain a good lookout and situational awareness
- Protect the tail by maintaining a minimum of 5 ft AGL skid height hover
- Maintain a safe distance from obstacles (minimum 10 metres)
- Monitor temperatures and pressures of the engine and transmission
- Memorise some of the aircraft power and wind limits
- Memorise the slope limits for the particular helicopter type
- Manage power and RRPM
- Talk to the instructor and announce what you are doing
- Be aware of the effect of the rotor wash on the ground

Common faults

Common faults to guard against during the air exercises:

- No lookout; instead, focusing on the ground
- Over-controlling, especially with the cyclic
- Not identifying the slope
- Not recognising a rollover developing
- Launching off the ground with too much up collective
- Very tense on the pedals
- Concentrating too much on the instruments instead of looking outside
- Not anticipating what is coming up next

Air Exercise 5-1: Lift-off and Landing

Introduction

The instructor will demonstrate a lift-off and a landing. After the demonstration, the instructor will give the student all the controls and position the helicopter on a flat, hard, level surface (concrete, bitumen helipad, or runway).

Because the lift-off and landing sessions can be very intense, they may be broken up with some transitions and circuit training to allow the student to relax, get some fresh air in the cabin and have a small break.

Work cycle

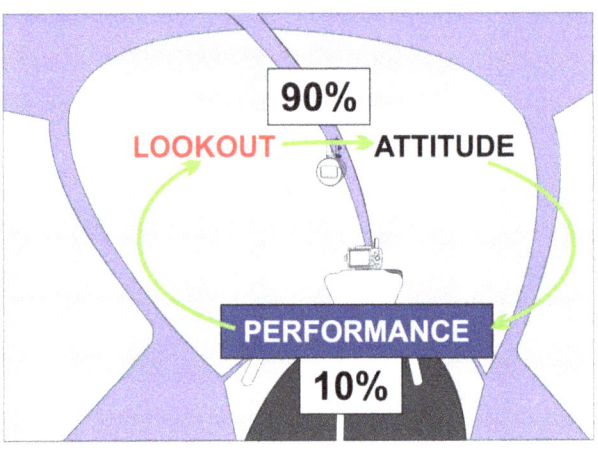

Process

Move the head and eyes to achieve the following work cycle.

Look	At...	And...
Outside	**Far** distances	Select hover reference points: - Raise the collective until the skids feel light on the ground - Look inside and confirm RPM and Power - Look outside again and continue to raise the collective to lift-off the ground (hover off the ground) - Use the cyclic to maintain or adjust the attitude as required and manage the ground position. Hold it there - Use pedals to control the yaw and keep the nose of the helicopter pointing at the nominated alignment point in the distance - ***Then***, continue to raise collective to lift vertically off the ground into the hover
Outside	**Middle** distance	Continually assess the nominated reference points so you can easily see if there is movement: - Fine-tune the ground position with cyclic. - Fine-tune the direction with pedals. - Use the collective to control the height AGL.
Outside	**Near** distance	Judge height AGL and use the collective to fine-tune the height
Inside	Instrument panel	Glance in quickly at the RPM, Power, T's and Ps and adjust if necessary.
	REPEAT	

Chapter 5 Lift-off, Transition and Landing

Lift-off

Lift-off Lift the helicopter off the ground to a stabilised 5 ft skid height hover.

Step	Action	Discussion
1	Preparation	Conduct the HEFFR Check.
2	Lookout	Look outside and consider: - The area is clear, and it is safe to lift-off - Slope and wind - Selection of hover reference points - The effects of the downwash on the surface
3	Announce	*"Lifting"*
4	CYCLIC	Position the cyclic slightly into wind and estimate its position based on past experience given the conditions. Allow for any slope.
5	PEDALs	Position the pedals ready for the introduction of power, anticipating some left pedal requirement.
6	COLLECTIVE	Alternate the scan between looking outside to inside on a continuous 50/50 split. Gently raise the collective until the skids feel light on the ground but not yet lifting off. Look inside and confirm: - RRPM top of the green - Power margin (should only be pulling approximately 55-60% Tq or 18-20" MAP) - Ts and Ps in the green
7	PEDALS	Continue to adjust pedals as the collective is raised to keep straight.
8	CYCLIC	Continue to adjust cyclic, starting to get a "feel" for the configuration of the helicopter and preparing the disc to maintain ground position.
9	SCAN	Look outside. Your lookout and scan should now be focused 99% outside to the middle and far distance. Do not be tempted to look inside to check the instruments until you are stabilised at the hover.
10	COLLECTIVE / CYCLIC / PEDALS	The three controls will now have to be manipulated together to lift the helicopter vertically up to a 5 ft skid height hover by: - raise collective slowly to lift the helicopter vertically off the ground. - move the cyclic as required to maintain a constant ground position with no sideways or backwards movement. A small amount of forward movement is acceptable as long as it is stopped when at the hover. - adjust the pedals as required to keep straight.
11	Hover	Stabilise the helicopter at the hover.
12	Hover checks	Complete the "Hover checks".

Landing

Landing — Land the helicopter from the hover back onto the ground.

Step	Action	Discussion
1	Preparation	Ensure the crew are prepared for a landing.
2	Lookout	Look outside and consider: - The area is clear and it is safe to land - Slope and wind - Selection of hover reference points - The effects of the downwash on the surface
3	Announce	*"Landing"*
4	CYCLIC	Maintain the hover position over the ground.
5	PEDALs	Maintain a constant heading. Anticipate some right pedal as collective is lowered.
6	SCAN	Look outside. Your lookout and scan should now be focused 99% outside to the middle and far distance. Occasional small glances into the near distance are permissible, but the primary focus is the middle and far distance. Do not be tempted to look inside to check the instruments until you are firmly on the ground.
7	COLLECTIVE	Gently lower the collective to commence a vertical descent to the ground. Giving a 5 count from 5 feet will assist. 5,4,3,2,1 skid touches the ground. At that point, lower the collective another physical inch (1") so that both skids are evenly and firmly placed on the ground and then pause. Hold the collective and all controls in place and allow the helicopter to settle without the full weight on the ground.
8	COLLECTIVE	Continue to lower the collective slowly and smoothly all the way to secure the landing.
9	CYCLIC	Centre the cyclic to the neutral position so that the rotor disc is parallel with the ground.
10	PEDALS	Pedals can return to neutral.
11	THROTTLE	IDLE
12	Frictions and Locks	ON

Air Exercise 5-2: Transition

Introduction — From the hover, the instructor will direct the student through a transition into forward flight up to a speed of approximately 50 kts, then from forward flight, a deceleration back to the hover.

This will require changes in the helicopter's attitude and power while maintaining alignment of the skids with the direction of travel with pedals. This exercise requires a lot of coordination and is a combination of all the flying skills learnt so far.

Transition into forward flight

Transition into forward flight — A transition from the hover into forward flight can also be called a take-off.

Step	Action	Discussion
1	Preparation	PWPTEM
2	Lookout	Area clear
3	Announce	*"Transitioning forward"*
4	CYCLIC	Use some forward cyclic to start the helicopter moving forward. As speed increases, move the cyclic further forward to keep the helicopter accelerating and counter any flapback of the disc.
5	COLLECTIVE	Use collective to control the height of the skids above the ground. The helicopter will want to sink in light winds, so raise the collective to prevent this as required.
6	PEDALS	Use the pedals to maintain the skids aligned with the direction of travel. Balance is not considered until passing through translational lift and above 50 ft AGL. Before this, keep the skids aligned with the direction of travel
7	TRANSLATIONAL LIFT	As the helicopter passes through translational lift, there will be a small shudder (normal), and the disc will want to flapback. Use more forward cyclic to counter the flapback. There will be a change in power required.
8	COLLECTIVE	Raise to 85% Tq or 25" MAP to set climb power.
9	CYCLIC	Set an attitude for 60 kts.
10	PEDALS	To maintain balance.

Transition back to the hover

Transition back to the hover A transition from forward flight back to the hover can also be referred to as short finals from the approach.

Step	Action	Discussion
1	**Preparation**	Short Finals Check: **ARP**
2	**Lookout**	Area clear
3	**Announce**	*"Transitioning to the hover"*
4	**CYCLIC**	Use some aft cyclic to place the helicopter in a decelerating attitude.
		Cyclic is used to control the closure rate of the helicopter to the nominated hover point.
5	**COLLECTIVE**	As speed decreases, use the collective to control the helicopter's rate of descent to the nominated hover point.
6	**PEDALS**	Use the pedals to maintain the skids aligned with the direction of travel.
		Once reaching 30 kts and approximately 50 ft AGL, balance is not considered.
7	**TRANSLATIONAL LIFT**	As the helicopter loses translational lift, in light winds, there will be a small sink and a sudden demand for more power (normal).
		As collective is introduced, the cyclic will have to be moved further forward to allow the helicopter to continue to move forward and down towards the nominated hover point.
8	**COLLECTIVE**	Raise collective to control the height AGL as the helicopter arrives at the 5 ft skid height hover.
9	**CYCLIC**	As required to maintain the ground position.
10	**PEDALS**	To keep straight.

Air Exercise 5-3: Sloping Ground Take-offs and Landings

Introduction | In this air exercise, the instructor will demonstrate how to identify the limits of a slope, and you will practice cross-slope take-offs and landings, up-slope take-offs and landings, and down-slope take-offs and landings.

Sloping Ground Landing

Sloping ground landing | Landing on sloping ground is described below.

Step	Action	Discussion
1	Preparation	Ensure the crew are prepared for a landing.
2	Lookout	Look outside and consider: ■ The area is clear and it is safe to land ■ Slope and wind ■ Selection of hover reference points ■ The effects of the downwash on the surface
3	Announce	"Landing"
4	CYCLIC	Maintain the hover position over the ground.
5	PEDALs	Maintain a constant heading. Anticipate some right pedal as collective is lowered.
6	SCAN	Look outside. Your lookout and scan should be focused 99% outside to the middle and far distance. Occasional small glances into the near distance are permissible, but the primary focus is the middle and far distance. Do not be tempted to look inside to check the instruments until firmly on the ground.
7	COLLECTIVE	Initiate landing as for a normal landing by lowering the helicopter toward the ground. If cross slope, one skid will touch the ground before the other, or If up or down slope, the heel or toe of the skids will touch first.
8	Initial Ground Contact	As the skid gear touches the ground, continue the descent. As the descent is continued, the aircraft will pivot and roll about the point of ground contact.
9	PEDALS	Use the anti-torque pedals to maintain direction and resist any tendency to pivot in the yawing plane after initial ground contact.
10	CYCLIC	To prevent the helicopter from rolling or sliding down the slope, apply sufficient cyclic in towards the slope to hold it there while continuing to lower the collective until both skids are on the ground.

Step	Action	Discussion	
11	Both skids on the ground	When both skids have full contact with the ground, stop lowering the collective and determine the stability of the helicopter.	
		If there...	Then...
		is **NO** tendency to slide or roll down slope	centre the cyclic and continue to lower the collective to complete the landing
		IS a tendency to slide or roll down slope	maintain the cyclic input and lift-off by raising collective and repositioning to an area with less slope

Sloping Ground Take-Off

Sloping ground take-off Take-off is accomplished in a reverse sequence to landing, as described below.

Step	Action	Discussion
1	**Preparation**	Ensure the crew is prepared for a lift-off.
2	**Lookout**	Look outside and consider: ■ The area is clear, and it is safe to lift-off ■ Slope and wind ■ Selection of hover reference points ■ The effects of the downwash on the surface
3	**Announce**	"Lifting"
4	**CYCLIC**	Move the cyclic to an estimated position into the slope in preparation for lifting the downslope skid off the ground.
5	**PEDALs**	Maintain a constant heading. Anticipate some left pedal as collective is raised.
6	**SCAN**	Look outside. Your lookout and scan should now be focused 99% outside to the middle and far distance. Occasional small glances into the near distance are permissible, but the primary focus is the middle and far distance. Do not be tempted to look inside to check the instruments until you are firmly on the ground.
7	**COLLECTIVE**	Initiate the lift-off by raising the collective. **Result:** If cross slope, the downslope skid should start to lift off the ground while the upslope skid now acts as a pivot point, or If up or down slope, the heel or toe of the skids will lift first.

Step	Action	Discussion
8	CYCLIC	As the downslope skid lifts, adjust the cyclic to bring the disc attitude level with the horizon. This is a fine balance between cyclic and collective to prevent any roll into the slope about the pivot point against not enough cyclic that the helicopter starts to slide down the slope. The helicopter can feel very sensitive at this point.
9	PEDALS	Use the anti-torque pedals to maintain direction and resist any tendency to pivot in the yawing plane after initial ground contact.
10	Once level	**Once the helicopter is in a level hover position with one skid on the ground and the other off the ground, then:**
11	COLLECTIVE	Raise more collective to lift the helicopter up away from the ground.
12	CYCLIC	Use some cyclic so that as the skids leave the ground, the helicopter is moved gently away from the slope (sideways).
13	PEDALS	Use pedals to keep straight.
14	Turns	Any turns required close to the slope should always be done to move the tail away from the slope.

6

The Circuit

Aim	To conduct a standard pattern for flying around, making an approach to, and taking off from a selected helicopter landing site (HLS).
Objectives	On completion of this lesson, the student will be able to: ■ explain the Height Velocity Graph ■ list the two types of circuit patterns used by helicopters ■ list the legs of a circuit ■ state the three (3) standard circuit heights ■ describe the circuit direction and how to determine it ■ describe the active and non-active sides of a circuit and what they are used for ■ discuss how wind may affect the circuit ■ explain the lookout in the circuit, and ■ apply the various checks required.
Motivation	You will encounter other aircraft when operating at, approaching or leaving an airport. It is important that everyone is flying in a standard, predictable manner to maintain separation and safety. Afterall, if we all came to an airport and just decided to make an approach at different heights, from different directions, we would have no idea what to expect, who has the right of way and just what that other guy is doing. Learning a standard circuit or traffic pattern allows for the smooth and orderly flow of aircraft traffic around a landing area.

Preparation: The Circuit

What is a circuit? The circuit is a standard path that pilots follow when taking off or landing at an airport. It is designed to put order into how aircraft approach, depart and operate to and from a runway.

Circuit height The standard fixed-wing circuit height is set at 1000 ft AGL following a reasonably large, rectangular shape. A helicopter circuit is usually set below and inside the fixed-wing circuit because helicopters:

- are slower and more manoeuvrable
- often do not use the runway but a small helipad located beside the runway for safety and to avoid adding unnecessary traffic on the runway, and
- usually do smaller circuits in a racetrack pattern instead of a rectangular pattern.

During training, it is not unusual to vary the circuit height to give the student experience managing the differing heights AGL. Initially, the instructor will use 1000 ft because it gives the student more time to experience each leg and experience the full circuit pattern. Over time the circuit height will be reduced so each circuit can be done quicker, and there will be less conflict with fixed wing traffic.

Also to be considered is local procedures at individual airports. There may be a documented requirement for a helicopter to operate at a certain circuit height inside a specific helicopter training area. All this information will be given to you by your Flight School.

Height Velocity Graph or Dead Man's Curve

Combinations of height and speed to avoid Due to the nature of a helicopter, there are combinations of height and speed to be avoided, if possible. If the engine and systems are operating correctly, these height and speed combinations represent no threat. However, a safe autorotational landing may be impossible if the pilot experiences a malfunction and operates at one of these combinations of heights and speeds.

Each helicopter will have a Height/Velocity graph (commonly referred to as the Deadman's Curve) in its flight manual, showing those areas of flight to avoid.

Bell 47 Height Velocity Graph

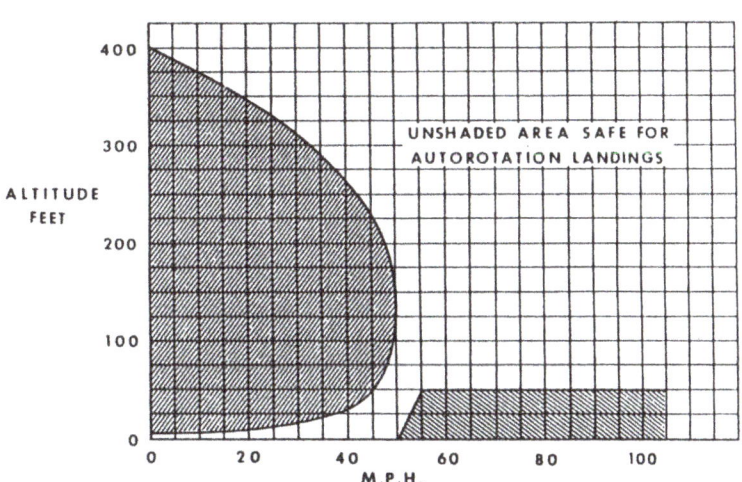

Chapter 6 The Circuit

Example

For example, using the Height/Velocity graph for the Bell 47 shown previously, a helicopter at 10 kts and 200 ft AGL experiencing an engine failure would be unable to enter autorotation safely, and the helicopter would sustain damage on touchdown.

Relevance in the circuit

The dead man's curve is relevant to the circuit as it gives you your take-off and approach profiles. As long as you stay out of the shaded areas as you complete the circuit, a safe autorotational landing should be possible from any part of the circuit.

R44 Height / Velocity Graph

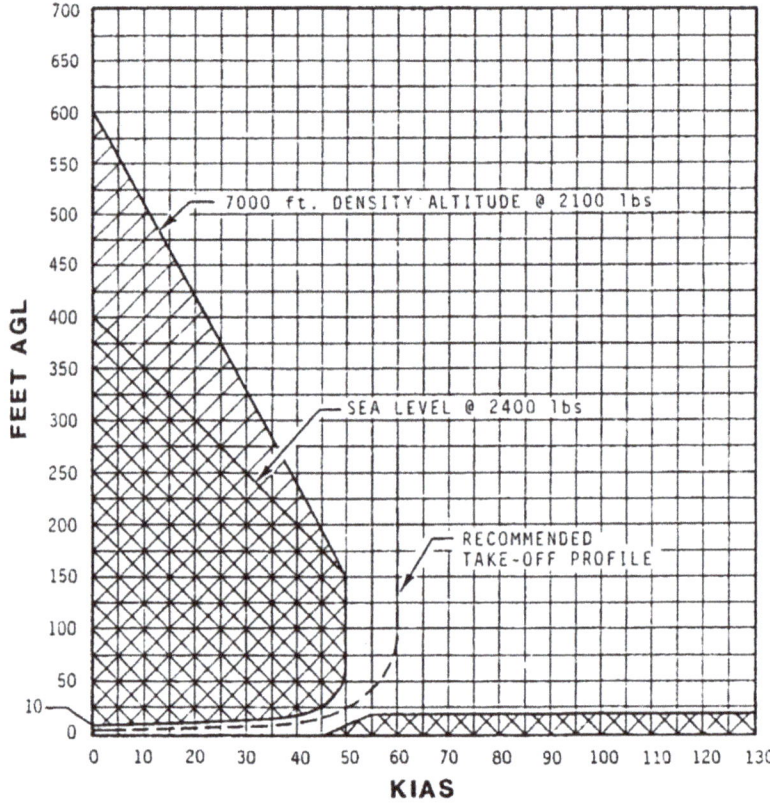

Example

The R44 Height/Velocity (H/V) curve above allows this helicopter to go from the hover to 55 kts below 50 ft AGL; then, the helicopter can climb at 60 kts to any altitude and stay out of the deadman's curve.

If, however, the helicopter continued to accelerate to, say, 80 kts at 20 ft AGL, then it would enter that part of the deadman's curve where if there were an engine failure, the pilot response time would not be quick enough to prevent the helicopter from hitting the ground. A descent at 60 kts would be good during an approach, with speed progressively reducing to zero once below 100 ft AGL.

Types of Circuits

Circuits can be of different shapes and sizes and be described as standard or non-standard. A standard circuit will be described below and will consist of either the:

- rectangular circuit, or the
- racetrack circuit.

Initially, you start with a rectangular circuit because it's useful as a teaching tool. It provides a little more time for the pilot to maintain situational awareness, perform relevant checks and fly the aircraft. As your coordination, judgement, and confidence improve, you will progress to a racetrack circuit, the most common helicopter circuit.

A non-standard circuit is anything that is not rectangular or racetrack.

What is the difference?   The term Rectangular and Racetrack pattern refers to the shape and turns conducted on each leg, not the overall ground track of the helicopter.

Rectangular circuit   A Rectangular circuit (also called a square circuit) will have the helicopter turn through 90 degrees on each leg, in effect "squaring" up each leg of the circuit.

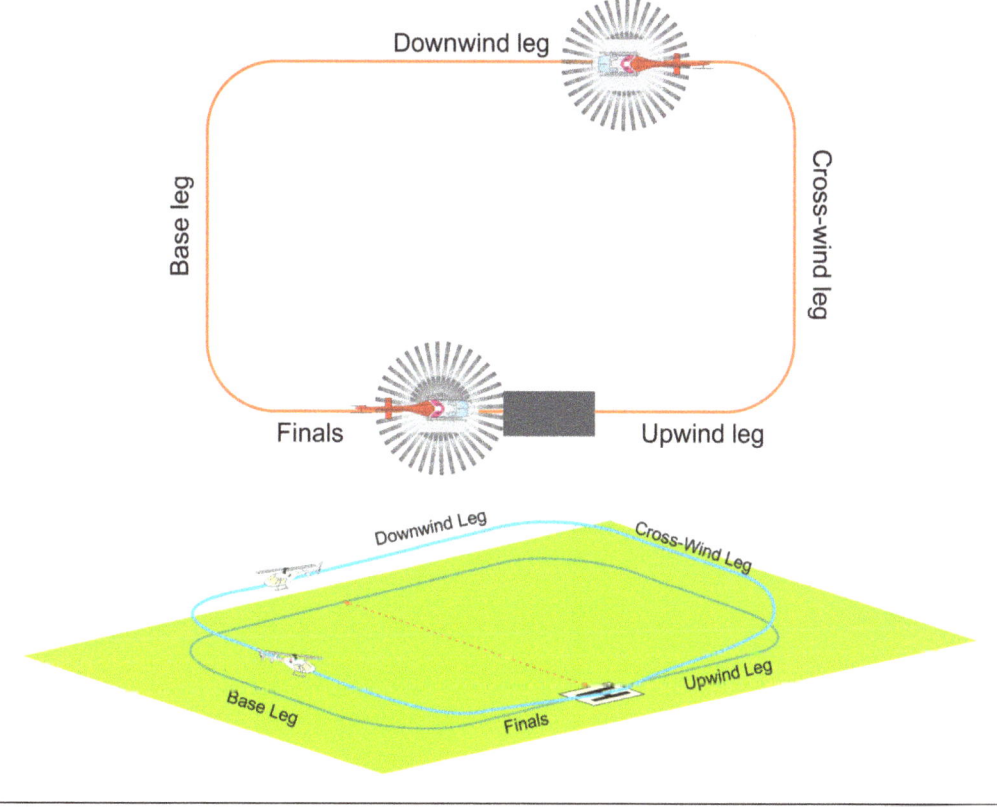

Chapter 6 The Circuit

Racetrack circuit A racetrack circuit will have the helicopter conduct a constant turn through 180 degrees on the cross-wind and base legs, rounding out the circuit to look more like a racetrack pattern.

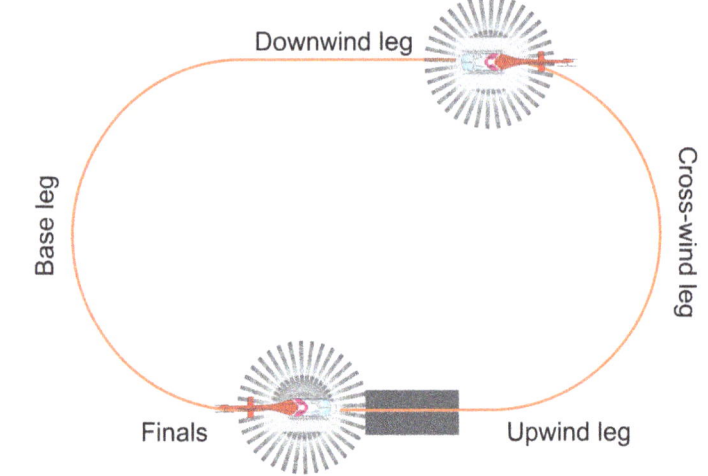

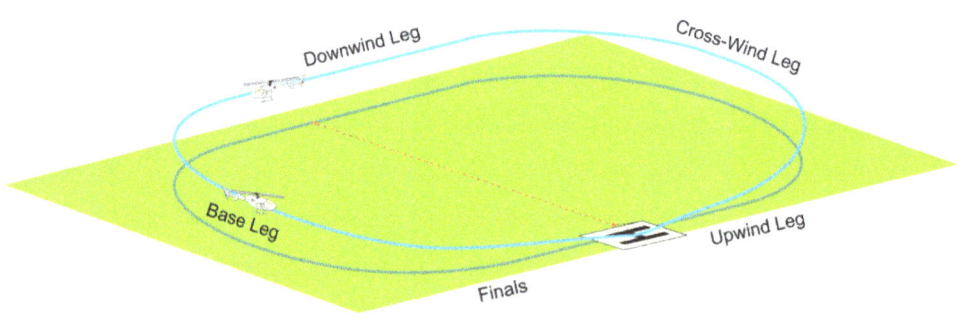

Which one Which one is flown will depend on several factors:

- How familiar the pilot is with the local circuit pattern and the aircraft being flown.
- How good the visibility is and whether it is day or night.
- How fast the helicopter is going and how fast other aircraft are going within the circuit, and how many of them are in there.
- What height is the circuit being flown at, and if fixed wing and helicopters are trying to operate within the same areas?
- What is the prevailing wind strength?

The more conservative you need to fly, the bigger and the squarer the circuit.

The less conservative you need to fly, the smaller and tighter the circuit can be in a racetrack pattern.

Student pilots typically start learning the rectangular circuit and progress into the racetrack circuit as they become more competent.

Circuit Height

Circuit heights

Three (3) standard circuit height options are documented in the Visual Flight Rules (VFRs) depending on the aircraft's performance. It will be up to the pilot to determine which category to use as follows:

Type of aircraft	Circuit speed range	Standard circuit height
Low performance Includes ultra-lights and helicopters	Up to 55 kts indicated airspeed	500 ft AGL above the aerodrome elevation
Medium performance Normal fixed-wing circuit and helicopters may also decide to operate in this category	Between 55 kts and 150 kts indicated airspeed	1000 ft AGL above the aerodrome elevation
High performance Includes jets and turboprops	Greater than 150 kts indicated airspeed	1500 ft AGL above aerodrome elevation

Diagram

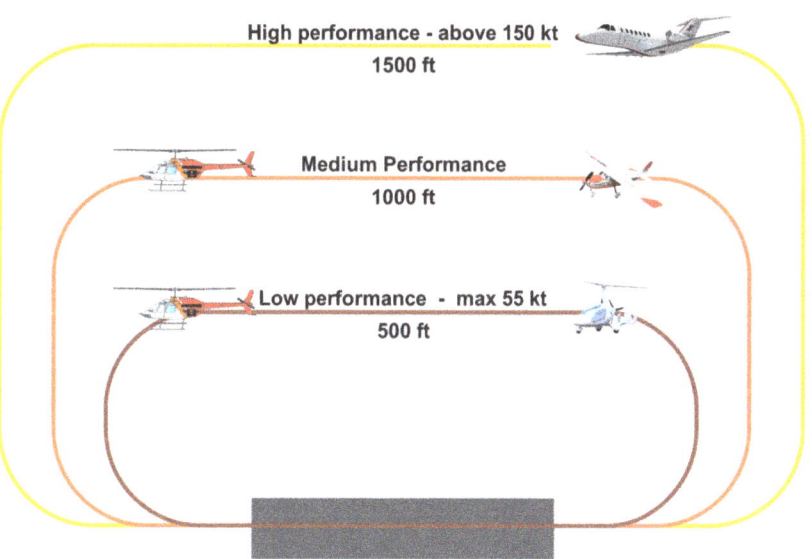

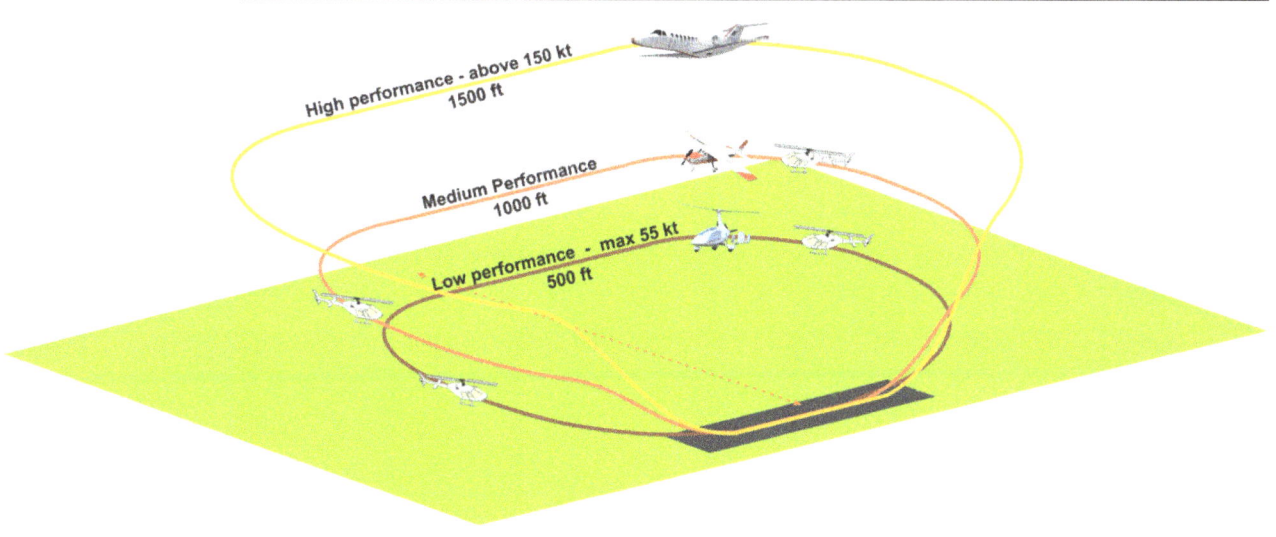

Circuit heights

Chapter 6 The Circuit

The circuit area The circuit area is generally considered anything that is within 3 NM from the aerodrome reference point up to 2000 ft AGL.

Any operations outside these limits are considered outside the circuit area.

If helicopters are transiting through an area that will be close to an aerodrome, they should remain at least 3 NM away from the runway, or if within 3 NM, should plan to overfly at not less than 1500 ft AGL or 2000 ft AGL if there are high-performance aircraft in the area.

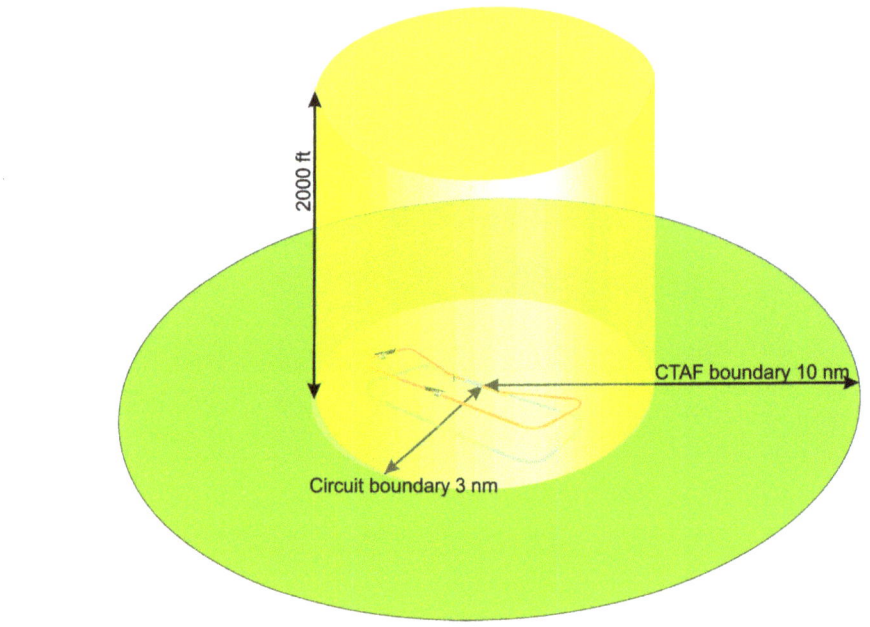

Circuit Direction

Circuit direction: Left-hand circuit Unless otherwise advised, all circuits are to be left-hand in direction. This means the cross-wind turn shall be made to the left as a standard rule. The left-hand turn is most common because all fixed-wing captains sit on the left-hand side of the cockpit, so it is assumed they will have better visibility in a turn, making it safer.

Circuit direction: exceptions There are exceptions to this. If the left-hand circuit takes the aircraft over populated areas or into an area that is unsafe, then a right-hand circuit will be used.

Helicopters can make left or right-hand circuits at their discretion. This is because our circuit is typically half the size of a fixed-wing circuit simply because we perform differently. Additionally, helicopters are allowed to make contra circuits (circuits on the other side of the active circuit area) to have them stay out of the way of normal fixed-wing operations.

Non-Standard Circuit:

If doing a circuit in a different direction to that published for the runway it is considered a non-standard circuit.

Publications

To know the circuit direction for a particular runway, the pilot needs to refer to the documentation provided by the regulator. Each Country will publish landing plates, and for VFR pilots, these can be found in a Visual Flight Guide (VFG), Enroute Supplement (ERSA) or similar document.

If the runway has a standard left-hand circuit, there will typically be no mention of circuit direction in the publication because a left-hand circuit is considered standard.

If the circuit direction is to the right, then this is something different from the standard, and there will be a statement in the publication advising this.

The extract below for Caloundra Airport in Australia states that right-hand circuits are required when operating on runways 12 and 23. Left-hand circuits are still required if operating on runways 30 and 05 but are not mentioned.

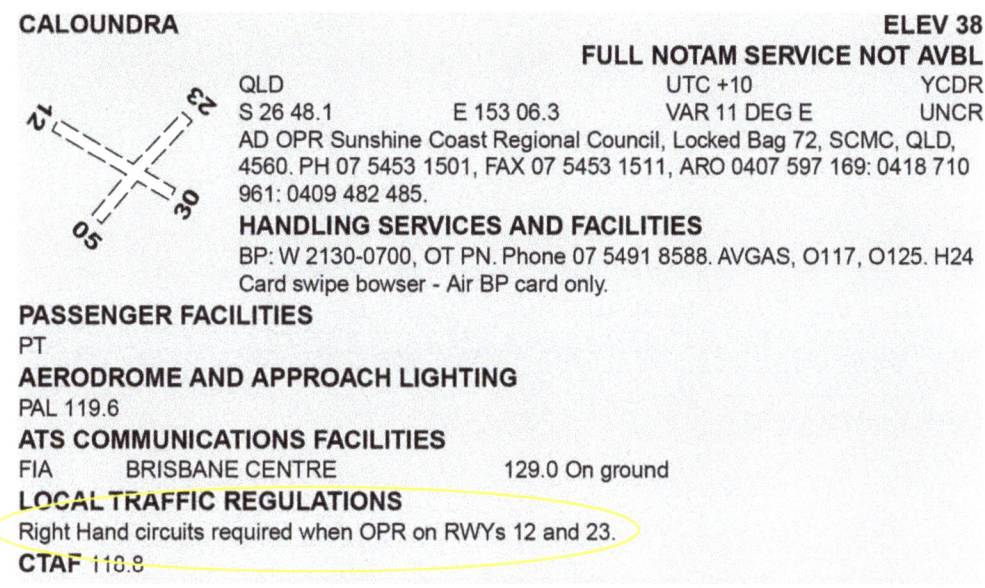

Active and non-active side

When reviewing a circuit, whichever way the pilot has to turn cross-wind will determine the circuit direction, but it will also determine which side of the runway is referred to as the *"active"* and *"non-active"* side.

The active side is the side of the runway where all the aircraft traffic is expected to be if conducting the correct circuit pattern for that runway.

The active side is where all the circuit traffic should be when at the correct circuit height; therefore, the other side of the runway should not have any circuit traffic and will be referred to as the *"non-active"* side.

The non-active side is where aircraft can manoeuvre and descend prior to joining the circuit pattern. Helicopters may conduct contra circuits (circuits in the opposite direction to that required) at a lower height (say 800 ft AGL) and stay out of the way of all the other traffic on the active side.

Chapter 6 The Circuit

Alternative terminology Depending on where a pilot has trained, alternative terminology may be used to describe the active and non-active side as follows:

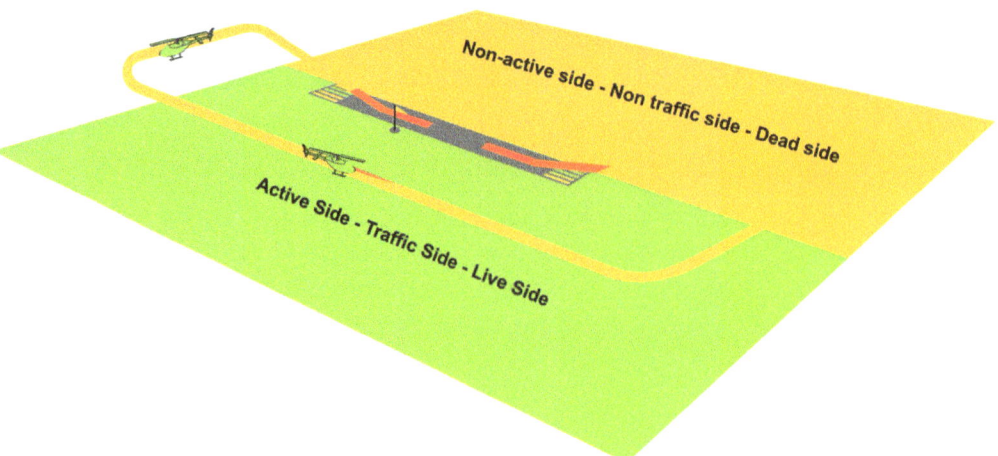

It does not matter which words you use; they all mean the same thing. In general, whichever way the cross-wind turn is done will determine the circuit direction and the active side.

Legs to a Circuit

Legs The circuit is divided into five (5) separate legs (as in a leg of a journey). When all the legs are combined, they form a standard circuit pattern.

The five (5) legs of a standard circuit pattern are:

1. Upwind leg (which is also the initial take-off path)
2. Cross-wind leg
3. Downwind leg
4. Base leg
5. Final leg

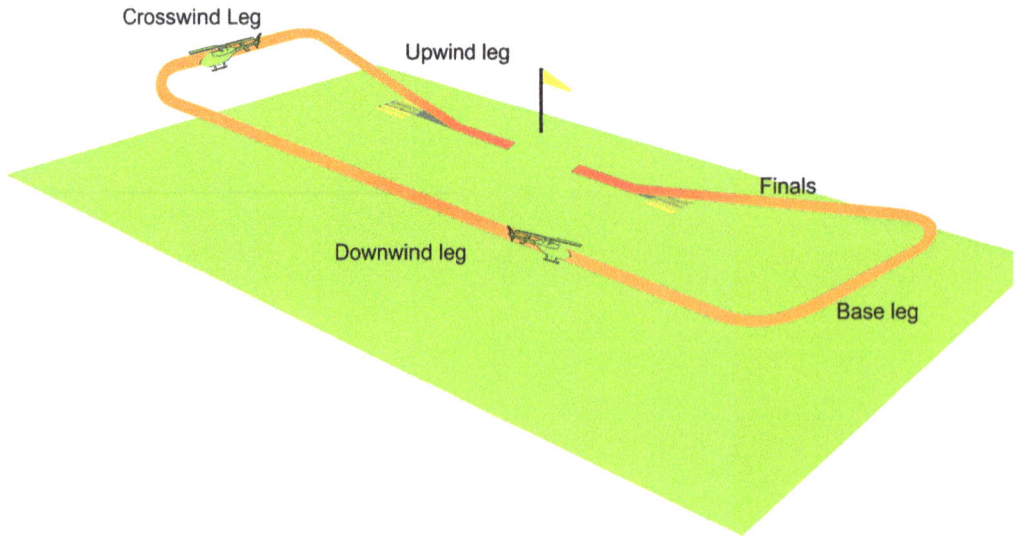

Additional legs

These legs can be further divided depending on where the helicopter is on each leg at the time in relation to the standard circuit pattern:

- Extended upwind
- Extended cross-wind
- Early downwind, mid downwind, late downwind
- Wide base, mid base, close base, and
- Long final, mid final or short final.

If departing or arriving to join a circuit, these legs can also be used by the pilot to identify on the radio their location and then intentions.

For example:

Consider a helicopter just taking off and wanting to depart straight ahead. On the radio, they can advise all traffic that they are departing the circuit from the upwind leg.

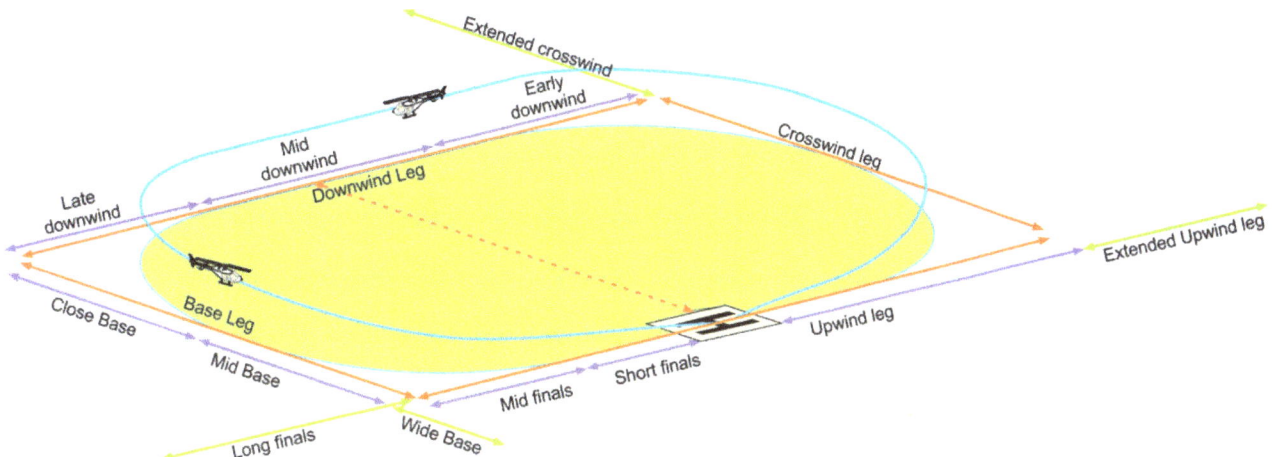

Effects of Wind

Wind

The wind can affect the circuit. The stronger the wind, the more the helicopter will be affected, and the pilot will have to compensate to maintain a standard circuit pattern shape. Remember, wind will not affect the airspeed but significantly affect the ground speed and ground track.

Strong headwind

With a strong headwind, the upwind leg may be very short and the downwind leg very fast and long across the ground, so the pilot may decide to extend the upwind leg to give more time and room in the circuit.

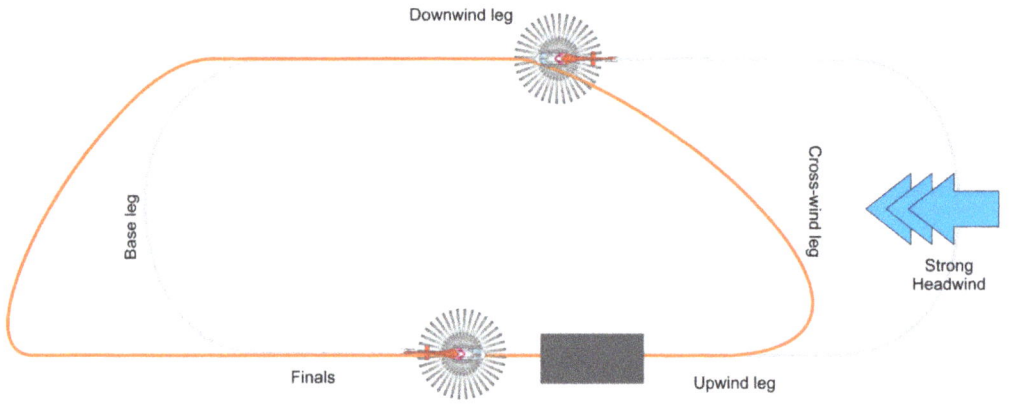

Headwind

In general, if the take-off wind is on the nose and:

- **below 15 kts** then fly a standard circuit.
- **above 15 kts** then:
    - extend the upwind leg, and
    - continue to climb to the circuit height before turning cross-wind and continuing with the normal circuit.

Headwind less than 15 kts

Standard circuit – headwind less than 15 kts

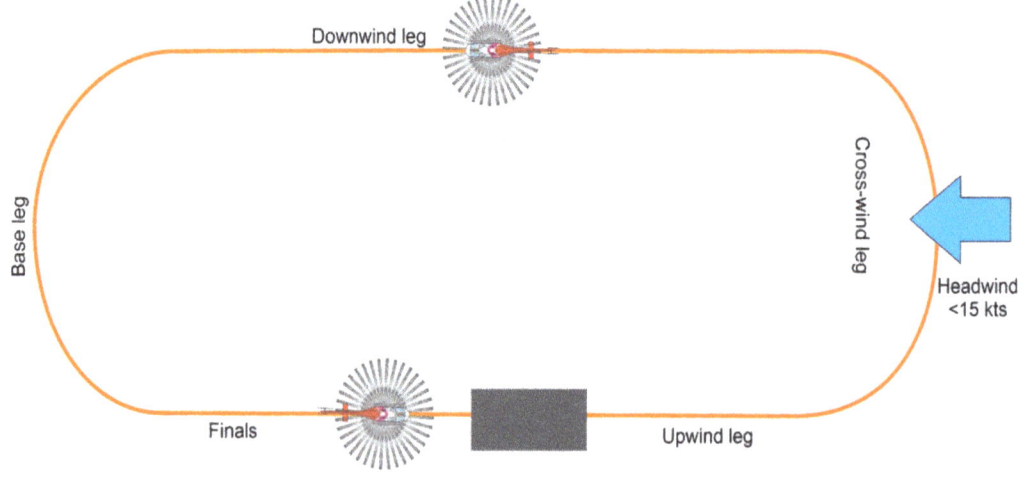

Headwind greater than 15 kts	**Extended time in the Upwind leg – headwind greater than 15 kts**
	Extending the time in the upwind leg until you reach the full circuit height before turning cross-wind allows the helicopter to reach its normal position over the ground within the circuit.

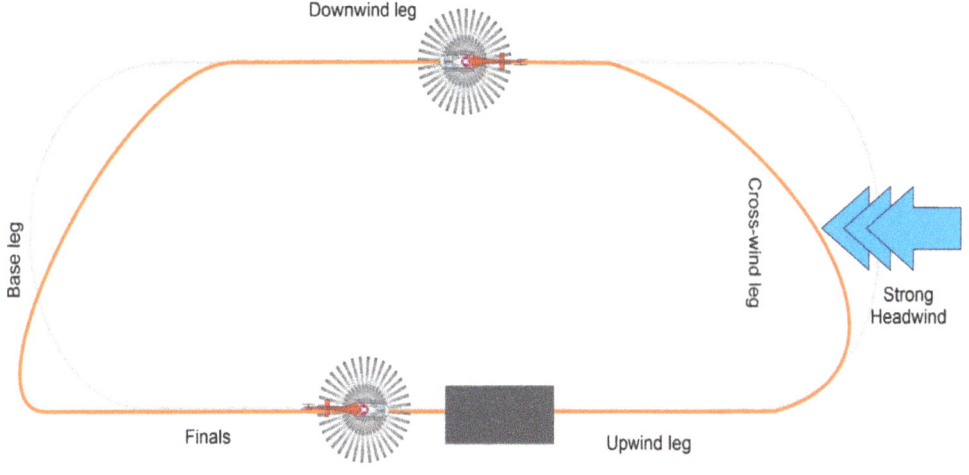

Strong cross-wind	With a strong cross-wind the helicopter may be blown closer or further away from the circuit area, so the pilot will have to adjust the heading to compensate for the drift caused by the wind.

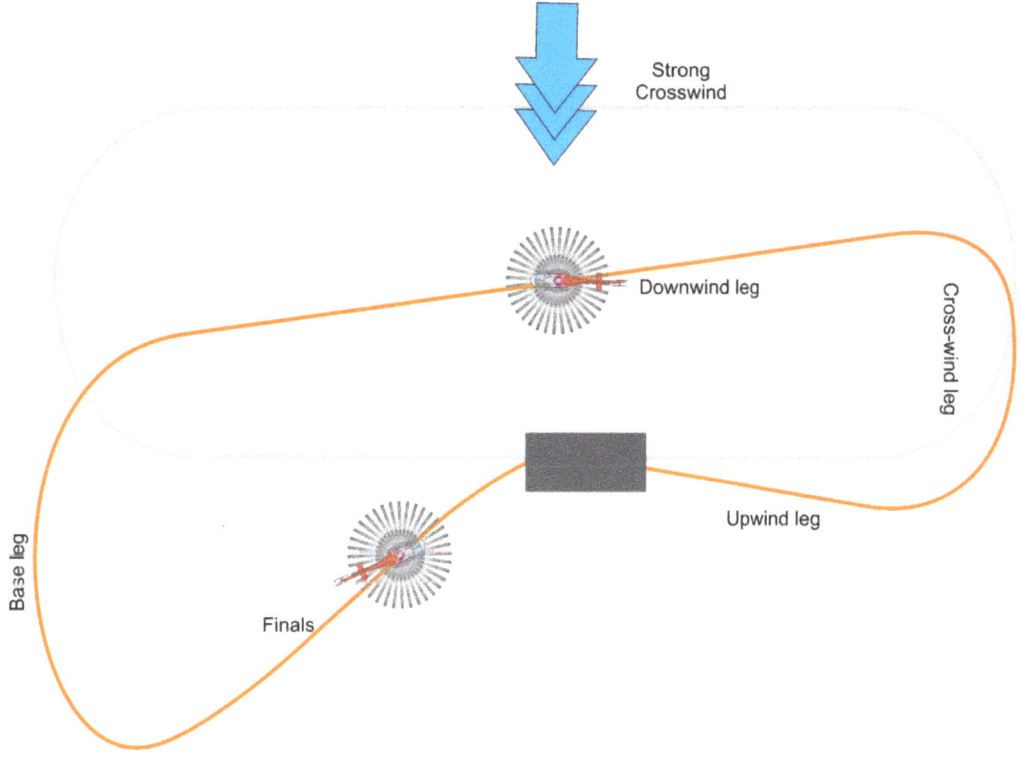

Chapter 6 The Circuit

Cross-wind   In general, if the take-off wind is coming from the left or right:

- **below 15 kts** then:
    - fly a standard circuit, but
    - ensure the heading is adjusted and reference points are selected on the ground to help maintain a constant circuit pattern.
- **above 15 kts**, the pilot will have to compensate for the wind to maintain the correct spacing from the runway or HLS.

If the pilot has a choice, it is often easier to turn cross-wind with the wind coming from behind (downwind), which means the base leg will be into wind, and the pilot can better control the descent and fly to the finals leg rather than be clown past it.

Cross-wind   Whichever way you decide to turn, if the wind is above 15 kts, extend the time on the upwind leg and continue to climb to the circuit height before turning cross-wind and continuing with the normal circuit and making allowance for the wind.

Lookout in the Circuit

Scanning technique   Because the circuit pattern is taught consistently to pilots, it becomes very predictable where to look for traffic, and the pilot can focus on those areas of particular interest where traffic would normally be expected to be coming from.

Of course, the normal lookout prior to any turn is essential, but the following areas can be focused on:

Upwind	Lookout ahead and above for any traffic that may be crossing overhead at 90 degrees to your take-off. Other aircraft joining the circuit may do so at 1000 ft AGL over the runway threshold.
Cross-wind	When turning cross-wind, the main area of focus for the lookout is for other aircraft that may be joining the circuit from a downwind position. If conducting a left-hand circuit, look for aircraft on the right. If conducting a right-hand circuit, look for aircraft on the left.
Downwind	The downwind leg can have traffic coming from several different areas. There may be traffic ahead, behind, outside or inside your circuit leg. There may also be traffic running at 90 degrees to you as they join overhead or at the same level joining the circuit pattern. The downwind leg can be very busy and will require good situational awareness.
Base	Before turning on to the base leg, check for other aircraft on a wide base or final. If conducting a left-hand circuit, look to the right, along the extended final leg and in the direction of your turn. If conducting a right-hand circuit, look to the left, along the extended final leg and in the direction of your turn. Particularly look along your intended path below.
Final	Before turning final, look back along the extended final path and ensure you will not turn in front of another aircraft. If conducting a left-hand circuit, look to the right. If conducting a right-hand circuit, look to the left.

| Diagram: scanning | The diagram below illustrates the lookout focus in a circuit.

RED is the most significant area to focus on where conflicting traffic may come from.

ORANGE is the next area of focus on where conflicting traffic may be coming from.

YELLOW is the last area where we may see traffic, but the possibilities of a collision are less as they are in front, moving in the same direction (hopefully). |
|---|---|

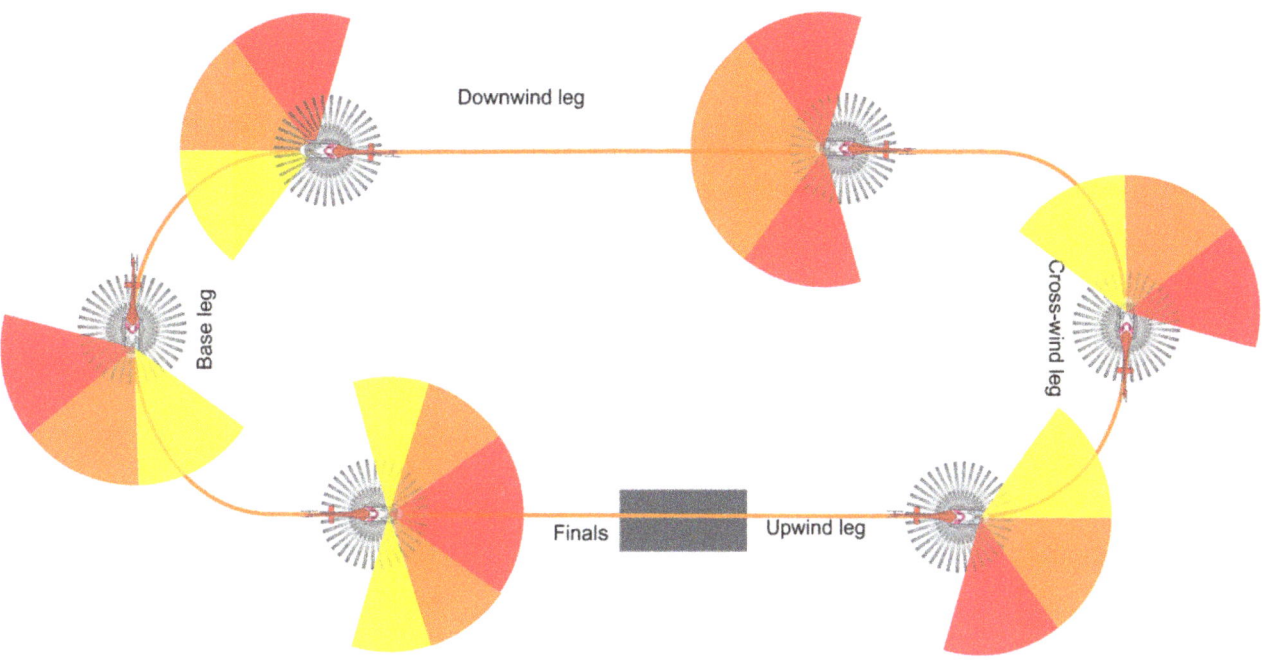

| Right-of-way rules | There are standard right-of-way rules that apply in a circuit, as follows:

1. Aircraft in front has the right of way
2. Give way to aircraft on your right
3. Aircraft below has the right of way
4. Aircraft established on final has the right of way
5. Aircraft already established in the circuit and on one of the legs of the circuit has the right of way over an aircraft joining the circuit
6. The aircraft landing has the right of way over an aircraft wanting to take-off
7. If two aircraft are head-on, then both shall turn right
8. Power-driven heavier-than-air aircraft (helicopters) shall give way to airships, gliders, balloons and aircraft towing other aircraft or objects regardless of the position in the circuit and regardless of the other right-of-way rules above. |
|---|---|

Chapter 6 The Circuit

Short Finals Check

The Short Finals Check is designed to confirm to the pilot that the helicopter is under control and able to continue and make a smooth transition to the hover. If any of the parameters explained below are not as they should be, then the pilot is to abort the landing or, if experienced enough, be aware of and plan for a more difficult approach and transition.

The Short Finals Check consists of:

- Airspeed over ground speed
- ROD is under control and
- Power in hand.

Airspeed vs Groundspeed

Comparing the airspeed to the ground speed checks that the airspeed is greater than the ground speed, which confirms a headwind.

This can be done visually by estimating the rate of closure (the combined speed across the ground and the rate of descent, which takes some past experience to comprehend) against the current airspeed.

Alternatively, referencing the airspeed against the groundspeed displayed on the GPS.For example

Look at the ASI and the GPS to compare the two readings.

Headwind example

If the ASI is higher than the GPS groundspeed, then there is a headwind. Continue with the approach.

The example below indicates a 10 kts headwind. *Continue with the approach.*

ASI   GPS

 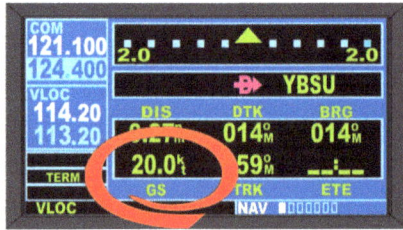

Tailwind example

If the ASI is less than the GPS ground speed, then there is a tailwind, and the approach should be aborted.

If the ASI is reading zero and the helicopter is still moving forward across the ground, then there is definitely a tailwind and the approach should be aborted.

The example below indicates a 10 kts tailwind. *Abort the approach.*

ASI   GPS

 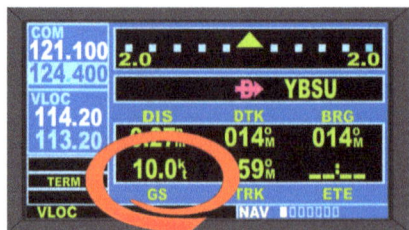

ROD is under control	The Rate of Descent (ROD) is under control check takes some practice for a trainee pilot to fully understand. The pilot is required to make a judgement based on the helicopter's ROD (vertical movement) and forward speed (horizontal movement) along the desired glideslope (shallow, normal, steep or vertical). These combined are referred to as the "Closure rate". The closure rate is how fast the helicopter seems to be closing in on (approaching) the landing area.

For example:

Consider going down a slide (representing the glideslope) with a brick wall at the end of it.

Too fast
- If the ROD and speed are **too high** (high closure rate), you will hit the wall (and it will hurt).

Too slow
- If the ROD and speed are **too low** (low closure rate), you may never get to the bottom.

Best ROD
- The **best closure rate** would be one where the ROD and speed down the slide resulted in stopping just before the wall.

Chapter 6 The Circuit

Helicopter feels too fast

If the pilot "feels" that the helicopter is too fast and the collective needs to be lowered to make the termination point, then the ROD is not yet under control. The closure rate is too high.

The best way to imagine this is the feeling you get when you try to put on the brakes of a bike in the rain, and it does not respond. No matter what you do, it does not want to stop; so you crash.

Helicopter feels too slow

If the pilot "feels" that the helicopter is slightly too slow and the collective is being raised so that the pilot is now making the helicopter move forward into the confined area, then the ROD is under control, and the closure rate is under control.

The best way to imagine this is if you "feel" as if you have picked the helicopter up and are carrying it into the landing area. You are taking the helicopter there; it is not taking you there.

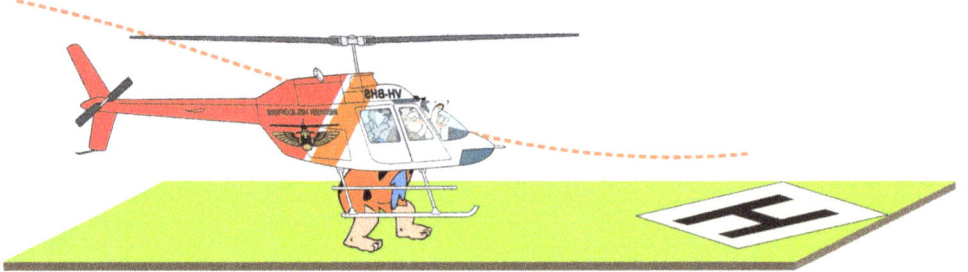

ROD is under control

On short finals, if the ROD is under control, then a small increase in collective will see the helicopter respond, and the ROD reduce. If a small increase in collective has no effect, then slow down until it does.

If the pilot cannot control the helicopter's ROD, the approach should be aborted.

| **Power in hand** | Cross-check the power currently being used compared to what will be required in the landing area at the hover based on the power check and the power margin. |

For example in a Jet Ranger:

If the current power being used is 50%, hover power is 85%, and maximum power available is 100% this will give a power margin of 15% at the hover.

If the power currently required is less than hover power and is what would be expected on approach at a low airspeed, then there is power in hand and the approach may continue.

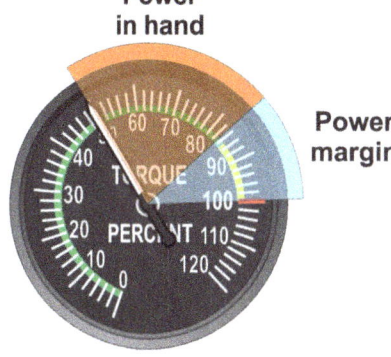

If the power currently required is too high or is at or above the hover power, and if the helicopter is still moving forward between 15-30 kts, there will not be enough power in hand to come to the hover, and the approach should be aborted.

| **Summary** | The finals check is an important tool that should be habitually used on every single approach and transition to the hover. |

Air Exercises: The Circuit

Introduction

This air exercise involves flying a circuit or traffic pattern and includes:

- lift-off to the hover
- transitioning from the hover to forward flight (the take-off)
- making a circuit that combines climbing and descending turns, levelling off and straight and level
- transitioning from forward flight back to the hover (the approach), and
- landing from the hover.

Common faults

Common faults to guard against during the air exercises:

- Poor lookout
- Not managing the shape of the circuit or compensating for wind
- Not managing the base and final leg
- Not managing power and altitude
- Not being prepared
- Not memorising the required checks

Air Exercise 6-1: The Circuit

Take-off profile

During translational lift, a large height gain is not desired, and a slight power reduction may be necessary to remain on the desired take-off profile.

At the same time, an excessive nose-low attitude is undesirable, so cyclic should not be used to convert the full effect of translational lift into forward speed.

The take-off profile is important because it keeps the helicopter in the best flight envelope in case of engine failure. From a pilot and safety point of view, if you keep the tip path plane (as seen in the sight picture when looking forward) above the horizon line during your take-off, then this will give the best compromise between acceleration and attitude so if something does go wrong you should still be able to either flare or at the very least get the skids level with the ground prior to ground contact.

A steeper take-off altitude not only requires a lot more power but can be the difference in you hitting the ground hard if the engine fails or not.

Air Exercise

The following diagram illustrates a left-hand circuit pattern at 1000 ft AGL.

Each helicopter type will use different power settings and speeds which you will know by now. The heights can either be standard or non-standard.

Mike Becker's Helicopter Handbook

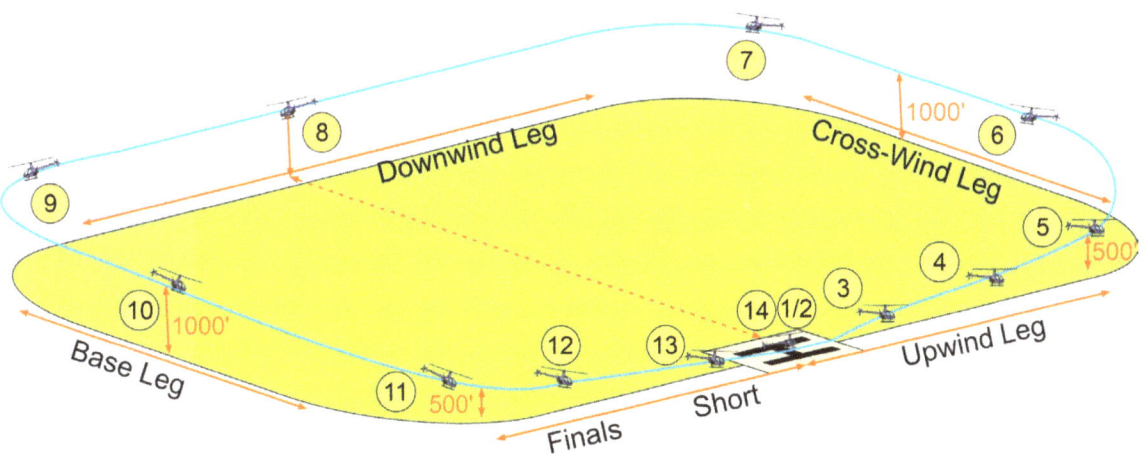

Step	Description	Discussion
1	Lift-off	Complete the Pre-liftoff checks (HEFFR), then lift off the ground to a stabilised 5 ft skid height hover.
2	Hover	Maintain a stabilised 5 ft skid height hover and conduct the pre-take-off checks (PWP) before completing a 90-degree clearing turn to check for traffic.
3	Transition	Ease the cyclic forward and transition from the hover to forward flight.
4	Climb	Accelerate straight ahead. On reaching the target climb speed, let the nose come away (up) and continue a climb straight ahead to 500 ft AGL.
5	Climbing turn Cross-wind	At 500 ft AGL' (or, as a general rule, ½ the desired circuit height), have a good lookout and commence a climbing turn.
6	Cross-wind leg	Level out at 1000 ft AGL.
7	Turn downwind	Lookout and then turn downwind. Once established on the downwind leg, check the height, heading, speed and spacing within the circuit.
8	Downwind leg	Continued downwind at: ■ Cruise power ■ 100% RRPM ■ Cruise speed, and ■ Conduct the pre-landing checks (HEFFR) and state a PWP for landing.
9	Turn into Base Leg	Lookout, carb heat ON and then commence a descending turn onto the base leg.
10	Base leg	Begin your descent with: ■ looking outside, straight ahead ■ decreasing power by lowering collective as required for the descent ■ maintain RRPM mid-green ■ maintain altitude with aft cyclic, and ■ pedal for balance. **Note**: Power - Attitude - Balance.
11	Turn into finals	Lookout, and then while descending and before passing 500 ft AGL, turn onto finals.
12	Finals	Rolling out onto finals: ■ carb heat OFF, and ■ progressively reduce speed using power to control glide slope.
13	Hover	Transition to hover.
14	Land	From the hover, land.

Chapter 6 The Circuit

Approach

The approach begins with a descending turn downwind until you are lined up for a straight approach at approximately 500 ft and 60 kts into wind. This is known as finals. The procedure for the approach is described in the table below.

Step	Description	Discussion
1	**Begin descent**	Initiate descent by: ■ looking outside ■ lowering some collective, as required, normally 10-17" MAP or 35% Tq ■ using aft cyclic to maintain attitude, and ■ using pedals for balance. *Straight and level Attitude* *Deceleration Attitude* **Note:** Power - Attitude - Balance.
2	**Begin descending turn**	Begin a descending turn by easing cyclic left/right to commence the turn. In the descending turn, lose 500 ft and speed so that you are at 500 ft and 60 kts on final approach.
3	**Adjusting the approach**	Adjust the approach, if required. | If the approach is too...	Then
---	---
High	■ lower collective to regain the approach angle, and ■ use aft cyclic to counter the effects of flap forward and maintain attitude and approach speed.
Low	■ raise collective to regain approach angle, and ■ use forward cyclic to counter the effects of flapback and maintain attitude and approach speed.
Fast	■ use aft cyclic to control speed, and ■ lower collective to maintain the approach angle.
Slow	■ use forward cyclic to control speed, and ■ raise collective to maintain the approach angle.	 **Result:** As the helicopter comes closer to the touchdown point, several things will happen: ■ the touchdown point will start moving down the bubble as you approach the hover ■ speed will steadily decrease if the helicopter is maintained in a flat attitude, and ■ power will need to be increased as the helicopter passes back through translational lift and comes to a hover.

Step	Description	Discussion
4	**Effect of ground cushion**	As the helicopter nears the ground, the intensity of the ground cushion will build up to slow the rate of descent, so you may need to make slight adjustments using the collective and pedals to continue the descent.

12-15 knots
Maintain 10 to 12 degree approach angle.

Translational Lift
12-15 knots.
Translational lift felt as a vibration.
Increase throttle.

10 knots
As you enter ground cushion, descent will slow.

Hover

Step	Description	Discussion
5	**Touchdown**	Once the skids touch the ground, lower the collective 1", then pause. With both skids on the ground and the helicopter feeling secure, gently continue to lower the collective until fully down.
		Note: Consider yourself always in the 'hover' until the collective is fully down and frictioned on.

Don't rush

Actions should not be rushed, and the lift-off and landing should be completed slowly. It is a common fault to hurry the lift-off and touchdown to avoid the prolonged and precise cyclic control the student must develop.

Lifting off slowly also provides the pilot with the opportunity to:

- recognise tail rotor drift and roll
- recognise dynamic roll-over
- recognise any helicopter malfunction
- feel any lack of control in cyclic or pedal due to high wind velocity, and
- feel that the centre of gravity is within limits, and there is adequate cyclic control.

Into wind

Where possible lift offs and touchdowns are made into wind. In strong wind conditions, cyclic pitch and tail rotor pitch may not be adequate when landing or lifting off out of wind. Actions taken in such circumstances will be covered later.

Most common errors

The most common errors when making the approach are as follows:

- On finals, the most common error is coming to a hover with your touchdown point still in front of you. Requiring you to taxi to the pad. In other words, you fall short of the pad.
- On finals, trying to control glide slope with the cyclic instead of collective and, therefore, coming in either too fast or too slow. If too fast, you will overshoot the touchdown point; if too slow, you will stop before it or undershoot.
- During the transition from forward flight to the hover, failing to anticipate the increasing requirement for power and instead of a nice steady increase culminating in a hover, the student does one big pitch pull, which results in large torque changes and attitude changes.

Chapter 6 The Circuit

Air Exercise 6-2: The Racetrack Circuit

Air Exercise The diagram below indicates racetrack circuit. Each helicopter type and airfield will use different power settings, speeds and heights for your particular circuit and helicopter. The circuit below is not type-specific; it is merely an example of what you can expect in your training.

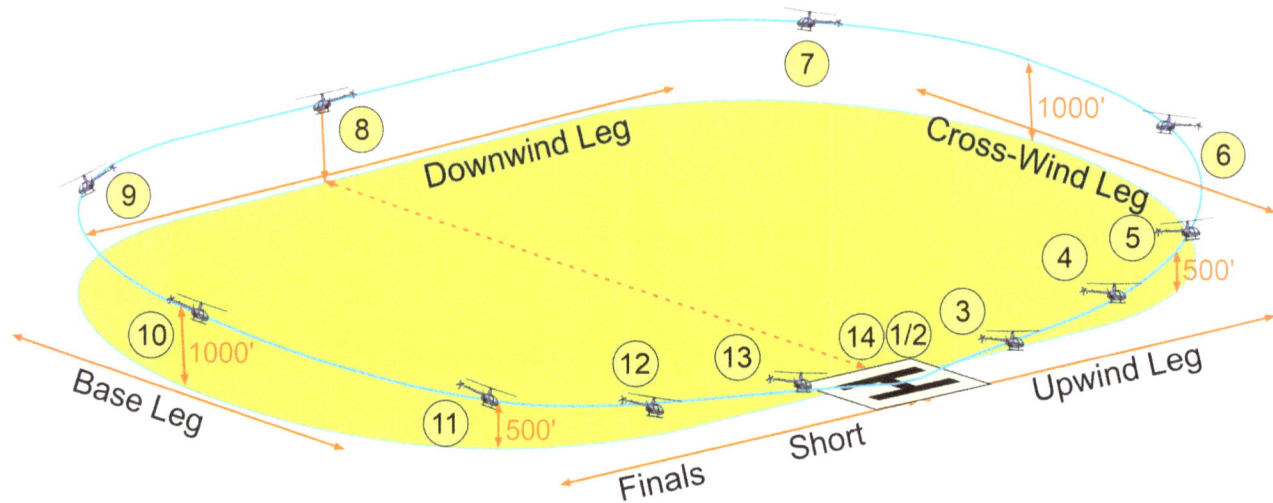

7

Basic Autorotation

Aim To conduct autorotations.

Objectives On completion of this lesson, the student will be able to:

- state at least three (3) symptoms of an engine failure
- explain the difference between a freewheeling unit and an engaging clutch
- state how to increase or decrease Rotor RPM during autorotation
- list the three (3) flare effects and what they do
- enter autorotation
- make a stabilised autorotative descent
- conduct a go-around, and
- conduct a flare and power termination to the hover.

Motivation During flight, there may come a time when the pilot experiences an emergency, such as an engine failure, that requires an autorotation.

An autorotation is the only flight condition that will keep the rotor blades turning after the engine failure and allow the pilot to maintain control of the helicopter to make an emergency landing. For this reason, it is essential to know how to do one.

Practising autorotations is designed to help the student to:

- respond automatically to an engine failure and, therefore, save valuable time in maintaining full control of the helicopter, and
- practice making decisions so that in a real emergency, the responses, drills, actions and decisions have been practised and become automatic and instinctive.

It will take time and repetition for the student to get a good feel and understanding of how the helicopter flies and responds to the flight control inputs during the transition from powered flight to autorotation, so the entry to autorotation will be practised often.

Preparation: Autorotation

Introduction

Contrary to popular belief, especially amongst our fixed-wing cousins (and your mum), the helicopter does not fall like a rock ending up in a heap of mangled bits on the ground at the bottom if the engine fails. Yes, it loses altitude quickly, and its glide ratio (the difference between height lost versus distance travelled across the ground) is not fantastic. However, unlike a fixed wing, helicopters can dramatically reduce the rate of descent and forward speed just before reaching the ground, which is a definite advantage when crashing into an unprepared landing zone.

Why don't the blades stop when the engine stops?

The question often asked is:

"Why do the rotor blades continue turning if no engine drives them?"

In powered flight, rotor drag is overcome by engine power. When the engine fails (or is deliberately disengaged from the rotor system for training), another force is required to maintain Rotor RPM.

Upon experiencing an engine failure, the pilot immediately lowers the collective to reduce drag on the rotor blades. This allows:

- inertia to keep the blades turning (for a limited time), and
- the helicopter to descend.

This descent creates an upwards airflow through the still-turning rotor blades. Lowering the collective pitch lever (to the minimum) allows the upwards airflow to strike the blades at an angle sufficient to produce:

- a positive angle of attack, and
- a driving force to maintain Rotor RPM.

When a helicopter is in autorotation, the **airflow from descending becomes the power driving the rotor blades**, just like a windmill.

Practising Autorotations

Autorotative speeds

Each helicopter will have its specific autorotative speeds and Rotor RPM limits specified by the manufacturer. Therefore, refer to the RFM for your helicopter's autorotation limitations. In this chapter, for descriptive purposes, we will assume an average speed of 60 kts indicated airspeed (KIAS) for practice autorotation.

Training altitude

An engine failure can occur during any phase of flight; however, for this initial stage of training, the basic autorotation will be commenced from a minimum height of 1000 ft AGL to an open, clear area where an emergency landing can still be made (such as over a runway at an airport) in case the engine really did fail during the practice.

School Limitations

When conducting autorotations, the crew must be fully aware of the limitations put on any practice scenarios by the Flight School so that inadvertent damage does not occur to the helicopter by taking the practice autorotation unnecessarily too far.

The most common training accident is during a practice autorotation, so take this seriously.

Chapter 7 Basic Autorotation

Sterile cockpit

When conducting autorotations, the crew must remain focused on the task and not discuss or bring up irrelevant subjects. This is referred to as maintaining a "*sterile cockpit*".

Any administrative or not-so-important items can be conducted before entering the autorotation or after recovering from the practice. Put the phone away, make any radio calls early and focus on the exercise at hand.

Airmanship

When conducting a practice autorotation, it is good airmanship to brief the crew before commencing the practice. This allows all crew members to be mentally prepared and ready to respond. It displays good Crew Resource Management (CRM).

The lookout prior is essential. An autorotation in a helicopter involves a high rate of descent at a very steep angle, so making sure the area is clear below the helicopter before commencing the practice autorotation is essential.

Emergency checklists

During an autorotation, the emergency checklist will not be able to be used as there will be no time. Nor will the pilot have any available hands or cognitive capacity (mental energy) to open a checklist and try to read it. All pilot actions during an autorotation are required to be committed to **memory.**

Remember that if an autorotation commences at 1000 ft AGL, with the helicopter descending at 2000 ft per minute, the entire sequence is completed in approximately 30 seconds.

HASEL Check: Before Practising Autorotation

HASEL Check

Before conducting autorotation practice, the crew shall conduct the HASEL check which consists of:

H	**Hatches** are secured, **harness** are done up and firm
A	The **area** is **approved** and **appropriate** to progress to either: ■ a Go Around or ■ a Power termination to the hover or ■ a full touch down to the ground
S	**Security** in the cockpit. All loose items are stowed. There is nothing in the cabin or cockpit that will make the situation worse.
E	**Engine** temperatures and pressures are normal, and instruments and radios are all set. In piston-engine helicopters, carb heat is used throughout, and consideration should be given to rapid cooling and heating of the engine.
L	**Look out:** above, below, left and right. Conduct a 360-degree turn to clear the area before commencing the practice autorotation. Make any radio calls that may be required to advise traffic of your intentions now.

| Multiple practices | If multiple practice autorotations are conducted in the same area on the same sortie, then the **HASEL** check only needs to be done once or at any other time the crew think it is appropriate. |

Symptoms of an Engine Failure

In a single-engine helicopter, having an engine failure can be quite obvious.

Because the instructor will simulate the engine failure by slowly and smoothly rolling the throttle to IDLE, the engine will not be turned OFF or failed but will no longer be driving the main rotor blades because the Engine RPM will drop well below the Rotor RPM. Therefore, the symptoms may be less obvious, and some things do not happen as dramatically or at all during a practice engine failure compared to a real one.

Depending on the model of the helicopter, symptoms of a real or practice engine failure may include the following:

Symptom	When it is obvious
Change in engine noise (sound)	Both real and practice autorotation
Yaw left (in anticlockwise rotating main rotor blades) Yaw right (in clockwise rotating main rotor blades)	Both real and practice autorotation
LOW RPM warning light and horn	Both real and practice autorotation
ENGINE OUT warning light	Real engine failure only
Generator Fail warning light	Real engine failure only
Change in engine instruments	Both real and practice autorotation

| Change in engine noise | Depending on the type of engine (piston or turbine) and what caused the failure in the first place, the engine may: |

- start to run rough, so the engine hum seems lumpy (often a sign of a dropped exhaust valve)
- start to make strange noises and a high-pitched wine (bearing failure)
- cough and/or splutter (low or dirty fuel or carb ice)
- fail gently, which is the case in a turbine that has run out of fuel, and it starts to spool down (turbine)
- suffer from stalling or surging (turbine), or
- suffer a catastrophic failure with any associated noise (bang) and/or vibrations (piston and turbine).

Whichever way the engine fails, the pilot will notice a change in engine noise and experience some changes in the torque reaction, making the tail feel twitchy.

Clutch vs Freewheeling Unit

Introduction When an engine fails, it has to automatically be disengaged from the rotor system so that the dead engine does not drag the Rotor RPM down. This is the job of the freewheeling unit.

Freewheeling unit vs the clutch There is some confusion about the difference between the freewheeling unit and the engaging clutch. Many pilots think they have one or the other, while others think they are the same thing.

Clutch An engaging clutch connects the **piston** engine to the main transmission and, therefore, the rotor blades. The clutch allows engine starts without the added drag of the drive system (transmission and blades).

For example, the B47 has a centrifugal drum clutch built into the transmission base. As the engine is started and Engine RPM increases, the clutch shoes are thrown outwards by centrifugal force, which then starts the rotation of the transmission, and in turn, the rotor blades through gears and shafts etc. The H300 and Robinson have a clutch actuator, an electrical motor that turns a worm drive or screw that, when activated by the pilot, pulls on a wire cable and a spring or moves an actuator arm that pulls or pushes a pulley out onto the rubber belts which then tighten. When these belts are tight, engine power turns the transmission. When they are loose, the engine is not connected to the transmission. Once the clutch is engaged, it stays engaged, even during autorotation.

A turbine-powered helicopter typically does not have a clutch because the engine can freely spool up without the added drag of the rotor system during start. As the engine accelerates, it takes the rotor system with it.

B47 example

B47 drum clutch

H300 clutch actuator

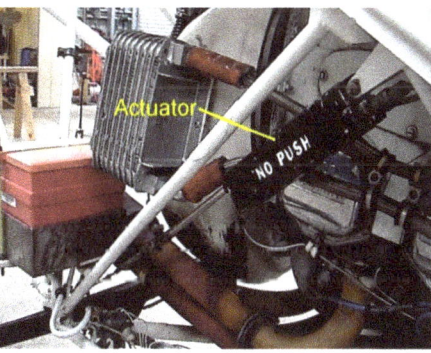

H300 example

H300 idler pulley lose

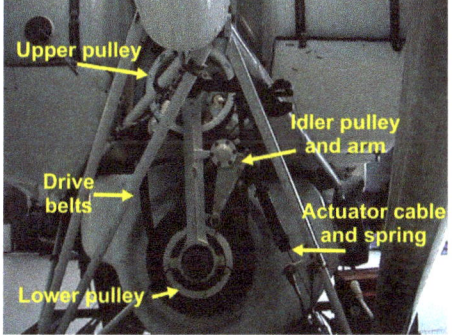

H300 idler pulley engaged

Automatic or manual clutch switch	Engaging clutches may be: - automatic, such as the B47, or - manual, such as the R22/R44, H300, Enstrom and Cabri, which require the pilot to activate a switch or engage a lever.

Freewheeling unit

When the engine fails or during training when the instructor rolls off the throttle (or in modern helicopters by merely lowering the collective and letting the correlator do the work), the Engine RPM will want to become lower than the Rotor RPM.

To prevent the lower Engine RPM from dragging the Rotor RPM down with it, the manufacturer has built in an overriding clutch or freewheeling unit (positioned somewhere between the engine and the rotor transmission system) that automatically disengages the engine from the rotor system.

It usually consists of rollers and pins that allow a driveshaft to turn freely one way but not the other. This is technically referred to as a *Sprague Clutch.*

Both piston and turbine-powered helicopters will have a Sprague Clutch (freewheeling unit) to allow autorotation without the added drag of the engine.

Most freewheeling units in small piston helicopters form part of the upper pully's inner bearing, which powers the rubber belts. If the engine fails, the blades can keep turning even though the engine has stopped.

B47 freewheeling unit   **H300 Sprague clutch**

Energy Management

Introduction
Once the engine has failed, only a limited amount of energy is available to control an approach and landing. Therefore, a basic understanding of what energy is available and how to best use that energy during autorotation is useful.

Types of energy
The types of energy stored in a helicopter during autorotation include:

- Potential energy stored as altitude
- Kinetic energy stored as forward speed, and
- Kinetic (rotational energy) stored in the momentum of the still-spinning rotor blades.

Using the energy
The pilot can use this energy to:

- In a limited capacity control the rate of descent, and
- manoeuvre to a landing spot (adjusting speed, conducting turns and adjusting rotor thrust).

Inertia and mass
Helicopter rotor systems are considered either low or high inertia based on the mass of the blades.

Low inertia rotor systems:

- take less energy to get moving, but
- stop more quickly, losing kinetic rotational energy faster.

High-inertia rotor systems:

- take more energy to get moving, but
- are harder to stop, losing kinetic rotational energy slower.

For example: A Bell 206 Jet Ranger has heavy metal blades and is considered a high-inertia rotor system. An R22 has very light metal blades and is considered a low-inertia rotor system.

Controlling Rotor RPM
The collective is the primary control for managing Rotor RPM; however, Rotor RPM can also be influenced by:

- the helicopters AUW
- cyclic movement and the manoeuvre
- changes in airspeed, and
- rates of descent.

Converting altitude into energy
The potential energy available from altitude can be converted into kinetic rotational energy:

- As the helicopter descends, the potential energy from altitude is converted into kinetic rotational energy by lowering the collective and allowing the resulting upwards airflow to drive the rotor blades, and
- kinetic energy for forward flight.

Zero airspeed

Autorotating at zero airspeed also reduces potential energy by increasing the rate of descent (and losing altitude). The faster the rotor blades turn, the greater the store of rotational kinetic energy. The amount of kinetic energy stored in the rotor system may increase due to the increased rate of descent; however, this is limited by the maximum Rotor RPM allowable.

In a low or zero-airspeed autorotation, the rotational kinetic energy stored in the rotor system is the only energy available to reduce the rate of descent before landing. In this case, a high-inertia rotor system will be more effective in arresting the rate of descent when compared to a low-inertia rotor system.

High forward speeds

Autorotating at high forward speeds reduces potential energy quickly by increasing the descent rate and wasting energy through increased drag.

However, the amount of kinetic rotational energy stored in the rotor system may increase but is limited by the maximum Rotor RPM allowable.

This forward speed may be converted into a flare to assist the rotational kinetic energy in the rotor system and alter the airflows onto the blades.

Controlling Rotor RPM

Maintaining rotor RPM

To maintain the rotational kinetic energy stored in the rotor system at engine failure, the pilot must lower the collective fully and then use the collective to keep the Rotor RPM within its autorotative design limits to:

- reduce the amount of drag produced by the main rotor blades
- create an upwards relative airflow from the rate of descent, and
- produce a positive angle of attack on the blades, allowing the blades to produce some rotor thrust.

During autorotation, the rotor blades still produce thrust equal to the aircraft weight, which allows a stable rate of descent rather than an increasing rate of descent (it's not falling).

The diagram below shows the aerodynamics of the blade within the shaded blue area.

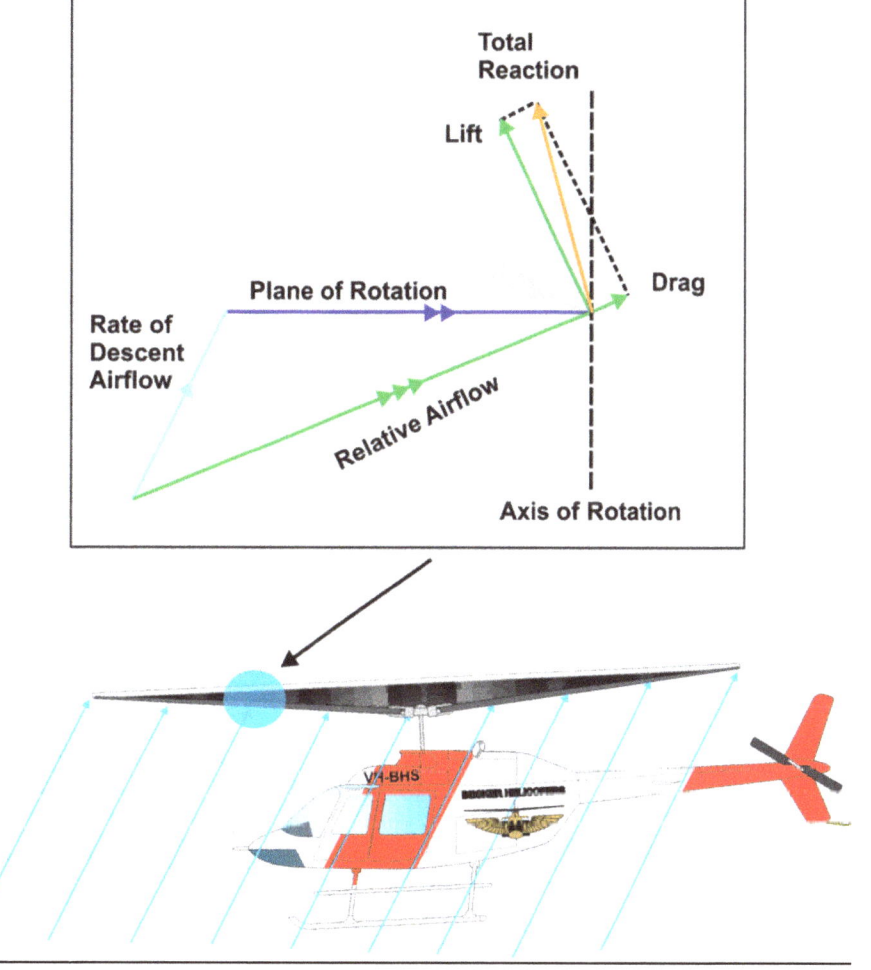

Autorotation limits

The manufacturer will have found (via testing) the optimum Rotor RPM and speeds for various autorotation configurations (endurance, normal and range). These numbers and autorotative limits are in the RFM and should be memorised by the pilot. These limits will also be marked on the airspeed gauge (in some helicopters) and Rotor RPM gauge (in all helicopters).

The optimum autorotative speed giving the lowest rate of descent usually corresponds to the minimum drag speed depicted on the drag curve.

Rotor RPM limits extended	When Engine RPM is no longer in consideration (because it has failed), the pilot can manipulate the Rotor RPM within a larger operating band. This allows the pilot to utilise the collective by raising or lowering it to control Rotor RPM and pitch, which can influence rotor thrust and rate of descent.

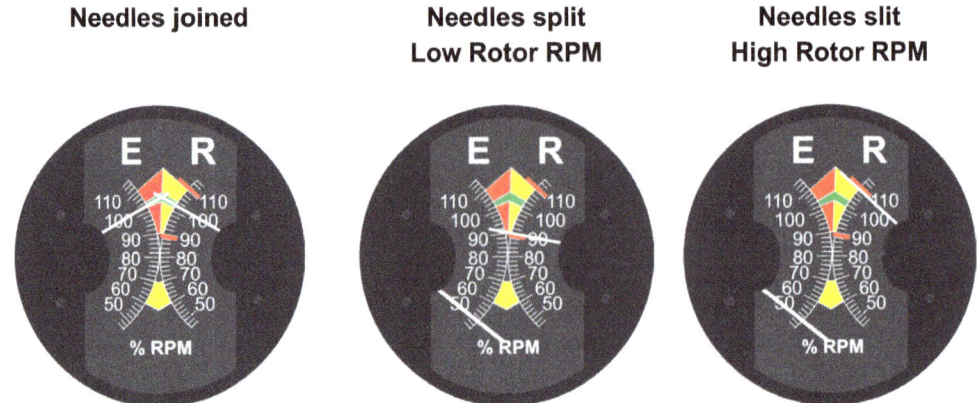

The pilot must remain within the Rotor RPM limits depicted on the RRPM gauge.

Using the collective	Assuming that Rotor RPM is kept within limits, then:

When collective is...	Rotor RPM will...	Rotor thrust will...
Raised	Decrease	Increase
Lowered	Increase	Decrease

Operating outside of the manufacturer's Rotor RPM limits can cause the opposite effect to that desired.

Exceeding the Rotor RPM limits	If Rotor RPM is: ■ **too high** (exceeding the upper limit), there is minimal rotor thrust, the rate of descent will increase more, and the centrifugal stress on the blades and head will increase to the point of causing damage. ■ **too low** (exceeding the lower limit), the total rotor thrust will now start to decrease with the lower velocity of the blades leading to high coning angles, the rate of descent will increase, and the blades can experience very high angles of attack which can lead to an unrecoverable blade stall.
Using the cyclic	Rotor RPM can also be affected by the use of the cyclic. Harsh manoeuvring or changing the disc attitude relative to the airflow can change the rate of descent airflow passing up through the disc and alter the disc loading (G forces).

If Cyclic is moved...	This exposes...	Rotor RPM will...
Aft	the **underside** of the disc to more of the oncoming airflow and increases the disc loading	Increase
Forward	the **top** of the disc to more of the oncoming airflow and decreases the disc loading	Decrease

Altitude

At altitude, due to the less dense (thinner) air, rates of descent and autorotative Rotor RPM will initially be higher (because of the reduced drag) on entry and needs to be controlled by the pilot with collective.

As the helicopter descends, the air becomes denser (thicker), and Rotor RPM will want to reduce (because of the increased drag); therefore, the pilot will need to be lowering collective to decrease the pitch and decrease the drag to maintain the Rotor RPM.

All up weight

A higher all-up weight will increase Rotor RPM, as will the additional loads imposed in a turn due to disc loading.

Heavy AUW   **Light AUW**

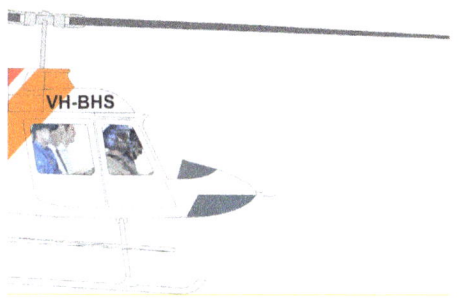 

High Rotor RPM   **Low Rotor RPM**

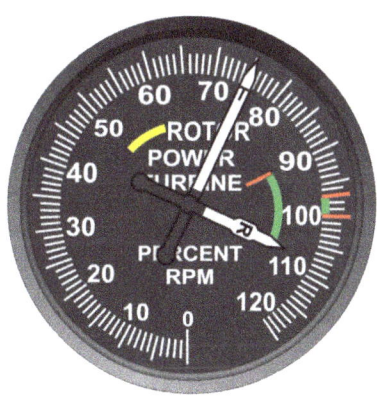 

Summary

When…	Rotor RPM will be…	To control Rotor RPM within limits, the collective is
High altitudes Hot air temperatures Heavy AUW	Higher	Raised
Low altitudes Low air temperatures Light AUW	Lower	Lowered or full down

Aerodynamics on the Blades during Autorotation

Vertical autorotation

Let's look closer at the aerodynamics of each blade during autorotation.

If we consider a vertical autorotation (because, initially, it is simpler to explain as each blade will have the same velocity), the rate of descent over each blade throughout its length will be the same, as the whole helicopter is descending at the same rate.

However, the aerodynamics vary along the length of the blade, including:

- rotational velocity, and
- pitch due to the washout incorporated into the blade design.

Autorotative regions on the blade

These different aerodynamic effects along the length of the blade result in three aerodynamic regions:

- Driven (propeller) region (where the blade's angle of attack is low)
- Autorotative (driving) region (where the blade's angle of attack is effective), and
- Stalled (dragging) region (where the blade's angle of attack is high).

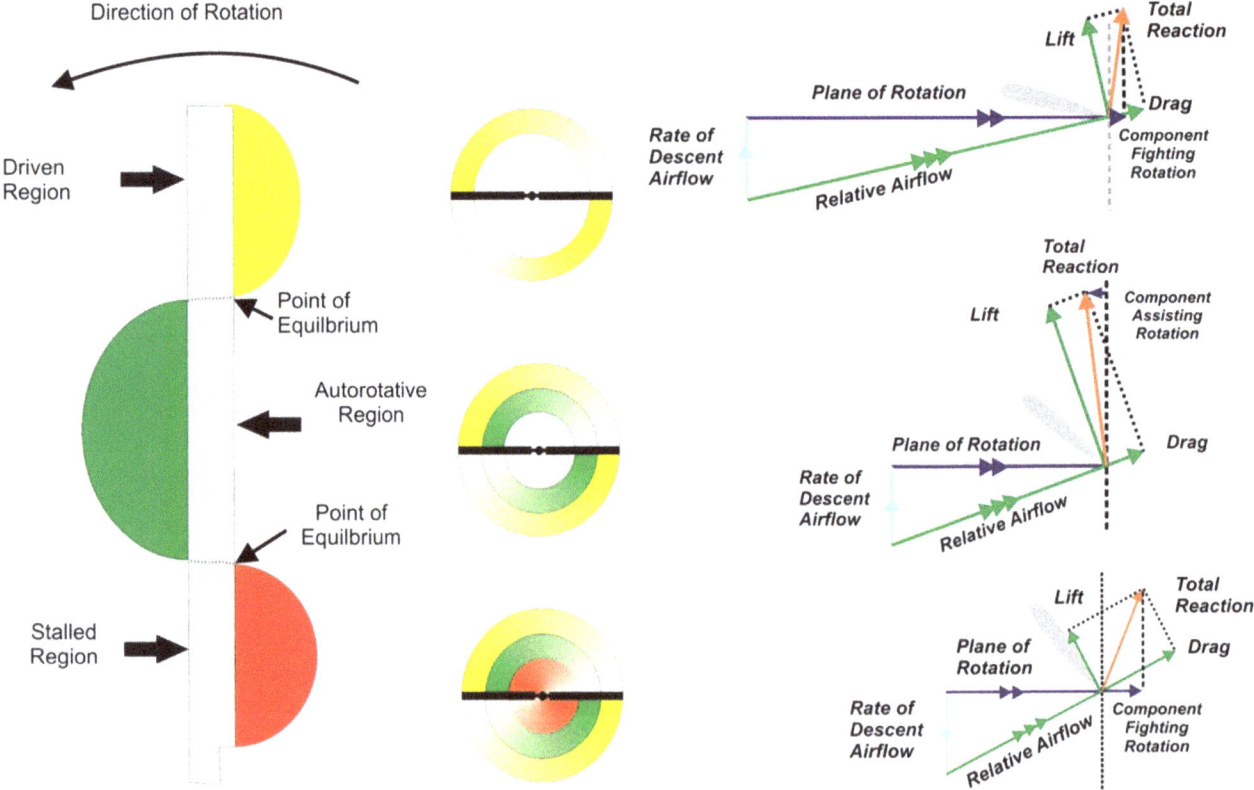

Driven region

The driven (or propeller) region at the blade's tip is the driven has a:

- high velocity
- low pitch angle (due to washout)
- same rate of descent as other regions of the blade, and
- the Total Reaction is tilted back, producing beneficial lift but acting backwards as drag.

Total Reaction will have a component fighting rotation, which will want to reduce Rotor RPM.

Chapter 7 Basic Autorotation

Autorotative region

The autorotative region at the mid-section of the blade has a:

- medium velocity
- medium pitch angle
- the same rate of descent as the other regions of the blade, and
- a productive angle of attack.

The autorotative region provides the driving force on the blade, and the Total Reaction is tilted forward, with a component assisting rotation, which will increase Rotor RPM.

Stalled region

The stalled region at the root of the blade has a:

- low velocity
- high pitch angle
- the same rate of descent as the other regions of the blade, and
- a high angle of attack.

The root section has stalled, producing drag. The Total Reaction will have a component fighting rotation, which will want to reduce Rotor RPM.

Note, though, that the drag acts upwards against weight!

Equilibrium

At the points of equilibrium, the Total Reaction is in line with the axis of rotation, and the blade is not driving forward or dragging back.

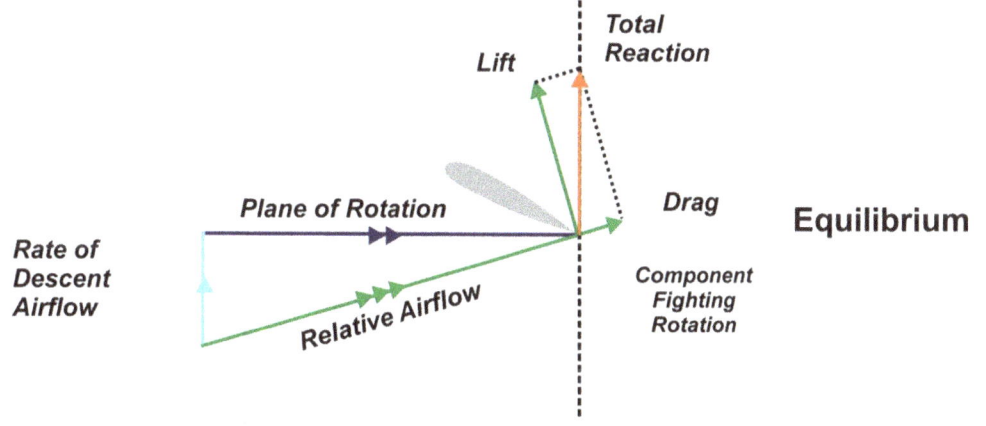

Summary

Of the total disc area, the:

- inner 25% (approx.) is operating at too great an angle of attack and is stalled, producing drag, and therefore trying to slow the rotor blades down
- central 45% (approx.) provides the autorotative force to drive the rotor system producing thrust and, therefore, trying to speed the rotor blades up, and
- outer 30% (approx.) is the propeller or driven region, which produces positive lift, but the Total Reaction is pointing backwards; therefore, the overall effect is a force trying to slow the blades down, while
- two points of equilibrium (neutral points) on the blade where the Total Reaction is in line with the axis of rotation, and the blade at that point is neither assisting or hindering the rotation of the blades.

Effect of forward speed

When autorotating at forward speed, the velocity of the advancing blade is greater; therefore, the relative airflow must change over the whole blade. The retreating blade is also affected.

The net result is the autorotative forces on the:

- advancing blade, move inwards, and
- retreating blade, move outwards.

In effect, all regions move to the retreating blade side.

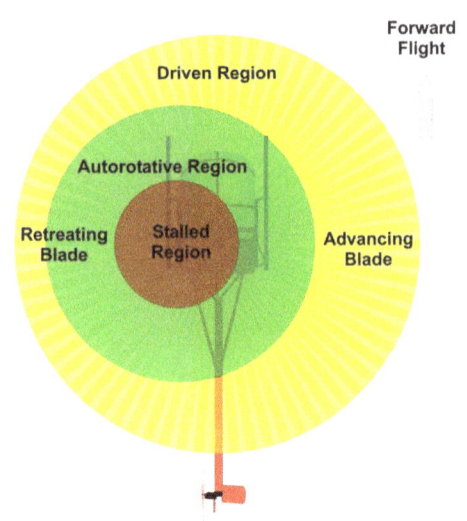

Flares during Autorotation

Introduction

The flare is designed to reduce forward speed and the rate of descent.

Part of the technique for an autorotative landing includes flaring the helicopter close to the ground by:

- applying aft cyclic, thereby raising the nose of the helicopter and reducing the rate of descent and speed before levelling out and touching down on the ground, or
- in the early stages of training, increasing power and levelling out to recover at the hover, referred to as a power termination autorotation.

Flare effects

Three distinct flare effects happen to the helicopter because of the action of the flare. They are:

- thrust reversal
- increase in total rotor thrust, and
- increase in Rotor RPM.

Trust Reversal

To initiate a flare, ease the cyclic aft, which will tilt the rotor disc backwards, and the fuselage will follow. This creates a rearward component of rotor thrust, working in the same direction as drag, which slows the helicopter down quickly. The changed fuselage attitude produces more drag, which also helps to slow the helicopter down.

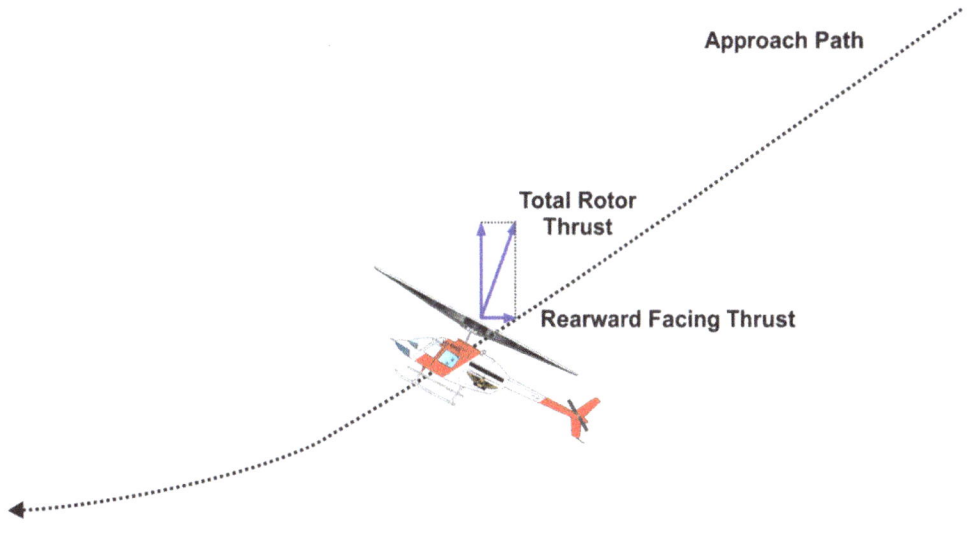

Increase in Total Rotor Thrust

Having used cyclic to initiate the flare and tilt the disc, the rate of descent airflow comes from a more obtuse (bigger) angle from below the disc, altering the angle of attack on the blades. As the angle of attack is increased on all the blades, the rotor system will experience an increase in Total Rotor Thrust even though the pilot has not raised the collective This will reduce the rate of descent and help slow the helicopter down further.

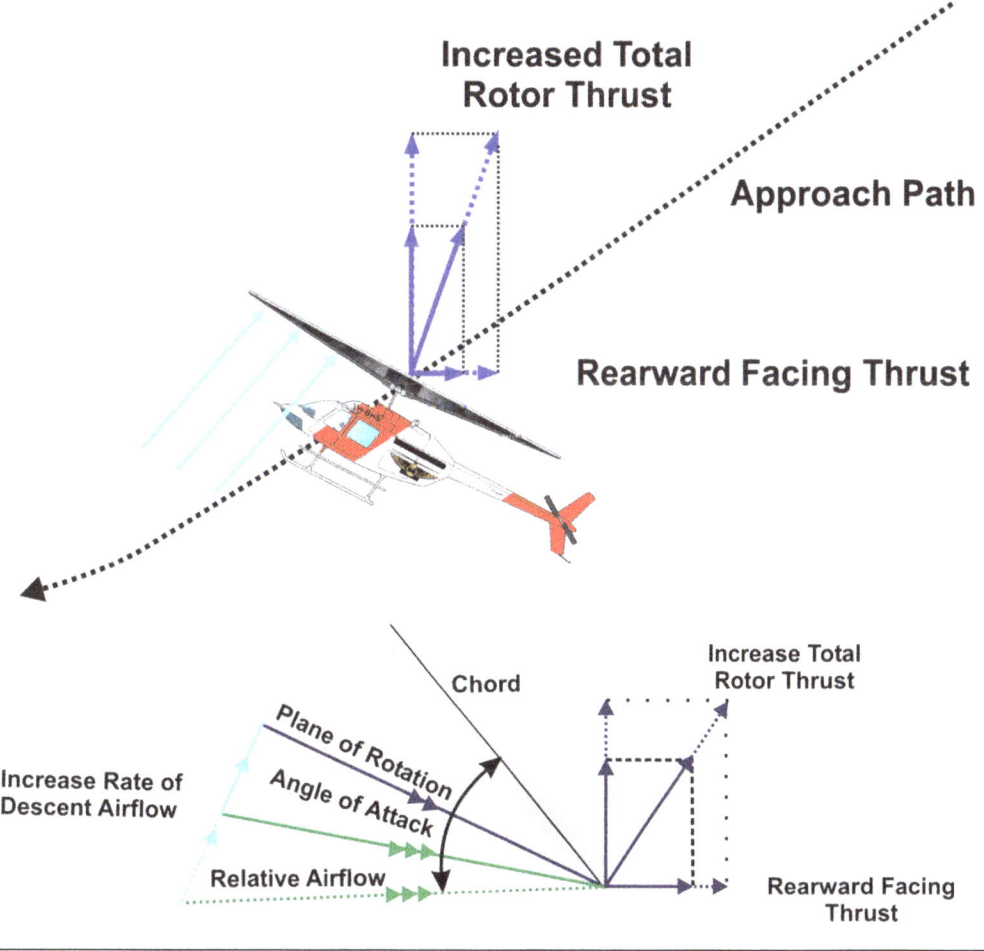

Increase in Rotor RPM

The extra lift produced on each blade by the increase in the angle of attack will affect the coning angle; the blades will begin to cone up. A greater coning angle will produce more 'Coriolis Effect', and as a result, the Rotor RPM will increase.

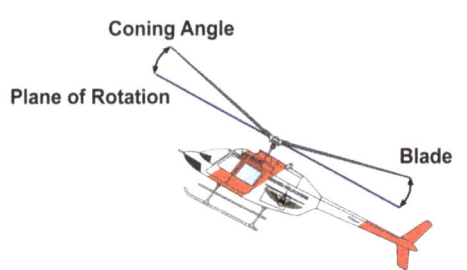

The collective can now be used briefly to maintain the angle of attack (and, therefore, rotor thrust) to cushion the helicopter onto the ground. In effect, the extra Rotor RPM is extra energy stored in the rotor system that the pilot can use.

A high-inertia rotor system is beneficial here as the effects of the Rotor RPM increase will last longer when compared to a low-inertia rotor system, where the effects disappear quickly when some pitch is applied.

Optimum speed

As the rate of descent and forward speed reduces rapidly, the effects of the flare will disappear rapidly, but if correctly used, the flare can facilitate a gentle low-speed touchdown without using the engine.

The effects of the flare will vary in direct proportion to forward speed. At very low forward speeds, flare effects are of little use in arresting a high rate of descent. A harsh flare from a high speed would cause the helicopter to balloon up in autorotation. There is, therefore, an optimum speed range in which the flare will have the desired effects. This varies with helicopter types and will be stipulated in the RFM, but it is generally between 45 – 75 kts.

Hold cyclic neutral

Once the helicopter touches the ground, the pilot will instinctively want to use aft cyclic to stop the ground run. This is to be avoided. Using aft cyclic will only cause the disc to strike the tail boom. Therefore on touch down, the cyclic must be placed in the neutral (centre) position or forward of the neutral, and it takes some discipline to do this.

To help slow the helicopter down, the pilot may lower a **small** amount of collective so that the extra weight causes friction with the ground and the helicopter begins to slow down.

Chapter 7 Basic Autorotation

Summary The diagram below summarises the flare in autorotation.

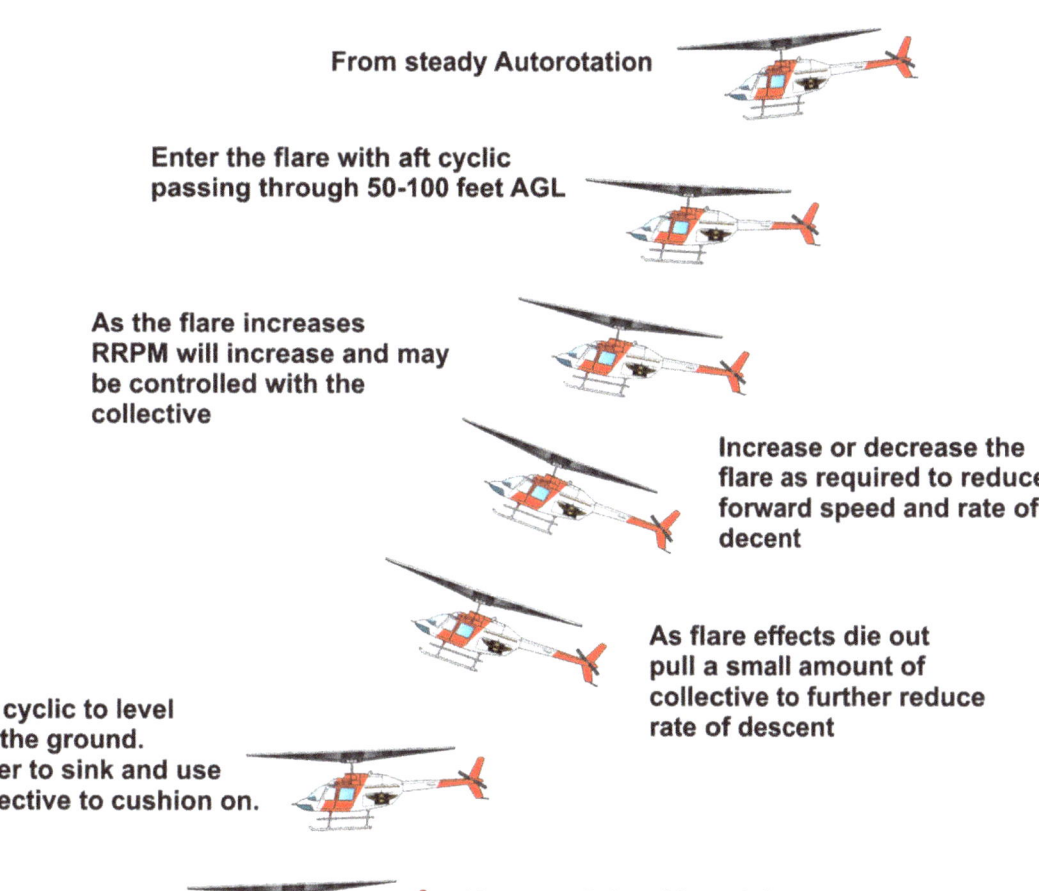

From steady Autorotation

Enter the flare with aft cyclic passing through 50-100 feet AGL

As the flare increases RRPM will increase and may be controlled with the collective

Increase or decrease the flare as required to reduce forward speed and rate of decent

As flare effects die out pull a small amount of collective to further reduce rate of descent

Apply forward cyclic to level the skids with the ground. Allow helicopter to sink and use remaining collective to cushion on.

Keep straight with pedals
Hold the cyclic forward of neutral

Air Exercises: Autorotation

Introduction	This section describes the air exercises and demonstrations to practice autorotations.
Airmanship	■ Memorise the speed and Rotor RPM limits ■ Memorise the entry procedure ■ Memorise the descent checks ■ Memorise the flare procedure ■ Memorise the Go Around procedure ■ Develop a robust lookout and situational awareness ■ Look after your engine
Common faults	The following are common faults of students: ■ Eyes are in the cockpit instead of on the horizon where they should be. ■ Not lowering the collective fully on entry. ■ Not correcting for yaw with pedals. ■ Failure to maintain a 60 kts attitude with cyclic, instead chasing the Airspeed Indicator (ASI), which has a lot of lag. ■ When collective is lowered, the rotor disc will experience less flap back effect; therefore, the nose of the helicopter will want to pitch down dramatically. Often the student does not recognise this, resulting in a dive at the ground exceeding 80 kts. Remember, when lowering collective, use aft cyclic to maintain attitude. ■ Difficulty judging the timing of the flare. ■ When commencing a go-around, failing to confirm that the Rotor RPM is within Engine RPM limits before reintroducing engine power. ■ When conducting a go-around, not maintaining the attitude and allowing the helicopter to climb, instead of accelerating and holding level. ■ No lookout, especially below.

Air Exercise 7-1: The Entry, Descent and Go-Around procedure

Introduction The instructor will demonstrate the basic autorotation's entry, descent and go-around procedure.

After the demonstration, the instructor will give the trainee all of the controls and position the helicopter at a minimum of 1000 ft AGL in a suitable area before allowing the trainee to practice.

Limits Go-around to commence before 300 ft AGL or at any time the instructor determines.

Air exercise The air exercise is detailed below.

Entry and Descent

Step	Action	Detail
1	Preparation	Conduct the *HASEL* check.
2	CRM	Confirm who is at the controls and who is responsible for the throttle.
3	Announce	Instructor: **"3, 2, 1, Practice Engine Failure"** The instructor will slowly and smoothly roll the throttle to IDLE.
4	Lower Collective	Lower the collective fully within 2 seconds.
5	Look Outside	Look out towards the horizon to determine attitude and heading.
6	Attitude	Use the cyclic to set a 60 kts attitude.
7	Keep Straight	Use the pedals to keep straight and maintain the heading.
8	Pause	Do not rush. Take several seconds here to stabilise the helicopter and get it fully under control. Breath! When ready…
9	Turn into wind and pick a spot	Turn into wind and pick a spot on the ground to aim for. In the early stages of training, the basic autorotation will be commenced into wind at the aerodrome or commenced at a high altitude away from the airfield with a go-around. So this item may not be important to consider in the early stages, but it becomes more important the closer to the ground the autorotation is commenced.
10	Descent	Continue straight ahead and conduct the descent checks: **Check** / **Description** **Engine RPM** — Engine Idling **Rotor RPM** — Rotor RPM in the Green **Balance** — In balance **Speed** — Set for 60 kts

Decision and Go-Around

Step	Action	Detail
11	Decision Height	**INSTRUCTOR:** Between 700-500 ft, announce: *"Go Around"*.
12	Reintroduce Power	**INSTRUCTOR:** By 500 ft AGL, the instructor will: - Check that Rotor RPM is within the engine limits, then - In a piston, roll the throttle to join the engine and rotor needles and call "CONFIRM NEEDLES JOINED." - In a turbine, roll the throttle to full and call "CONFIRM THROTTLE FULL."
13	Confirm power	The student shall then also confirm the throttle position: - Physically check the throttle position - Check the needles are joined - In a piston, call "NEEDLES JOINED CONFIRMED." - In a turbine, call "FULL THROTTLE CONFIRMED." **If the student notices that the instructor has not done this step, they must speak up.**
14	Go-Around	Commence a Go-Around:

Check	Description
Power	Raise collective to climb power.
Attitude	Maintain an attitude for 60 kts (40-70 kts acceptable initially) with some forward cyclic. Allow the helicopter to commence a climb; do not try to accelerate level.
Balance	Maintain balance with pedals throughout.
Recovery	Commence a climb and a turn if required, and go around and set up for the next exercise.

Air Exercise 7-2: Power Termination procedure

Air exercise The entry and describe are described in the previous air exercise, this air exercise continues from step 10 to conduct a power termination rather than a go-around.

Decision and Power Termination

Step	Action	Detail
11	Decision Height	**INSTRUCTOR:** By 700 ft AGL, *consider* the options available, which are to conduct a: • Go-Around, or • Power Termination. **By 500 ft AGL, the *decision* and the appropriate actions must be taken.** Announce the decision. In this case 'CONTINUE FOR A POWER TERMINATION' Remember, though, that if the autorotation is not going as planned at any time, the instructor can change the plan and give the command to "Go Around".
12	Reintroduce Power	**INSTRUCTOR:** The instructor will: ■ Check that Rotor RPM is within the engine limits, then ■ In a piston, roll the throttle to join the engine and rotor needles and call "CONFIRM NEEDLES JOINED." ■ In a turbine, roll the throttle to full and call "CONFIRM THROTTLE FULL."
13	Confirm power	Student confirms throttle position: ■ Physically check the throttle position ■ Check the needles are joined ■ In a piston, call "NEEDLES JOINED CONFIRMED." ■ In a turbine, call "FULL THROTTLE CONFIRMED." *If the student notices that the instructor has not done this step, they must speak up.*
14	Manage descent	Manage descent:

Check	Description
Power	Use collective to control the Rotor RPM but remain in a steady state of autorotation.
Attitude	Maintain an attitude for 60 kts with cyclic throughout. (50-70 kts IAS acceptable)
Balance	Maintain balance with pedals throughout.
Recovery	Maintain a consistent autorotation configuration until passing 100-50 ft AGL.

Step	Action	Detail
15	**Flare**	Use aft cyclic as required to flare the helicopter and reduce the rate of descent and forward speed.
16	**Develop**	Develop the flare with more or less aft cyclic as required to make the touchdown point.
17	**Hold**	Once the helicopter is set in an attitude that the pilot feels will allow the helicopter to arrive at the ground with zero forward speed and zero rate of descent, then hold that attitude until approximately 10 ft AGL.
18	**Initial**	At 10 ft AGL, quickly pull the collective up a small amount (approximately two (2) inches of travel) and hold it there. This is called an **'initial pitch pull'** or a **'pop'** of the collective. It is a quick little 'jerk up' of the collective. This initial pitch pull is just enough collective so that the additional Rotor RPM (energy) can be converted into rotor thrust to further arrest (slow) the helicopter's rate of descent and forward speed.
19	**Level**	When doing the initial pitch pull, use some forward cyclic to level the skids with the ground.
20	**Pull collective to hover**	As the helicopter comes level with the ground, raise collective enough to return to the hover in a power termination. Because the helicopter is now returning to powered flight, the pilot may experience some excessive YAW as the engine power is reintroduced, and the Rotor RPM may droop. This is normal and will require the pilot to correct as required to keep straight, maintain Rotor RPM and come back to the hover.

Air Exercise 7-3: Full Touch Down Autorotation

Introduction The only difference between a good Power Termination and a Full Touch down autorotation is the introduction of power at the hover.

In a full touchdown auto, the engine power will not be available, so once the helicopter has levelled, the pilot must allow the helicopter to sink the last 10 ft to the ground and delay the last pitch pull to cushion on.

Air exercise The air exercise continues from step 18 of the previous air exercise.

Full Touch Down

Step	Action	Detail
19	Level	When doing the initial pitch pull, use some forward cyclic to level the skids with the ground.
20	Sink	**Allow the helicopter to then sink towards the ground. At this point, the helicopter could still be moving forward at a very slow taxi speed, leading to a run-on landing or the helicopter may actually be stopped and be sinking vertically towards the ground.**
21	Pull to cushion on	Once the helicopter passes approximately 5 ft AGL, raise the collective to use any remaining energy in the rotor system to cushion the helicopter onto the ground.
22	Keep Straight	Maintain heading with pedals. From a practical point of view, this is done by looking outside and keeping the skids aligned with the direction of movement. If the skids are side on to the direction of movement, the helicopter could roll over on ground contact, so keeping straight by looking outside is vital.
23	Cyclic	Use the cyclic to maintain the ground position and keep it slightly forward so the main rotor blades are kept away from the tail boom.
24	Lower collective	Once the helicopter comes to a stop, gently lower the collective fully. At that point, the exercise is completed.

Mike Becker's Helicopter Handbook

Summary Diagram: Autorotation to Power Recover

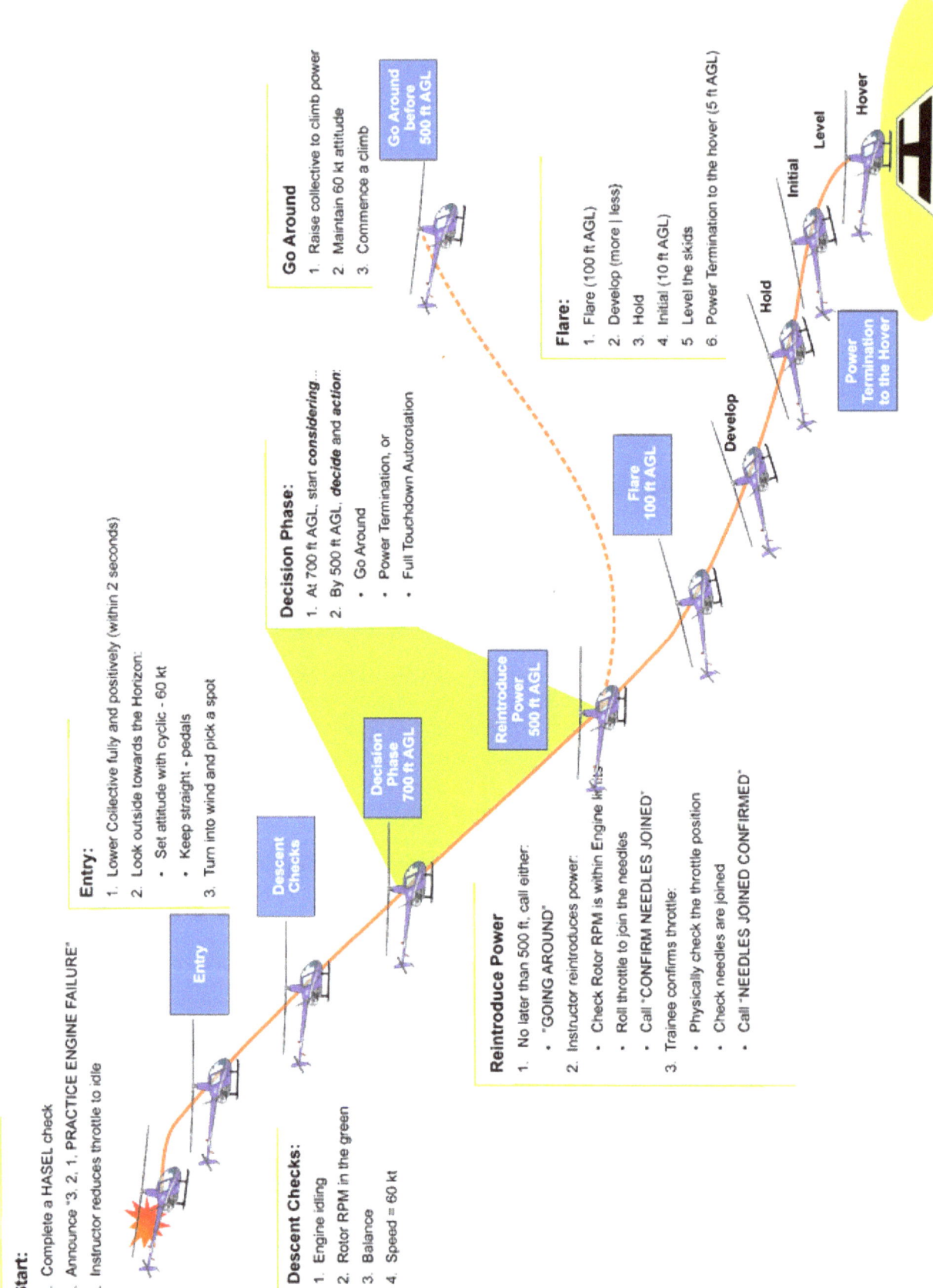

Manoeuvring during Autorotation

Aim
To manoeuvre the helicopter during autorotation to make a selected touchdown area.

Objectives
On completion of this lesson, the student will be able to:

- state how to configure the helicopter for autorotative range
- state how to configure the helicopter for autorotative endurance (a minimum rate of descent)
- state how to configure the helicopter for a Constant Attitude autorotation by adjusting speed
- turn the helicopter in autorotation, including, S-Turns, 180 and 360-degree turns, and
- identify suitable touchdown areas based on the helicopter's current height and speed and determine the appropriate method to make that area after an engine failure to complete a precision autorotation.

Motivation
Once the pilot has entered autorotation, they need to manoeuvre the helicopter to make a safe touchdown area. In real life, this could mean making some big manoeuvres and deciding early what to do as the areas may be limited, too close or too far away for a standard autorotation. Knowing how to extend the glide slope and go as far as possible (Range), stay in the air for as long as possible (Endurance), adjust speed and make turns to manage height and distance across the ground, and set the helicopter up to fall into a smaller area (constant attitude) are paramount to a safe landing.

Preparation: Manoeuvring during Autorotation

Descent Configurations

Angles of descent The angle of descent (glideslope) can be varied by altering collective pitch, increasing or decreasing the rate of descent, adjusting speed and turns or combinations of them together. This is done so a pilot can make a nominated touchdown area.

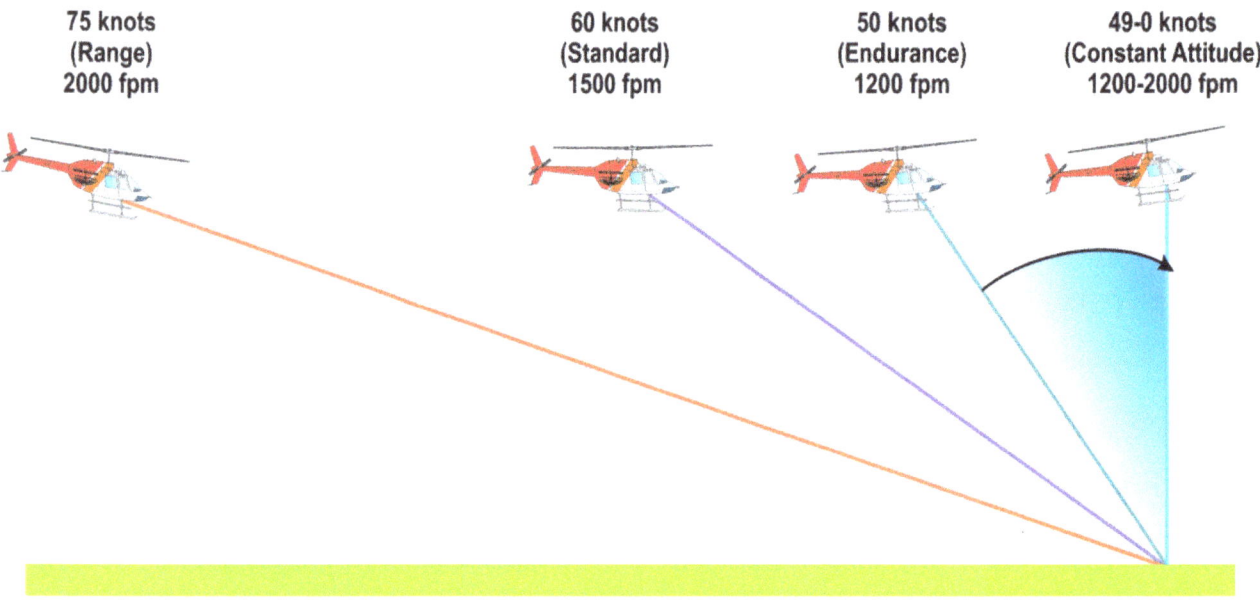

Autorotative range Autorotative range allows the pilot to **go as far as possible**. The best glide distance (range) is achieved by:

- using some forward cyclic to achieve a higher airspeed, and
- raising some collective to achieve the minimum allowable Rotor RPM and maximum Rotor Thrust.

For example, if the normal autorotative speed is 60 kts with Rotor RPM set in the mid-green, then for range, the speed is increased to 75 kts with Rotor RPM reduced to the bottom of the yellow.

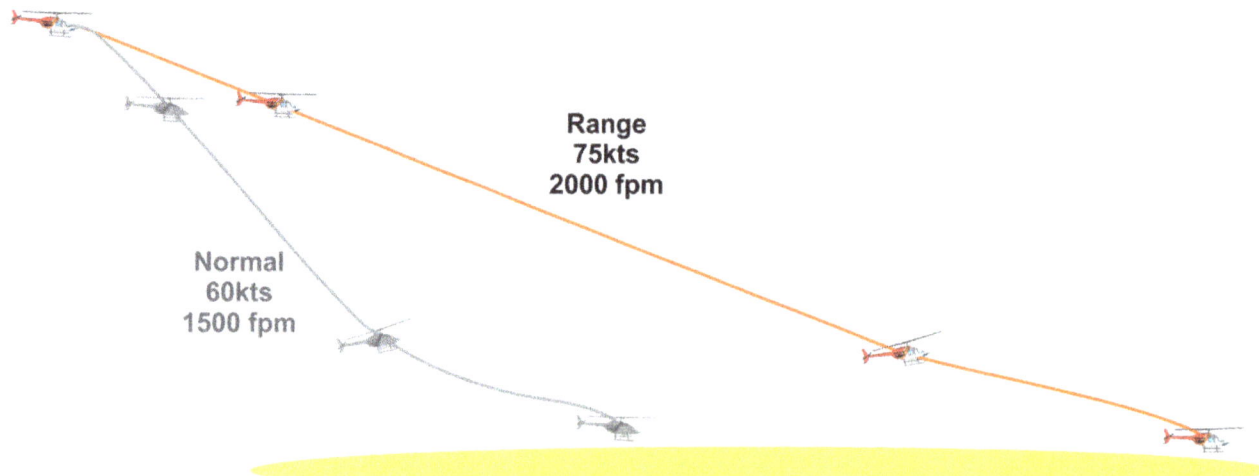

Although it seems counterintuitive, higher speeds and lower Rotor RPM will generate a higher rate of descent, but the extra speed will mean a better glide ratio, so the helicopter will cover a greater distance in a shorter time.

Chapter 8 Manoeuvring during Autorotation

Autorotative endurance

Autorotative endurance allows the pilot to stay in the air **for as long as possible** with **a lower rate of descent**.

This could be necessary for instrument meteorological conditions (IMC) or at night when the pilot cannot see how close the ground is. A minimum rate of descent autorotation is achieved at the minimum allowable rate of descent airspeed (as stated in the RFM and corresponds to the minimum drag speed for that helicopter) and minimum allowable Rotor RPM by raising some collective.

For example, if the normal autorotative speed is 60 kts with Rotor RPM set in the mid-green, then for endurance, the speed is reduced to 50 kts, and the Rotor RPM reduced to the bottom of the yellow range (to get maximum rotor thrust).

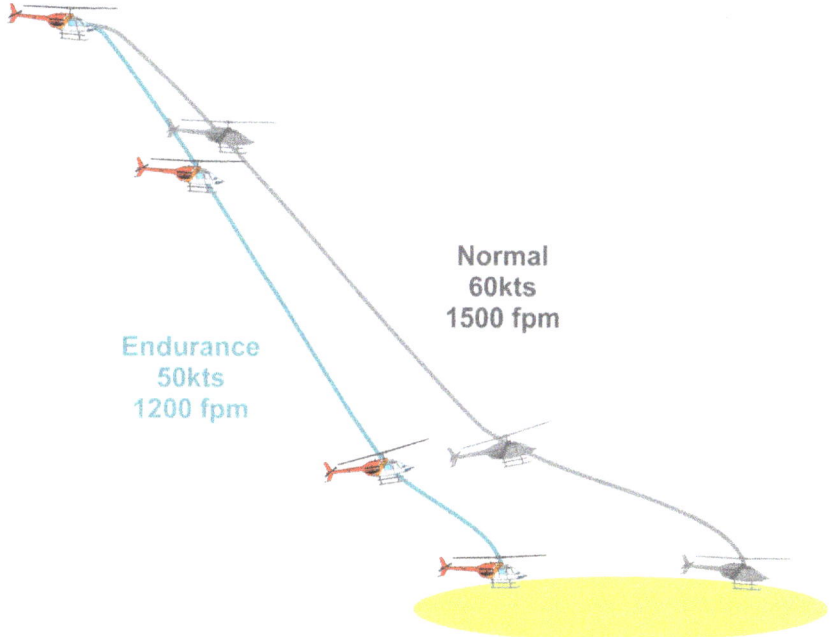

Lower speeds and lower Rotor RPM will generate a lower rate of descent (ROD). The lower speed will mean a worse glide ratio, but the helicopter will have more time in the air because the ROD is less. This may be crucial in certain circumstances to give the crew time to prepare for the pending landing (crash).

Constant Attitude autorotation

Autorotative constant attitude is normally required when:

- overshooting the selected landing area, or
- attempting to approach a small confined area.

Constant attitude autorotations allow the pilot to descend at a **constant angle** with a lower airspeed to make a tight landing area where a flare and forward speed may not be the best option.

Pilot technique and rotor design are going to be factors to consider before committing to a contant attitude autoration, given the lower speed, reduced ability to flare and arrest the rate of descent (ROD) at the bottom.

In a helicopter, the maximum angle would be straight down at zero airspeed; however, we also have the added advantage of being able to go backwards in autorotation, so the term maximum angle is used loosely.

The success of the landing relies solely on the judgment and skill of the pilot.

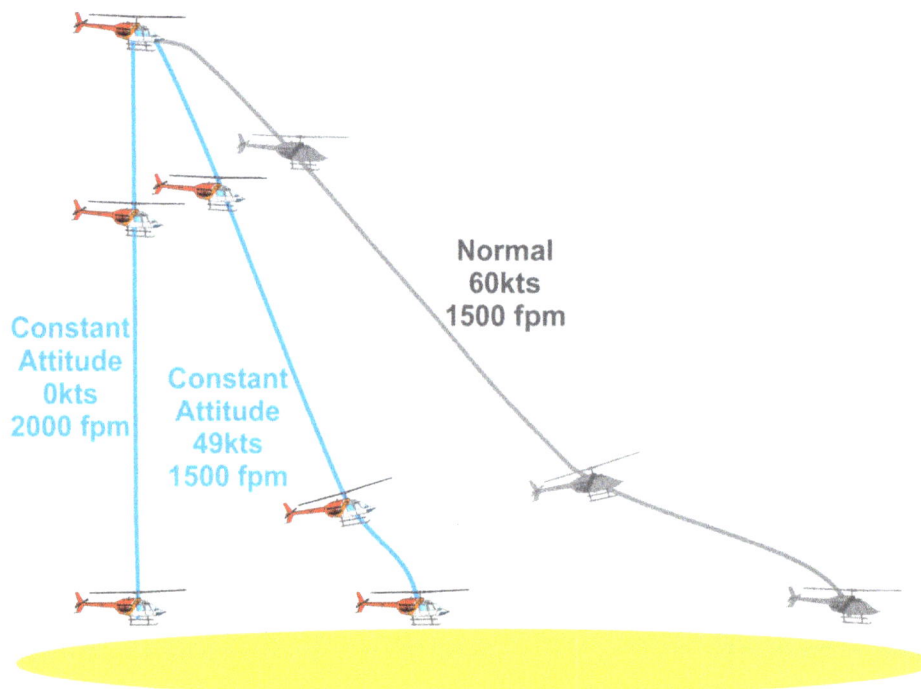

Turns during Autorotation

Introduction

During autorotation, the helicopter's flight characteristics remain the same as powered flight, except it is in a constant high rate of descent. Therefore, turning during the autorotation is done in the same manner as in powered flight, by using the cyclic for pitch and roll and pedals for balance.

Effect of turns

Turning will lead to:

- increased rate of descent (ROD)
- increased Rotor RPM due to an increased disc loading (felt by the pilot as a "G" force)
- reduced range (reduced distance across the ground), and
- reduced endurance (reduced time in the air).

Controlling Rotor RPM

When turning in autorotation, Rotor RPM must be controlled by raising or lowering the collective.

When entering a turn, the Rotor RPM will tend to increase because:

- disc loading will increase, and
- ROD will increase, forcing more air through the rotor disc.

Therefore, a small amount of collective needs to be raised to control Rotor RPM.

When exiting the turn, the Rotor RPM will tend to decrease because:

- the disc loading has been reduced, and
- ROD will reduce, and less air will be forced through the rotor disc.

Rolling out of the turn, the pilot will still have the collective raised, so it now must be lowered back to its original position to control Rotor RPM.

The inertia of the rotor system will have a significant effect on the changes in Rotor RPM during a turn. A high-inertia rotor system (such as the Bell Jet Ranger) will not experience big changes in Rotor RPM when in a turn compared to a helicopter with a low-inertia rotor system, such as the R22, which can experience big changes in Rotor RPM.

Use of turns

Turns during autorotation are useful in helping the pilot manoeuvre to a nominated touchdown area while maintaining a constant airspeed, compared to reducing speed.

The **disadvantage** is that the pilot may turn away from the nominated touchdown area and could lose sight of it as well as significantly increase the ROD.

Because turns are done in the same manner as in powered flight, they are easy to do and manage.

Types of turns during an autorotation

There are three (3) standard turns used in autorotation they are:

- 180-degree turn
- 360-degree turn, and
- S-Turns.

180-degree turns

The 180-degree turn is used when the helicopter is downwind and abeam the nominated touchdown area. If the pilot has a problem and enters autorotation, the wind will be directly behind at the time of the failure.

The correct sequence is for the pilot to enter autorotation and then immediately conduct a 180-degree turn into the wind so that the remainder of the autorotation is done into wind.

180-degree turns are usually done from any height above 500 ft AGL.

If lower than 500 ft AGL, the pilot may have to consider conducting the autorotation with a tailwind due to the lack of height and time to be able to do the 180-degree turn. Pilot experience, the circumstances and the terrain will all play a part in deciding to conduct a 180-degree turn below 500 ft AGL.

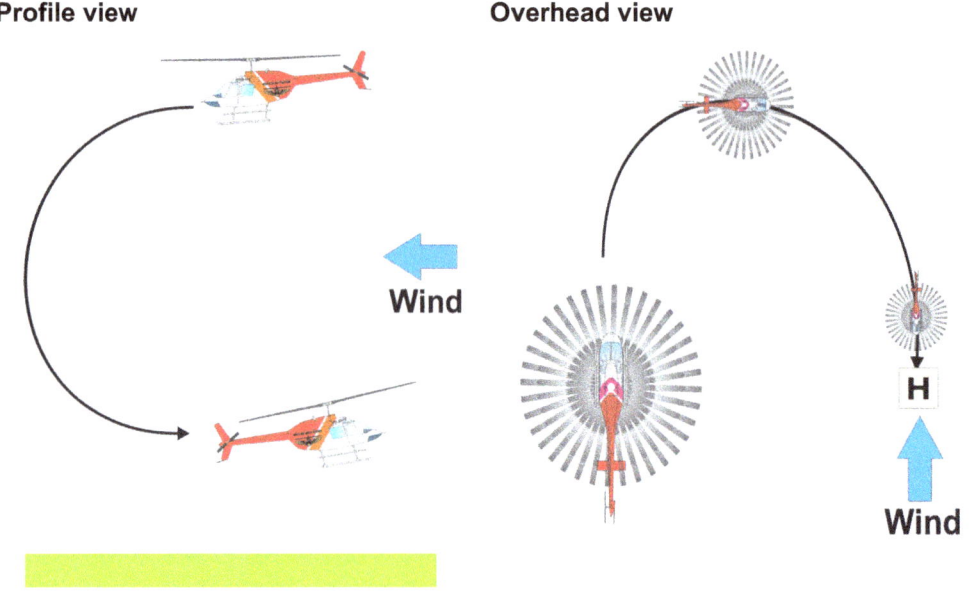

S-Turn autorotation

The S-Turn is used when the pilot needs to lose some height but does not want to travel across the ground to make a nominated touchdown area.

The correct sequence is for the pilot to enter autorotation and then immediately conduct one or more S-Turns as required.

S-Turns can be done from any height all the way to the flare height.

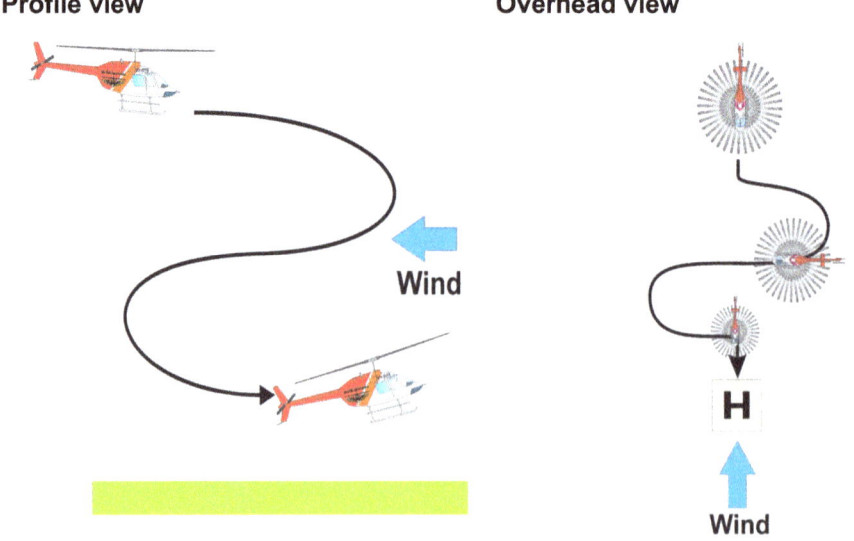

Chapter 8 Manoeuvring during Autorotation

360-degree turns

The 360-degree turn is used when the helicopter is very high and has a lot of height to lose, and the pilot wants to keep the nominated touchdown area in sight.

The correct sequence is for the pilot to enter autorotation and then conduct a 360-degree turn so that the helicopter is positioned the last 500 ft AGL into wind and able to make the nominated touchdown area.

360-degree turns are usually done from any height above 1000 ft AGL.

If lower than 1000 ft AGL, the pilot may have to consider other methods, such as S-Turns or a constant attitude autorotation, as the 360-degree turn can be difficult to judge and manage if commenced below 1000 ft AGL.

Pilot experience, the circumstances and the terrain will play a part in deciding to conduct a 360-degree turn below 1000 ft AGL.

Profile view   **Overhead view**

Effect of wind

Wind will play a role in managing a turn while in autorotation.

The stronger the wind, the easier it is to **miss** the nominated touchdown area if relying on turns to control the glide slope.

In general, winds up to 15 kts are very manageable and predictable. With winds greater than 15 kts, the pilot needs to ensure the nominated touchdown area remains in sight and close to the helicopter.

Wind and 180-degree Turns

When conducting a 180-degree autorotation with 15 kts or less wind, enter autorotation with the touchdown area abeam or slightly behind as viewed from the pilot's seat.

If the wind is very strong, the touchdown area needs to be in front of the helicopter on entry, or you will miss it.

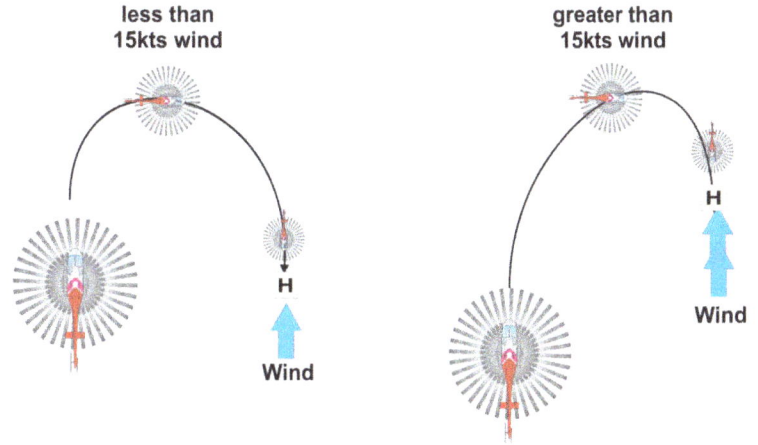

Wind and S-Turns When conducting S-Turns during autorotation and the wind is very strong, the touchdown area needs to be very close in front of the helicopter on entry, or you will miss it.

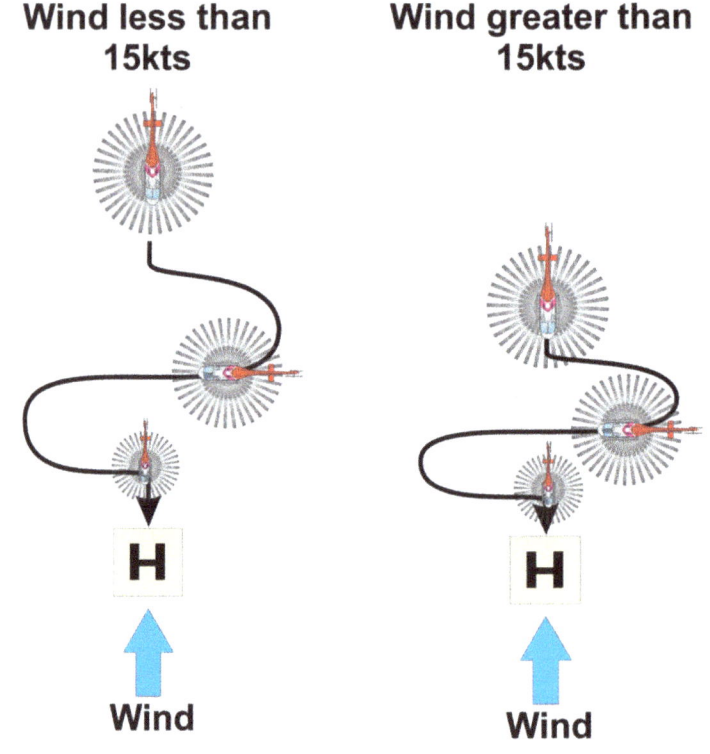

Wind and 360-degree turns When conducting a 360-degree autorotation with 15 kts or less wind, enter autorotation with the touchdown area in front or underneath the helicopter.

If the wind is very strong, the touchdown area needs to be directly underneath or, in some cases, behind the helicopter on entry to autorotation, or you will roll out with it way in front of you and miss it.

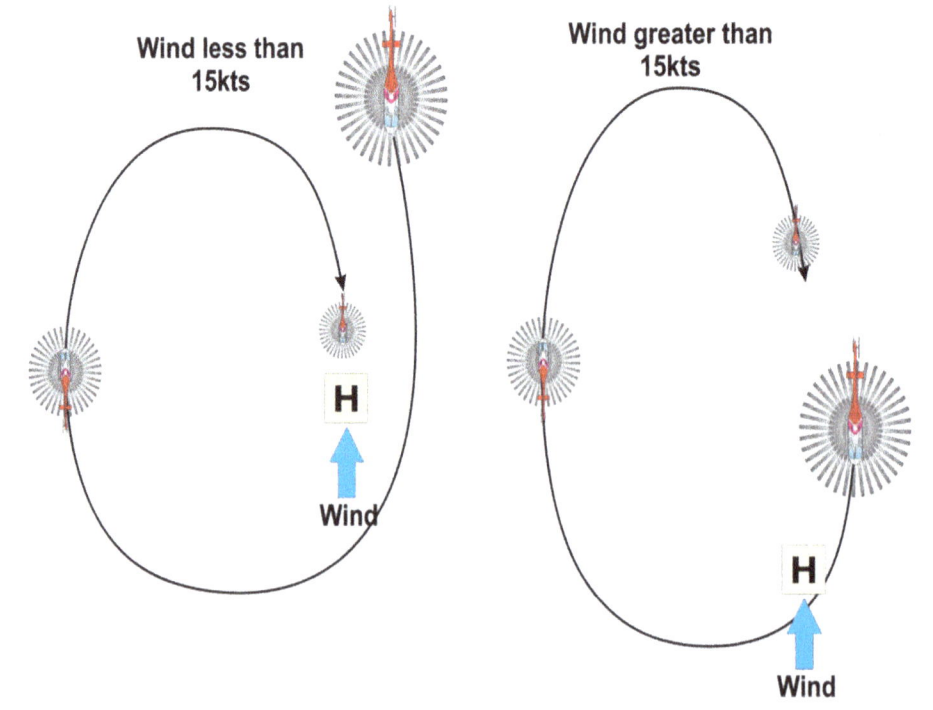

Chapter 8 Manoeuvring during Autorotation

Adjusting Speed during Autorotation

Adjusting speed Adjusting speed during an autorotation is a much more effective method in manoeuvring to a selected touchdown point because:

- rates of descent are controlled better and are usually lower
- Rotor RPM remains more constant
- range can be lengthened and shortened as required
- the pilot can keep the selected spot in sight at all times, and
- the pilot's workload in simply controlling the helicopter is much reduced.

Using speed The manufacturer will have found the optimum Rotor RPM and speed to conduct an autorotation, and basic training usually only centres around these pre-set numbers. However, flying with an experienced instructor and pushing the envelope in both speed and Rotor RPM is well worth the investment.

High forward speeds Autorotating at high forward speeds reduces potential energy quickly by increasing the descent rate and wasting energy through increased drag. However, the amount of kinetic energy stored in the rotor system may increase but is limited by the maximum Rotor RPM allowable.

The forward speed may be converted into a flare to assist the rotor system's kinetic energy and alter the blades' airflows.

Low forward speeds Autorotating at zero airspeed reduces potential energy by increasing the rate of descent. However, the amount of kinetic energy stored in the rotor system may increase but is limited by the maximum Rotor RPM allowable.

In the case of a low or zero-speed autorotation, the energy stored in the rotor system is the only thing available to reduce the rate of descent before landing. At zero speed, the pilot cannot flare. In general, the lower the speed, the less ability to flare at the bottom of an autorotation.

Using speed to manoeuvre When using speed to manoeuvre, it is desirable to re-establish the optimum autorotation speed when passing through 500 ft AGL ensuring sufficient speed for a flare.

However, if the terrain on landing is not conducive to an autorotative flare, the speed may be kept low, and a constant attitude is held to the ground (or at least 10 ft above the ground). This relies on Rotor RPM and, therefore, an increase in the angle of attack (and, therefore, rotor thrust) at the required time to instantly reduce the rate of descent.

This will require the pilot to have the timing of the collective pull just right, and if a mistake is made, there is no other option, and the landing will be hard.

In theory, it is possible to conduct a vertical autorotation with zero airspeed all the way to the ground. In practice, high-inertia rotor systems are much more capable of this type of autorotation, if flown by an experienced pilot. Low-inertia rotor systems can still do it, but the pilot needs to be perfect. However, it is much better to conduct a constant attitude auto to the ground at night or amongst tall trees instead of hitting those trees at 60 kts and 100 ft up in the canopy.

As with most things, practice and experimentation to see the various autorotation configurations that suit a particular helicopter type make for a safer and more confident pilot.

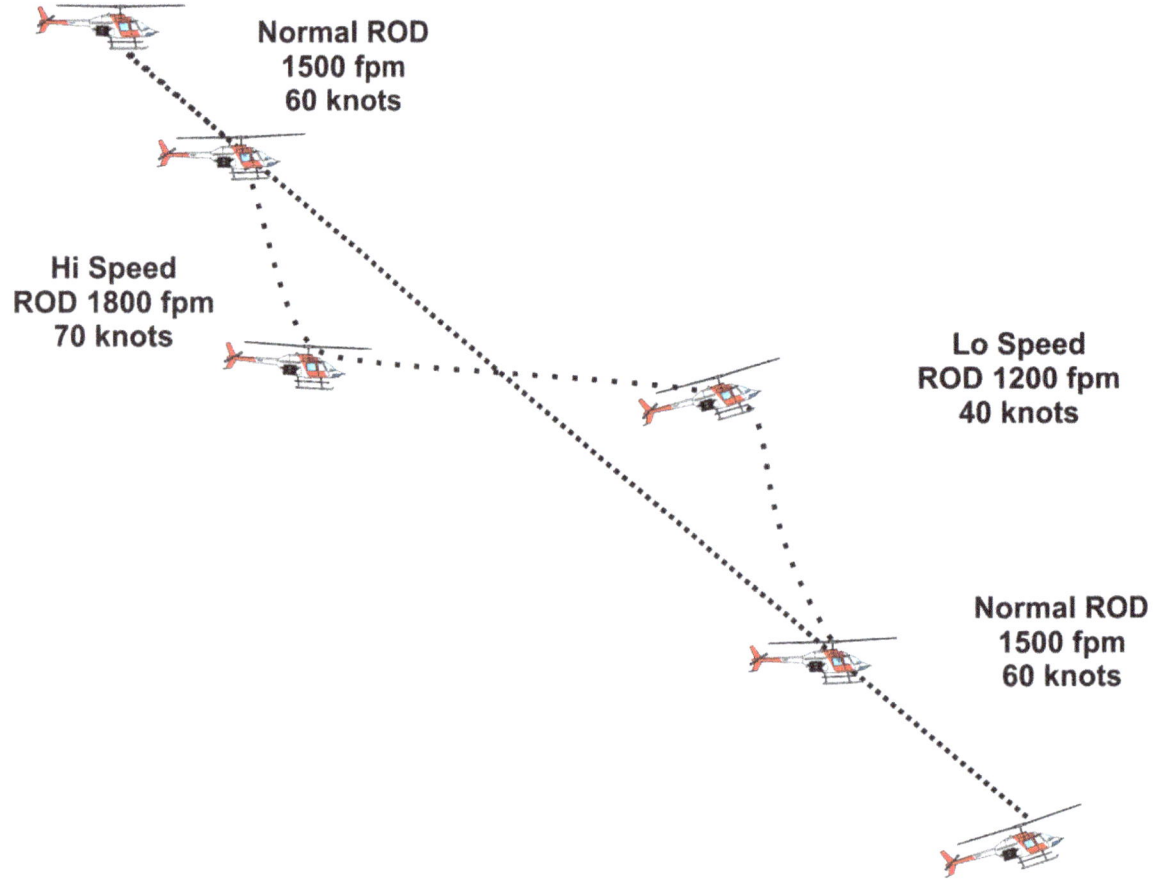

Precision Autorotation

Introduction

Once the pilot has learned the different methods to manoeuvre the helicopter while in autorotation, the next step is to start identifying suitable termination areas that the helicopter can make given the variations of height, speed, Rotor RPM, turning and wind and be able to decide which technique is best used to make the nominated touchdown area. Making a nominated touchdown area is referred to as a **Precision Autorotation**.

Arc of influence

A helicopter can only glide so far, much less than a fixed wing. How far it can glide will depend on the initial height AGL on entry to the autorotation, the wind and how the pilot decides to manoeuvre.

The pilot will have to choose the most suitable touchdown area that is within gliding distance of the helicopter and that will maintain a headwind component for the termination. This is referred to as the **"arc of influence"**.

The **higher** the helicopter is above the ground, the **more choices** the pilot will have to choose a suitable touchdown point within the arc of influence.

The **lower** the helicopter is above the ground, the **fewer options** the pilot will have to choose a suitable touchdown point within a smaller arc of influence.

Profile view

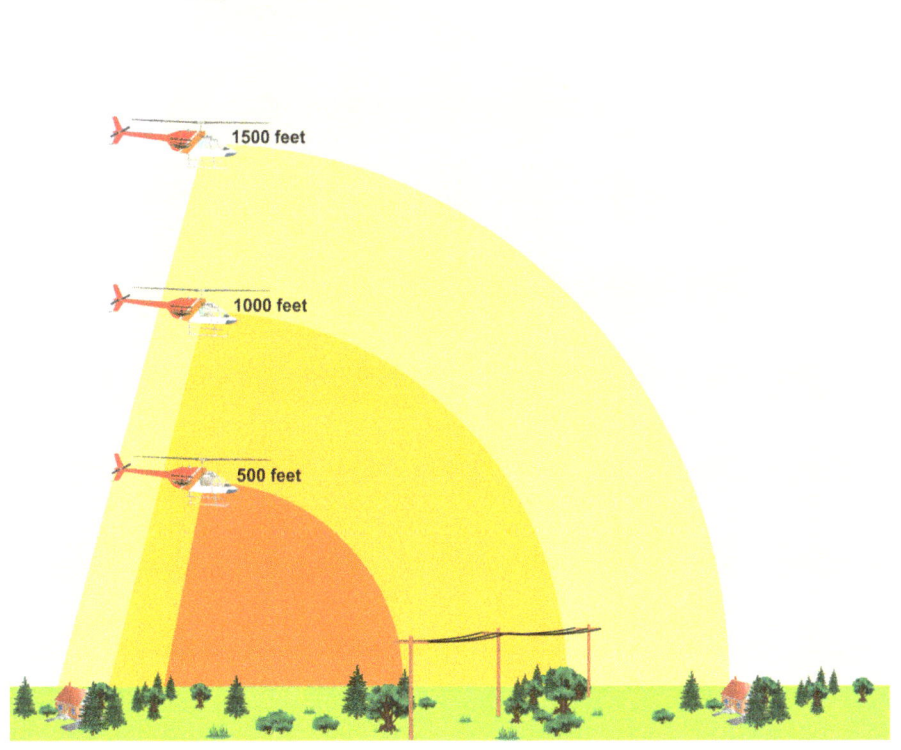

Overhead view

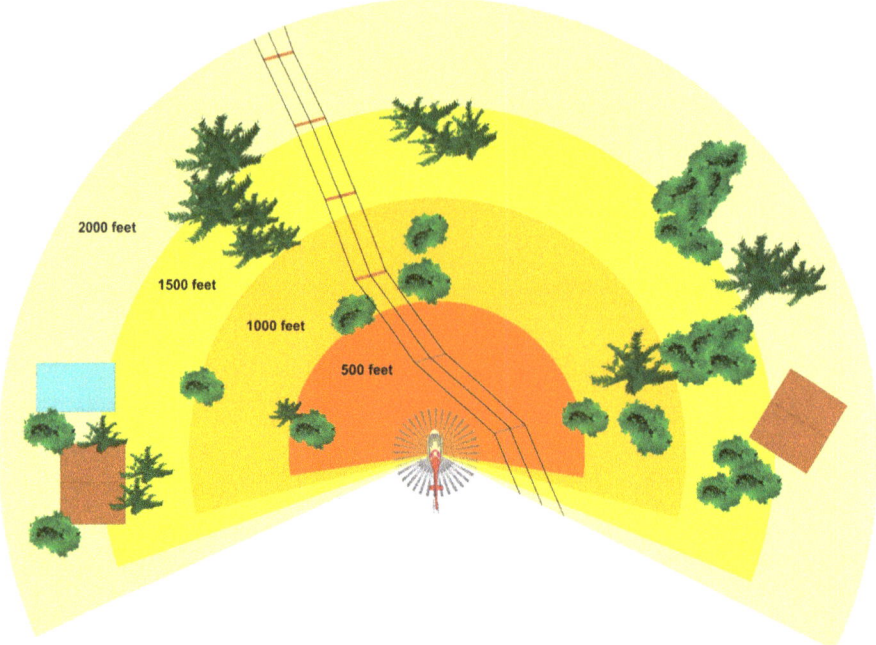

From the pilot's point of view, the outer edges of the arc of influence will always be just above the nose section of the helicopter. The inner edge of the arc of influence can be straight down or slightly behind.

Obviously, the higher the helicopter, the more ability the pilot has to turn and look in another direction, and the arc of influence will also change.

Cockpit view

Each helicopter type can have a different arc of influence and reference points in the cockpit for the pilot to consider and learn.

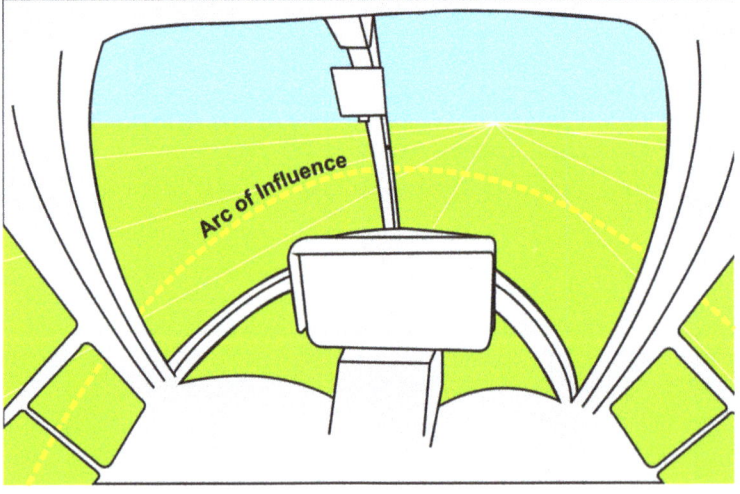

Chapter 8 Manoeuvring during Autorotation

Aiming point

The aiming point is the area the pilot has selected as the best place to land.

Selecting the aiming point is very important as it focuses the pilot's mind on how to manoeuvre the helicopter to make the aim point and also allows the pilot and crew to survey the surrounding area for obstacles and hazards.

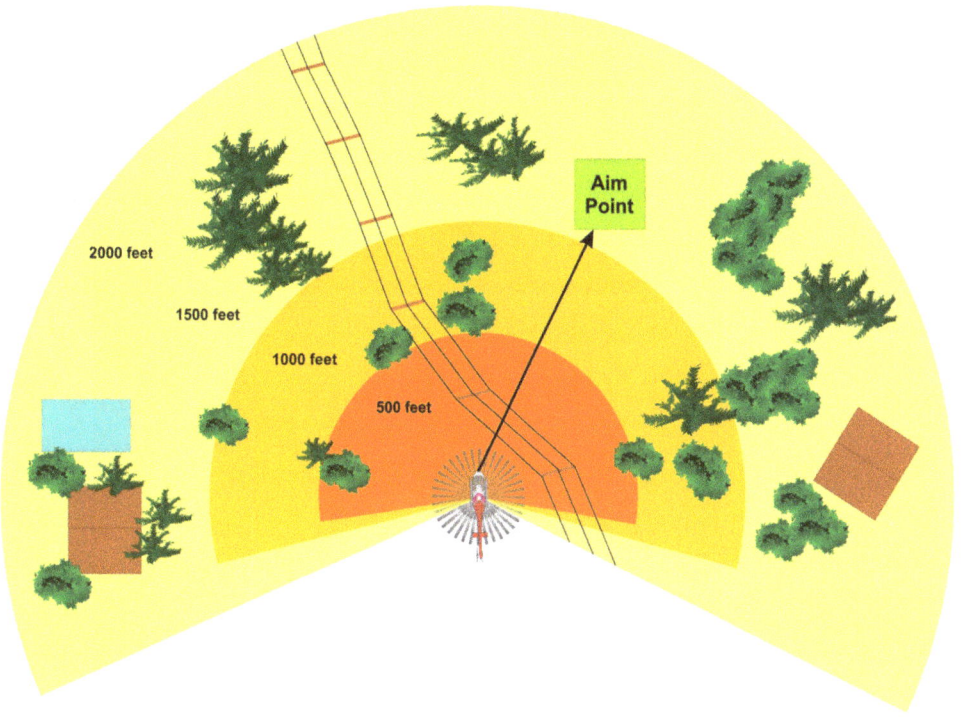

Making the aim point

From the pilot's point of view, if the correct aim point has been selected, it should be in a **relatively constant position** in the window; if conducting a constant attitude autorotation until the last 150 ft, then it may start to move down the window until it is under the helicopter at the termination.

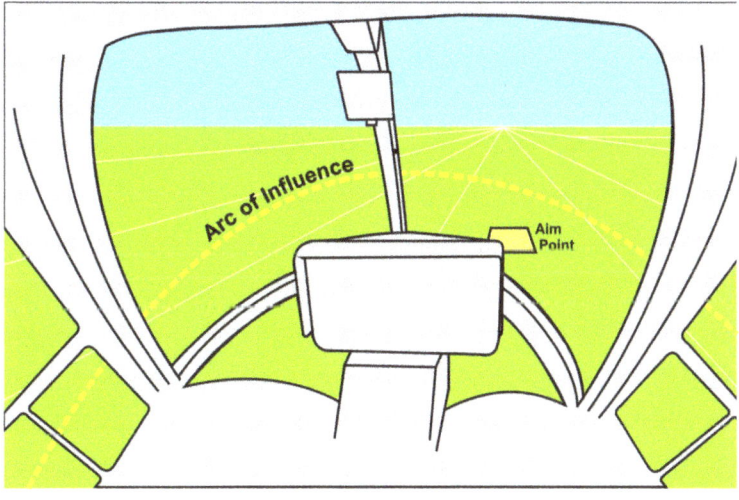

Under or Overshoot

If above 150 ft and the aim point starts to move:

- **down the window**, this indicates that the helicopter is **overshooting** (going past) the aim point.
- **up the window**, this is an indication that the helicopter is **undershooting** (falling short) the aim point.

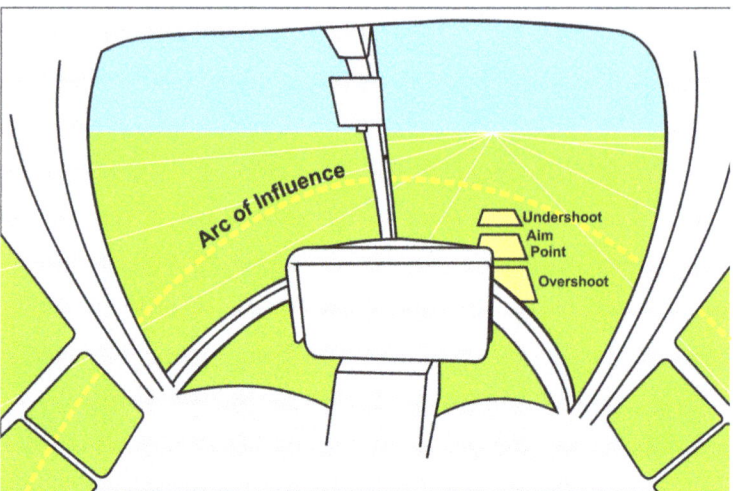

On recognising this, the pilot can adjust the manoeuvring technique to make the original aim point.

Air Exercises: Manoeuvring during Autorotation

Introduction
This section describes the air exercises and demonstrations to practice manoeuvring during autorotation.

Airmanship
- Memorise the speed and Rotor RPM limits
- Memorise the configuration for range
- Memorise the configuration for endurance
- Develop a robust lookout and situational awareness
- Always be looking for suitable termination areas in the event of having to conduct a real autorotation (Forced Landing)
- Always be aware of the position of the throttle.

Common faults
The following are common faults of students:
- Eyes are in the cockpit instead of on the horizon where they should be.
- Not lowering the collective fully on entry or controlling the Rotor RPM.
- Not correcting for yaw with pedals.
- Failure to control the attitude with cyclic
- Difficulty in judging the timing of the flare.
- When commencing a Go Around, failing to confirm that the Rotor RPM is within engine RPM limits.
- No lookout, especially below.
- Difficulty in judging how far the helicopter can glide.
- Difficulty in judging the amount of manoeuvring required to make the selected aiming point.
- Not making allowance for the wind.

Air Exercise 8-1: Range, Endurance, Constant Attitude

Introduction
The instructor will demonstrate range, endurance, constant attitude and adjusting speed when in autorotation to a power termination.

After the demonstration, the instructor will give the student all of the controls and position the helicopter at a minimum of 1500 ft AGL into wind in a suitable area before allowing the student to practice.

In the tables below, the items marked in **ORANGE** show the differences to the standard autorotation for that configuration.

Demonstration
To understand the difference in Range, Endurance, Constant Attitude and speed changes during autorotations, the instructor will first go through a demonstration sequence so that the student can appreciate the differences.

The helicopter will be placed over a position that can be identified by a reference point on the ground by the crew at a constant height (usually 1500 ft AGL) and at a constant airspeed (usually from the cruise speed for that helicopter).

The instructor will then do a Basic Autorotation. On returning to the starting point to demonstrate Range in autorotation, note the distance travelled across the ground from entry to power termination. Do the same for each configuration.

This is to show that from the same entry point, configuring the helicopter for range, endurance, constant attitude or speed changes, the helicopter will travel a different distance across the ground and have a different time in the air compared to a basic autorotation.

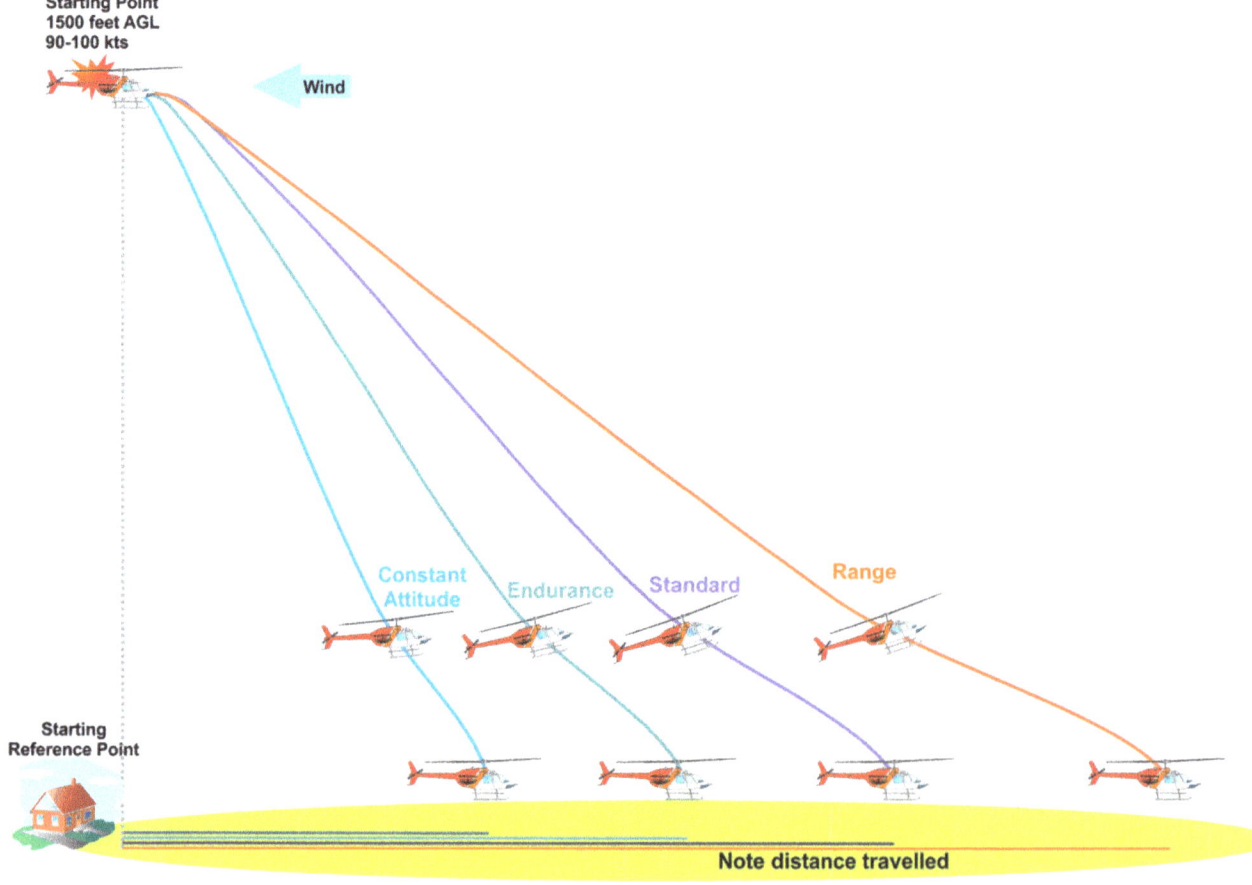

Range

Range — From a minimum of 1500 ft AGL at cruise speed (90-100 kts) into wind over a suitable area:

Step	Action	Detail
1	Preparation	Conduct the "*HASEL*" checks.
2	CRM	Confirm who is at the controls and which pilot is responsible for the throttle.
3	Announce	Instructor: **"3, 2, 1, Practice engine failure"** Instructor will slowly and smoothly roll the throttle to IDLE.
4	Lower Collective	Lower the collective fully within 2 seconds.
5	Look Outside	Look out towards the horizon to determine attitude and the heading.
6	Attitude	Use the cyclic to set a 75 kts attitude plus the estimated wind (the range speed for your helicopter). For example: If the wind is blowing at 15 kts, then the range speed will be 75+15 = 90 kts. This formula applies until the maximum autorotation speed is reached. This number is in the RFM and is usually depicted on the ASI as a blue line or a red and white barbers pole. Typically that number is 100 kts.
7	Keep Straight	Use the pedals to keep straight and maintain the heading.
8	Pause	Do not rush. Take several seconds here to stabilise the helicopter and get it fully under control. Breath! When ready…
9	Collective	Raise some collective until the Rotor RPM is sitting in the **lower autorotation range**.
10	Turn into wind and pick a spot	Turn into wind and pick an area to aim for on the ground.
11	Descent	While manoeuvring, conduct the descent checks: | Check | Description | |---|---| | **Engine RPM** | Engine Idling | | **Rotor RPM** | Rotor RPM in the lower range | | **Balance** | In balance | | **Speed** | Set for range |

Step	Action	Detail
12	Decision Height	**INSTRUCTOR:** By 700 ft AGL, *consider* the options available: • conduct a Go-Around • continue for a Power Termination The *decision* must be made by 500 ft AGL. **Announce decision.**
13	Reintroduce Power	**INSTRUCTOR:** By 500 ft AGL, the instructor will: ▪ Check that Rotor RPM is within the engine limits, then ▪ In a piston, roll the throttle to join the engine and rotor needles and call "CONFIRM NEEDLES JOINED." ▪ In a turbine, roll the throttle to full and call "CONFIRM THROTTLE FULL."
14	Confirm power	The student shall then also confirm the throttle position: ▪ Physically check the throttle position ▪ Check the needles are joined ▪ In a piston, call "NEEDLES JOINED CONFIRMED." ▪ In a turbine, call "FULL THROTTLE CONFIRMED." ▪ **If the student notices that the instructor has not done this step, they must speak up.**
15	Manage descent	At 500 ft AGL, manage descent: | Check	Description
---	---
Power	Lower some collective to allow the Rotor RPM to again increase and store more energy in preparation for the flare and termination but remain in a steady state of autorotation.
Attitude	If required, maintain an attitude for range with cyclic throughout, or if the area is going to be made, then speed may start to reduce to normal.
Balance	Maintain balance with pedals throughout.
Recovery	Maintain a consistent autorotation configuration until passing 100-50 ft AGL.	
16	Flare	Use aft cyclic, as required, to flare the helicopter and reduce the rate of descent and forward speed.
17	Develop	Develop the flare with more or less aft cyclic as the speed reduces to better make a predetermined touchdown point.
18	Hold	Once the helicopter is set in an attitude that the pilot feels will allow the helicopter to arrive at the ground with zero forward speed and zero rate of descent, then hold that attitude until approximately 10 ft AGL.
19	Initial	At 10 ft AGL, quickly pull the collective up a small amount (approximately two (2) inches of travel) and hold it there.

Step	Action	Detail
20	**Level**	Using some forward cyclic to level the skids with the ground when doing the initial pitch pull.
21	**Pull collective to hover**	As the helicopter comes level with the ground, raise collective enough to return to the hover in a power termination. Because the helicopter is now returning to powered flight, the pilot may experience some excessive YAW as the engine power is re-introduced and the Rotor RPM may droop. This is normal and will require the pilot to correct as required to keep straight, maintain Rotor RPM and come back to the hover.
22	**Note the distance**	On conducting a circuit back to the starting reference point, note the extra distance travelled across the ground during **range** compared to the standard autorotation.

Endurance

Endurance — From a minimum of 1500 ft AGL at cruise speed (90-100 kts) into wind over a suitable area:

Step	Action	Detail
1	Preparation	Conduct the "*HASEL*" checks.
2	CRM	Confirm who is at the controls and which pilot is responsible for the throttle.
3	Announce	Instructor: "**3, 2, 1, Practice engine failure**" Instructor will slowly and smoothly roll off the throttle.
4	Lower Collective	Lower the collective fully within 2 seconds.
5	Look Outside	Look out towards the horizon to determine attitude and the heading.
6	Attitude	Use the cyclic to set a 50 kts attitude ignoring the wind strength
7	Keep Straight	Use the pedals to keep straight and maintain the heading.
8	Pause	Do not rush. Take several seconds here to stabilise the helicopter and get it fully under control. Breath! When ready…
9	Collective	Raise some collective until the Rotor RPM is sitting in the lower autorotation range.
10	Turn into wind and pick a spot	Turn into wind and pick an area to aim for on the ground.
11	Descent	While manoeuvring conduct the descent checks:
12	Decision Height	**INSTRUCTOR:** By 700 ft AGL, *consider* the options available: • conduct a Go-Around • continue for a Power Termination The *decision* must be made by 500 ft AGL. **Announce decision.**

Check	Description
Engine RPM	Engine Idling
Rotor RPM	Rotor RPM in the lower range
Balance	In balance
Speed	Set for endurance

Chapter 8 Manoeuvring during Autorotation

Step	Action	Detail
13	Reintroduce Power	**INSTRUCTOR:** By 500 ft AGL, the instructor will: - Check that Rotor RPM is within the engine limits, then - In a piston, roll the throttle to join the engine and rotor needles and call "CONFIRM NEEDLES JOINED." - In a turbine, roll the throttle to full and call "CONFIRM THROTTLE FULL."
14	Confirm power	The student shall then also confirm the throttle position: - Physically check the throttle position - Check the needles are joined - In a piston, call "NEEDLES JOINED CONFIRMED." - In a turbine, call "FULL THROTTLE CONFIRMED." - **If the student notices that the instructor has not done this step, they must speak up.**
15	Manage descent	At 500 ft AGL, manage descent: | Check	Description
---	---
Power	Lower some collective to allow the Rotor RPM to again increase and store more energy in preparation for the flare and termination but remain in a steady state of autorotation.
Speed	If required, maintain an attitude for endurance with cyclic throughout, or if the area is going to be made, then speed may start to increase to normal.
Balance	Maintain balance with pedals throughout.
Recovery	Maintain the autorotation configuration until passing 100-50 ft AGL.	
16	Flare	Use aft cyclic as required to flare the helicopter and reduce the rate of descent and forward speed.
17	Develop	Develop the flare with more or less aft cyclic as the speed reduces to better make a predetermined touchdown point.
18	Hold	Once the helicopter is set in an attitude that the pilot feels will allow the helicopter to arrive at the ground with zero forward speed and zero rate of descent, then hold that attitude until approximately 10 ft AGL.
19	Initial	At 10 ft AGL, quickly pull the collective up a small amount (approximately two (2) inches of travel) and hold it there.
20	Level	At the point of doing the initial pitch pull, then use some forward cyclic to level the skids with the ground.

Step	Action	Detail
21	**Pull collective to hover**	As the helicopter comes level with the ground, raise collective enough to return to the hover in a power termination.
		Because the helicopter is now returning to powered flight, the pilot may experience some excessive YAW as the engine power is re-introduced and the Rotor RPM may droop. This is normal and will require the pilot to correct as required to keep straight, maintain Rotor RPM and come back to the hover.
22	**Note the distance**	On conducting a circuit back to the starting reference point, note the extra distance travelled across the ground during **endurance** compared to the standard autorotation.

Constant Attitude

Constant Attitude From a minimum of 1500 ft AGL at cruise speed (90-100 kts) into wind over a suitable area:

Step	Action	Detail
1	Preparation	Conduct the "*HASEL*" checks.
2	CRM	Confirm who is at the controls and which pilot is responsible for the throttle.
3	Announce	Instructor: **"3, 2, 1, Practice engine failure"** Instructor will roll off the throttle.
4	Lower Collective	Lower the collective fully within 2 seconds.
5	Look Outside	Look out towards the horizon to determine attitude and the heading.
6	**Attitude**	Use the cyclic to set a hover attitude ignoring the wind strength.
7	Keep Straight	Use the pedals to keep straight and maintain the heading.
8	Pause	Do not rush. Take several seconds here to stabilise the helicopter and get it fully under control. Breath! When ready…
9	**Collective**	Use some collective to set the Rotor RPM in the middle to upper autorotation range. Because we are not concerned with distance or time in the air, we are setting an angle to make a spot, so we want to maintain a good store of energy in the rotor system.
10	Turn into wind and pick a spot	Turn into wind and pick an area to aim for on the ground.
11	Descent	While manoeuvring, conduct the descent checks: | Check	Description
---	---
Engine RPM	Engine Idling
Rotor RPM	Rotor RPM in the mid to upper range.
Balance	In balance
Speed	Set for a constant attitude to make a spot.	
12	Decision Height	**INSTRUCTOR:** By 700 ft AGL, *consider* the options available: • conduct a Go-Around • continue for a Power Termination The *decision* must be made by 500 ft AGL. **Announce decision.**

Step	Action	Detail
13	Reintroduce Power	**INSTRUCTOR:** By 500 ft AGL, the instructor will: - Check that Rotor RPM is within the engine limits, then - In a piston, roll the throttle to join the engine and rotor needles and call "CONFIRM NEEDLES JOINED." - In a turbine, roll the throttle to full and call "CONFIRM THROTTLE FULL."
14	Confirm power	The student shall then also confirm the throttle position: - Physically check the throttle position - Check the needles are joined - In a piston, call "NEEDLES JOINED CONFIRMED." - In a turbine, call "FULL THROTTLE CONFIRMED." - **If the student notices that the instructor has not done this step, they must speak up.**
15	Manage descent	At 500 ft AGL, manage descent: | Check	Description
---	---
Collective	Use as required to manage Rotor RPM within the engine limits but remain in a steady state autorotation.
Speed	Some forward cyclic to accelerate and reset the attitude for basic autorotation (60 kts) speed.
Balance	Maintain balance with pedals throughout.
Recovery	Maintain the autorotation configuration until 100-50 ft AGL.	
16	Flare	Use aft cyclic as required to flare the helicopter and reduce the rate of descent and forward speed.
17	Develop	Develop the flare with more or less aft cyclic as the speed reduces to better make a predetermined touchdown point.
18	Hold	Once the helicopter is set in an attitude that the pilot feels will allow the helicopter to arrive at the ground with zero forward speed and zero rate of descent, then hold that attitude until approximately 10 ft AGL.
19	Initial	At 10 ft AGL, quickly pull the collective up a small amount (approximately two (2) inches of travel) and hold it there.
20	Level	At the point of doing the initial pitch pull, then use some forward cyclic to level the skids with the ground.
21	Pull collective to hover	As the helicopter comes level with the ground, raise collective enough to return to the hover in a power termination. Because the helicopter is now returning to powered flight, the pilot may experience some excessive YAW as the engine power is re-introduced and the Rotor RPM may droop. This is normal and will require the pilot to correct as required to keep straight, maintain Rotor RPM and come back to the hover.

Chapter 8 Manoeuvring during Autorotation

Step	Action	Detail
22	Note the distance	On conducting a circuit back to the starting reference point, note the extra distance travelled across the ground during **constant attitude** compared to the standard autorotation.

Air Exercise 8-2: Turns during Autorotation

Demonstration   The instructor will demonstrate 180, 360 and S-Turns during autorotation.

After the demonstration, the instructor will give the student all of the controls and position the helicopter at a minimum of 1500 ft AGL in a suitable area prior to allowing the student to practice.

Limits
- Commence from a minimum 1000 ft AGL
- Go-around to commence before 300 ft AGL or at any time the instructor determines.
- Remain within the Rotor RPM limits, and
- Remain within any airspeed limits set by the manufacturer.

Step	Action	Detail
1	Enter autorotation	Conduct a standard entry into autorotation.
2	**Turn into wind and pick a spot**	Turn into wind and pick an area to aim for on the ground.
3	**Turns**	Use cyclic to conduct the turns required. When turning, the pilot will have to use some collective to control Rotor RPM and some pedal in the direction of the turn to maintain balance.
4	Terminate autorotation	Conduct a standard autorotation termination to either a: - go-around, or - power termination.

Air Exercise 8-3: Adjusting Speed during Autorotation

Demonstration
: The instructor will demonstrate speed changes during autorotation.

  After the demonstration, the instructor will give the student all of the controls and position the helicopter at a minimum of 1500 ft AGL in a suitable area prior to allowing the student to practice.

Limits
: - Commence from a minimum 1000 ft AGL
  - Go-around to commence before 300 ft AGL or at any time the instructor determines.
  - Remain within the Rotor RPM limits
  - Remain within any airspeed limits set by the manufacturer

Step	Action	Detail
1	**Enter autorotation**	Conduct a standard entry into autorotation.
2	**Turn into wind and pick a spot**	Turn into wind and pick an area to aim for on the ground.
3	**Adjusting speed**	Use the cyclic to make speed changes to manipulate the glideslope. When adjusting speed, the pilot will have to use some collective to control Rotor RPM and some pedal in the direction of the turn to maintain balance.
4	**Terminate autorotation**	Conduct a standard autorotation termination to either a: - go-around, or - power termination.

Air Exercise 8-4: Precision Autorotation

Putting it all together

The Precision Autorotation is now putting all of this together and allowing the student to practice selecting suitable termination areas and deciding which technique to use to get there.

If undershooting, go for range.

If overshooting, reduce speed or make turns.

At 500 ft AGL, return to the basic autorotation configuration to conduct a normal power termination to the hover.

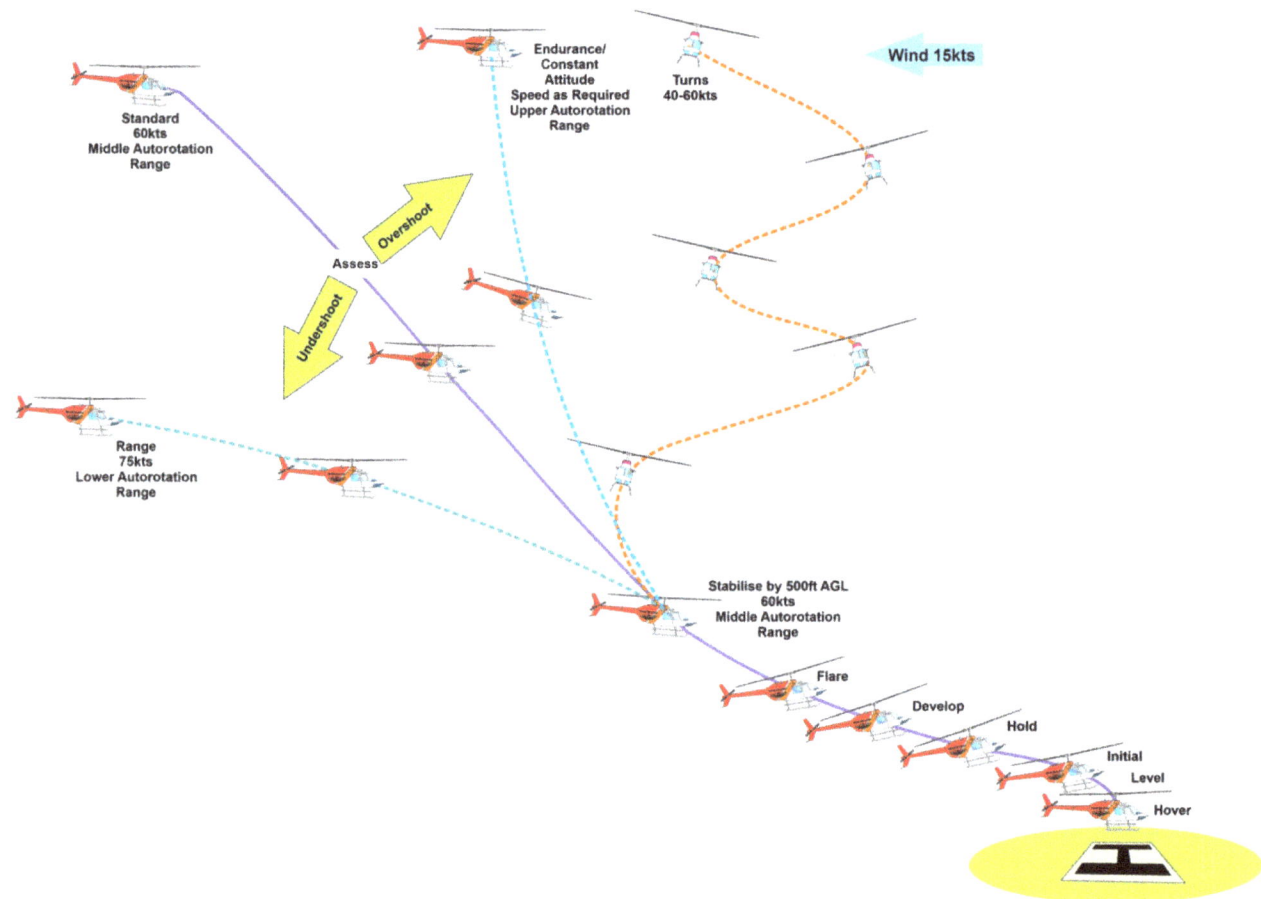

9

Flight Control Emergencies (FCEs)

Aim
To control the helicopter and land after a malfunction of one of the primary flight control systems (collective, pedals, throttle or cyclic).

Objectives
On completion of this lesson, the student will be able to:

- describe collective and throttle steering
- explain the difference between a control failure and a control jam
- identify a primary flight control jam in flight
- troubleshoot the jam and make a plan of action for landing
- make an approach, and
- land the helicopter.

Motivation
Emergencies involving a primary flight control becoming stuck are rare but can and have happened. Just because one control becomes stuck in its position does not necessarily mean that you lose control of the helicopter. Instead, the pilot needs to learn techniques that, if used correctly, can help the pilot safely land the helicopter by using the secondary effects of the other controls still operating correctly.

Although the combinations of FCEs are limitless, an instructor can teach principles that can be useful when applied properly within boundaries.

In this chapter, this author hopes to give you tools and awareness, not necessarily competency, on some of the options available.

Preparation: Flight Control Emergencies

Introduction

A Flight Control Emergency (FCE) involves one of the four primary flight controls (collective, cyclic, throttle or pedals) becoming stuck in its current position due to either:

- jamming, or
- a component in the flight control system failing, meaning it can no longer be moved and is similarly stuck in its current position as if it were a jam.

Any issue controlling the tail rotor is referred to as a Yaw Control Emergency (YCE), even though any issues with any flight control are collectively referred to as an FCE.

A failure of a flight control, therefore, is more about the system being completely out of the pilot's controllability. A component has broken apart (blade, rod, cable, bearing etc.) or a part of the system has fallen off the helicopter (e.g. tail rotor and gearbox). A complete failure of a system means the unit may not be stuck in its current position but be freely moving with no control or not moving at all, and the pilot may lose control. It is not possible to train for a failure in this sense.

Summary

An FCE can be summarised as follows:

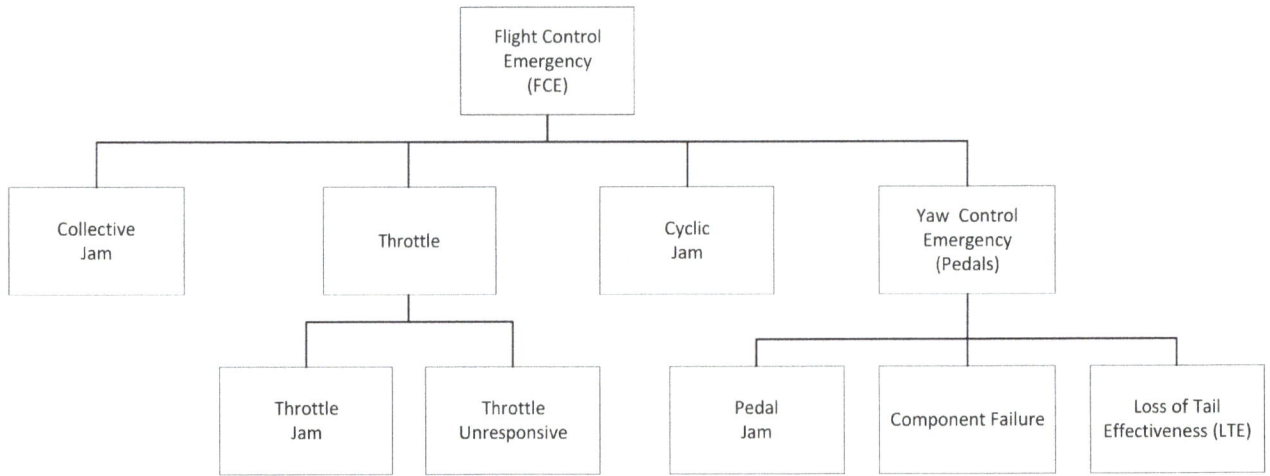

Anticlockwise rotating blades

Note: The emergencies covered in this chapter assume an anticlockwise rotating main rotor system. The techniques will be the same for clockwise rotating systems, but because the Torque Reaction is working in the opposite direction, the use of left and right pedals will be the reverse of that described.

Control Bind

Sometimes the flight control may not have jammed but has become very stiff due to something binding (getting tight or creating friction) in some part of the system. This is referred to as a control *"bind."*

In this case, the control may still be used but in a limited capacity.

Chapter 9 Flight Control Emergencies (FCEs)

In this chapter This chapter only considers control jams. The discussion and air exercise will revolve around:

- The **collective**, **throttle** or **cyclic** being stuck in a fixed position (referred to as a jam). To control the helicopter, the pilot will have to use the secondary effects of the controls that are working to compensate for the jammed one.
- When considering the **pedals** and a Yaw Control Emergency (YCE), in addition to the possibility of a jammed pedal, the pilot may also have to consider a:
  - failure of the tail rotor driveshaft or gearbox so that it no longer spins around and no longer provides thrust regardless of any pedal movement, or
  - situation where the air passing over the tail is disrupted to the point that the tail rotor can no longer provide tail rotor thrust and the effectiveness of the tail rotor is lost. This situation is referred to as Loss of Tail rotor Effectiveness or LTE, regardless of being able to move the pedals, or
  - component failure where parts of the tail rotor or tail rotor gearbox come away from the helicopter which may also change the fuselage's centre of gravity, affecting the ability to control the pitch attitude with cyclic.

Piston vs Turbine The collective, pedals and cyclic use is the same across most helicopters regardless of the type of engine installed. The throttle, however, may have similar functions but access and use can vary significantly between a piston and turbine engine.

All piston engine helicopters and some turbine helicopters have the throttle on the end of the collective so it can be manipulated by the pilot easily while in flight.

However, some turbine helicopters, especially bigger ones, do not have the throttle on the end of the collective. Instead, there is a separate throttle lever, located on the floor or the ceiling, that is moved independently of the collective and only used during start and shutdown. As a result, the pilot's hand cannot be on the collective and throttle at the same time. Therefore, in some helicopters, the throttle cannot be easily used in a FCE, and the pilots rely on the collective to manage yaw.

Troubleshooting When there is an FCE, it is essential the crew ask themselves questions because every jam may be different. Asking questions is part of the troubleshooting process, and allows the brain to start looking for answers. When a flight control jams, everything happens very quickly, so finding the answers is important.

Just answering the following questions allows the mind to settle and allows the crew to evaluate the situation before coming up with a plan of action.

No	Question
1	**Which control has jammed?** Identify it and then announce to the crew which one it is. Remember that the other controls are still working properly and are *not* jammed.
2	**Is the helicopter still controllable?** ■ If **yes**, the crew has time to troubleshoot, experiment by gently and slowly moving the other controls to help construct a solution and plan. ■ If **no**, then the crew may have a serious problem.
3	**Is the control jam greater, the same or less than the expected hover position or power required for that control?** This information allows the crew to determine the type of approach and landing that may ensue.

No	Question
4	**Can the helicopter fly to a safe area, or must the crew do something now?**
	If the helicopter is still flying reasonably normally, the crew may be able to fly to an airport where the landing area is open and flat, and emergency services can assist instead of landing immediately in a field with no support.

Collective and Throttle Steering

Tail rotor thrust is provided to counter the torque reaction of the main rotor blades when driven by the engine. If tail rotor thrust, for whatever reason, is no longer available, then the pilot can use airspeed, collective and/or the throttle to a small degree to control the yaw of the fuselage. This method is referred to as collective and throttle steering.

Collective Steering   When the **collective** in an anticlockwise rotating system is:

- **raised**, the torque reaction increases, and the fuselage wants to **turn right**.
- **lowered**, the torque reaction decreases, and because the pilot still has the same amount of left pedal forward, the fuselage will want to **turn left**.

Collective steering is typically used when in forward flight and can be remembered by the words:

RAISE RIGHT | LOWER LEFT

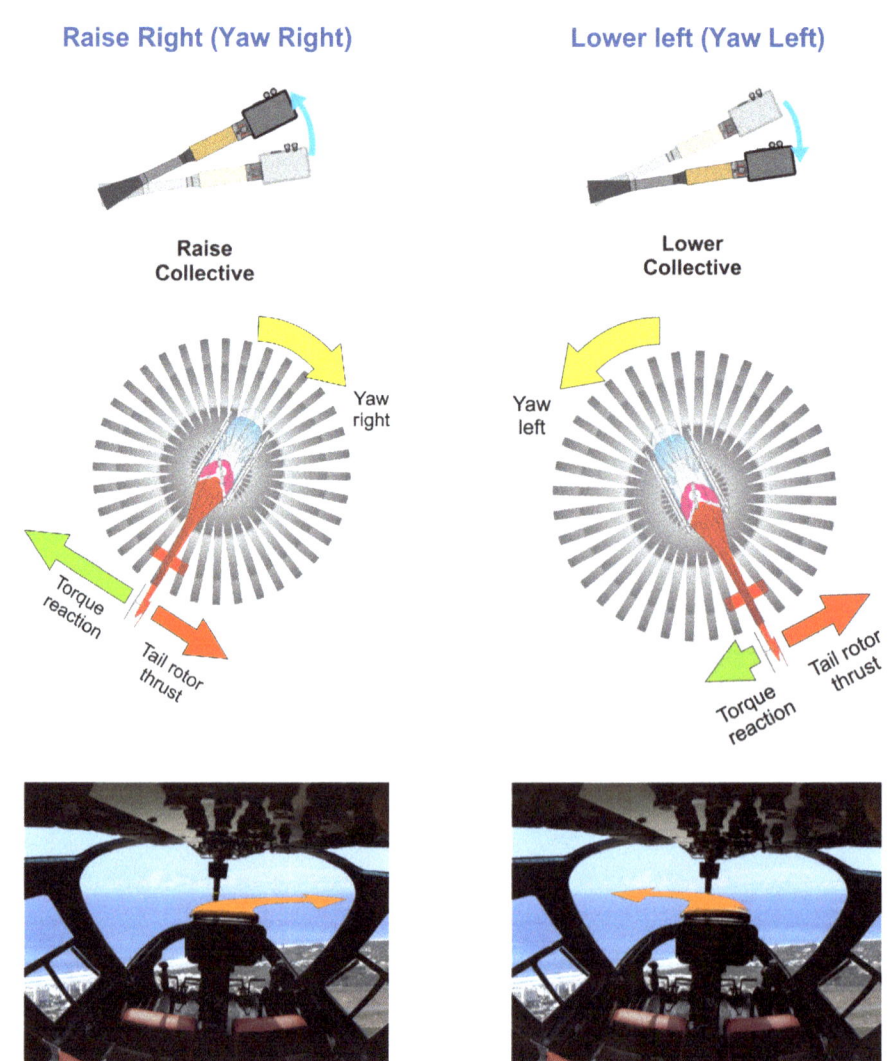

Chapter 9 Flight Control Emergencies (FCEs)

Throttle steering Throttle steering is only relevant in those helicopters where the throttle is located on the end of the collective and is accessible to the pilot. Additionally, throttle steering can only be used within the Rotor and Engine RPM limits, so it must be used sparingly.

When the throttle is:

- **rolled towards the IDLE position**, the torque will decrease, and the fuselage will want to **yaw left**.
- **rolled on towards the FULL position**, the torque will increase, and the fuselage will want to **yaw right**.

Throttle steering is normally used when:

- at the hover, or
- on short finals close to the ground.

Throttle steering can be remembered by the words:

RIGHT ON | LEFT OFF

Additionally, when using throttle steering, point the left index finger as you move the throttle; the finger will point to which way the nose will turn.

Rolling throttle on towards FULL the nose yaws right

Rolling throttle off towards IDLE the nose yaws left

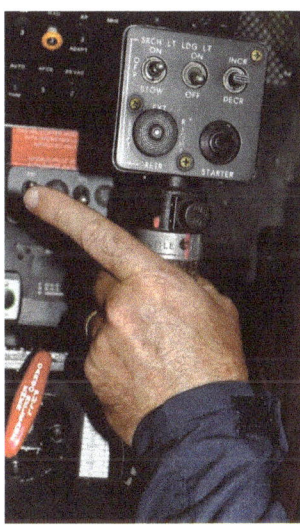

Jammed Controls at the Hover

Jammed Collective at the Hover

Introduction When the collective jams at the hover, the pilot cannot move the collective up or down. The pilot can no longer adjust the power and manage the height above the ground with the collective.

Cause Collective jams are usually a result of:

- something getting caught under the collective lever
- a passenger or other crew member sitting or leaning on the collective, or
- a binding of one of the control rods.

Scenarios

The three scenarios the pilot may be faced with are:

1. The collective jams at a **stable hover** IGE.
2. The collective jams at the hover, but the helicopter is already **descending** towards the ground.
3. The collective jams at the hover, but the helicopter is **climbing** away from the ground OGE.

Troubleshoot: stable hover

From a stable hover IGE

- If possible, check to see if anything is jamming the collective.
- If safely able to free the collective, then do so.
- If not able to free the collective, then:
  - slowly roll off a small amount of throttle and pause, waiting for the Rotor RPM to decay, thereby reducing the rotor thrust from the main blades
  - maintain ground position with cyclic throughout in the normal manner, and
  - maintain heading with pedals throughout in the normal manner.

The amount of throttle used will depend on the desired rate of descent to land.

- **Slowly** rolling off a **small** amount of throttle will allow for a **controlled** rate of descent.
- **Quickly** rolling off a **large** amount of throttle will cause a **high rate of descent** and possible damage on landing and should be avoided.
- If the rate of descent is too high, roll some throttle back on to increase RPM and rotor thrust.

The key to this emergency is to conduct it **slowly, patiently, and maintain control**.

Troubleshoot: descending hover

From a hover but the helicopter is already descending

Allow the helicopter to continue to descend until it touches the ground.

If it is coming **down too fast**, then:

- use some small forward cyclic so that on contact with the ground, the energy is divided between the vertical and horizontal by accepting some run-on.
- once on the ground, roll the throttle to IDLE.

Troubleshoot: climbing hover

From a hover but the helicopter is climbing OGE

Based on the situation, either:

- conduct the same steps as for a stable hover; or
- fly away and then conduct the jammed collective emergency from forward flight (see *Jammed Collective from Forward Flight*).

Jammed Throttle at the Hover

Introduction

In a turbine helicopter, the throttle is not usually moved in flight unless an emergency occurs. Therefore, any throttle failure is not apparent until it is moved on the ground at start-up or shutdown.

In a piston-engine helicopter, the throttle is constantly used in flight to fine-tune the RRPM. So, the pilot will notice when it jams or when the throttle is moved and nothing happens.

RRPM will now be affected by collective and pedal movements, which need to be used sparingly to avoid an overspeed or underspeed of the engine and rotor system.

Chapter 9 Flight Control Emergencies (FCEs)

Troubleshoot

When troubleshooting a jammed throttle at the hover, initially look for the obvious.

- Is the throttle friction on?
- Is there something binding the throttle?

Try to gently move the throttle but be mindful not to make the situation any worse.

It can be made worse by rolling the throttle too far towards the IDLE and not being able to roll it back on or too far towards FULL and not being able to roll it back off Maintaining RRPM is still important.

- If the current throttle and collective combination allows the Rotor RPM to remain within limits, then simply lower the collective to land.
- If the current throttle and collective combination cause the Rotor RPM to decrease, then continue to land, accepting a possible over-pitch situation and raising collective to cushion on.
- If the current throttle and collective combination cause the Rotor RPM to increase and exceed an upper limit and the helicopter is still not on the ground, then you need to find a way to reduce the RRPM without using the throttle.

Reducing RRPM without using the throttle

In a piston-engine helicopter, options include:

- Applying carb heat. This will reduce power by almost 2" MAP and RRPM will decay.
- Turning off one (1) magneto. This will reduce the power by almost 2" MAP and RRPM will decay.
- Hovering downwind. Hovering downwind requires more power which, if not applied, will allow the helicopter to sink towards the ground.
- As a last resort, turn the engine OFF and conduct an engine failure at the hover procedure.

In a turbine engine helicopter, there are fewer options to consider as the Fuel Control Unit (FCU) and Governor should still be functioning and managing the RRPM. If, however, the RRPM still wants to rise while at the hover, the best option may be to turn the engine OFF and conduct an engine failure at the hover procedure.

As soon as the helicopter touches the ground, continue to lower the collective to get some weight on the skids. If RRPM at that point still wants to increase and exceed a limit, then turn the engine off before lowering the collective fully. Yes, this will mean an abnormal shutdown and no cooldown, but this is OK in an emergency.

In a piston engine helicopter, if you take this option and have a passenger onboard, it is best to get them to pull the mixture or turn off the magnetos so that you are prepared for the instant changes in Torque Reaction as if it was an engine failure at the hover. Do not turn off the fuel valve, as this can lead to the engine coughing and spluttering with erratic changes in torque, making it more difficult to control.

In a turbine engine helicopter, the pilot can turn off the fuel valve, and it may take seconds or minutes for the fuel to starve at the Fuel Control Unit (FCU), but the engine stop should be relatively smooth.

Jammed Cyclic at the Hover

Introduction

A jammed cyclic is a serious event with little time to react. If the pilot cannot move the cyclic in any direction, then there is no direct control of pitch and roll with the cyclic.

This situation is usually due to something interfering with the cyclic flight control system, such as a crew seat moving or cockpit objects interfering with the movement.

If it is something:

- **simple**, fix it quickly.
- **critical**, then the crew will have a **serious** problem.

Land immediately

Landing the helicopter is a priority, as flying away **is not** an option.

A heavy landing or controlled crash is better than allowing the helicopter to transition into forward flight.

How the emergency is managed will depend on the severity of the jammed control and the helicopter's attitude.

If the situation occurred	Then
in a stable hover	A controlled landing is possible by gently lowering the collective.
whilst you are moving at the hover (sideways, backwards)	The likelihood of a successful landing is reduced. If the helicopter starts to **roll**, following the roll with pedal input will keep the skids relatively level for the touchdown and should align them with the direction of travel. If the helicopter starts to **pitch** (nose up or down), use abrupt and harsh collective movements to use flap back to effect the pitch before ground contact.

Troubleshoot

There is no time to troubleshoot unless the cause of the jam is evident and easily and quickly fixed. Instead, conduct the recovery actions before the pilot loses control.

Jammed Pedals at the Hover

Introduction

When the pedals jam at the hover, they may jam to various degrees in the:

- left pedal forward position, or
- the right pedal forward position.

The further forward the left or right pedal, the more difficult the situation. The closer to the hover pedal position, the easier the recovery.

Chapter 9 Flight Control Emergencies (FCEs)

Effect of wind

Wind will have a dramatic effect on any tail rotor control emergency. If the wind is on the nose, the wind will be equal on both sides of the fuselage.

If the helicopter is turning, then the wind will have maximum effect when it is 90 degrees to the fuselage based on whatever direction the jam has occurred (Left or right).

This can be an advantage if the tail is moving up into the wind as the wind will automatically be trying to slow the rate of turn.

The wind will also be a disadvantage if the tail is moving away from the wind as the wind will be trying to accelerate the rate of turn.

Left pedal jam

Consider a **left pedal jam** with the helicopter **turning left**

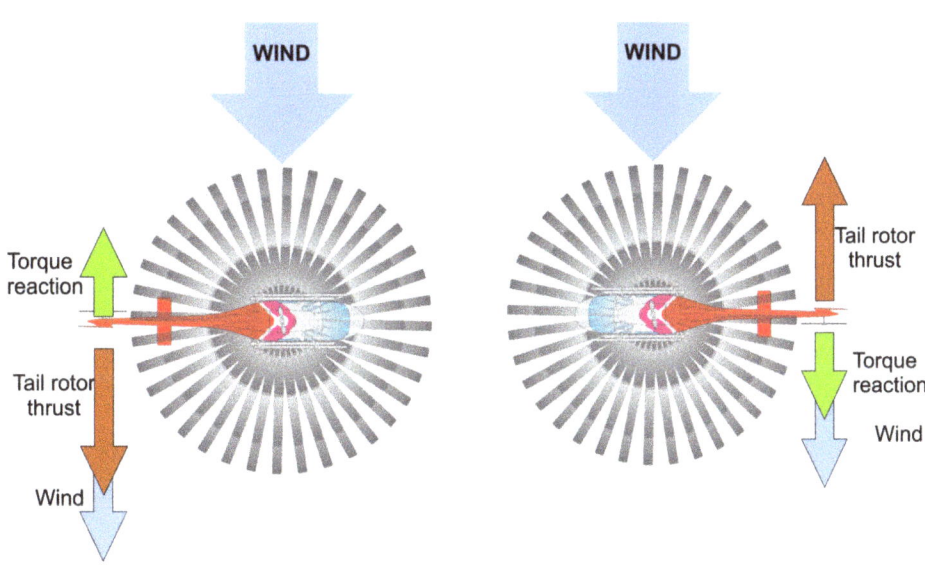

Right pedal jam

Consider a **right pedal jam** with the helicopter **turning right**

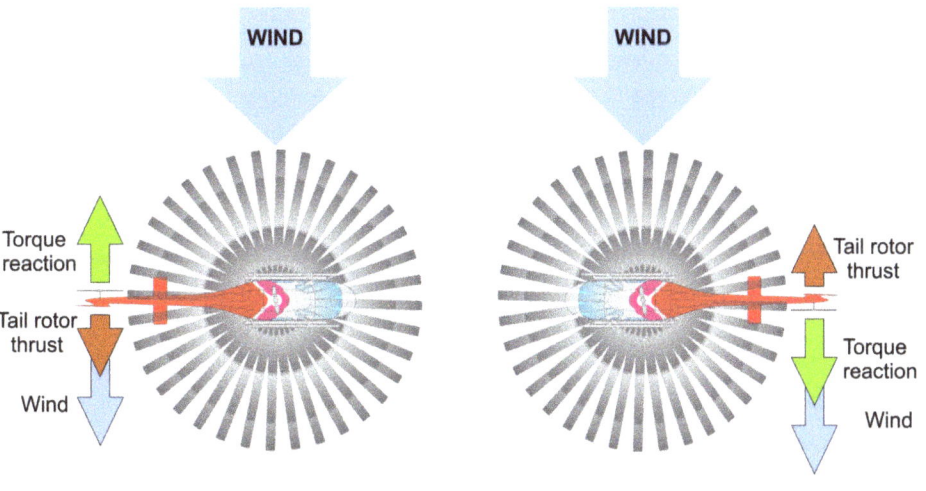

Ignore the wind

In theory, it is obvious to have the wind 90 degrees on the right with a left pedal jam and 90 degrees to the left with a right pedal jam, but because the helicopter is turning, it will constantly be changing and, in an emergency, to attempt to consider the wind will take mental energy and time. With the pilot already under stress, the wrong choices may be made.

Instead, ignore the wind and conduct the recovery actions and once on the ground, you will likely notice that the turn stopped 90 degrees to the wind without the pilot consciously deciding this.

For any emergency involving jammed controls, whether from forward flight or the hover, it is best to make any approaches directly into wind in the normal manner and simply conduct the recovery actions. The wind will automatically assist as the emergency and subsequent recovery unfolds.

Jammed hover pedal

When the pedals jam and the helicopter is in a stable hover with tail rotor thrust and torque reaction being equal, the helicopter will not be turning.

Typical hover pedal position

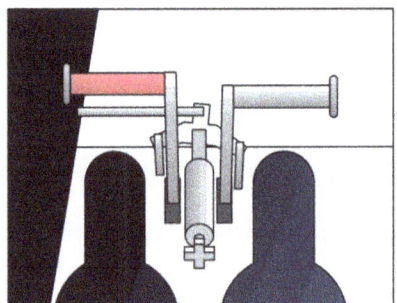

Forces in balance

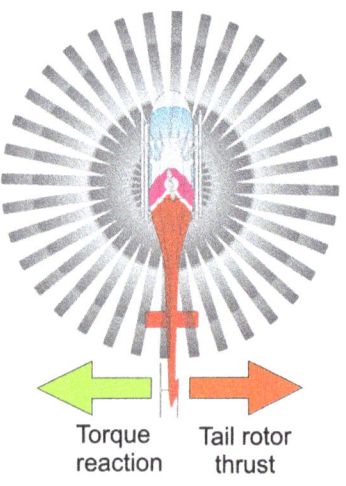

Torque reaction   Tail rotor thrust

Recovery

This is the easiest pedal jam to respond to, as the helicopter is already stable.

Simply lower the collective in the normal manner to conduct a landing.

When landing on a smooth flat surface, the skids will handle some turning motion.

If the fuselage wants to turn, use throttle steering to control the yaw, if applicable.

Use cyclic into the centre of the turn to prevent the centrifugal force from rolling the helicopter over in the opposite direction.

Troubleshoot

If the helicopter is stable, it may be appropriate to check the pedals to see if anything obvious is jamming them, and it is easy to fix. This is at the discretion of the pilot.

It is generally better to land and fix the problem on the ground. Trying to fix the problem while hovering may make it worse, and the pilot may lose concentration on flying.

Chapter 9 Flight Control Emergencies (FCEs)

Left pedal forward When the pedals jam with the left pedal forward of the hover position, tail rotor thrust will be greater than torque, and the helicopter will turn to the left.

Left pedal forward position   **Greater tail rotor thrust**

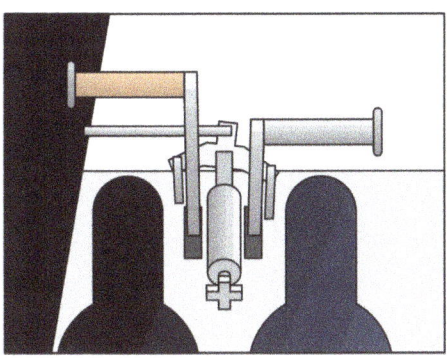
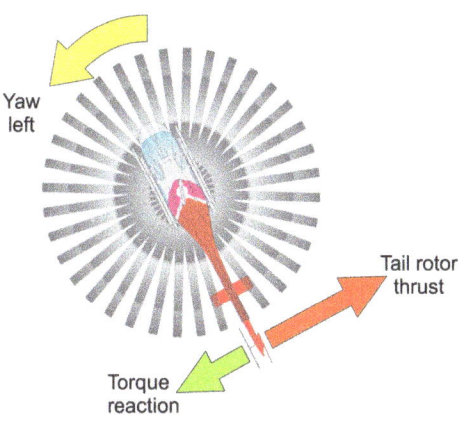

Bleed technique In this situation, a **bleed** technique is used. If the pilot slowly rolls off (bleeds) some throttle, then the RPM of the tail rotor blades decreases, and the tail rotor thrust will decrease.

At this stage, the helicopter will still be turning to the left, possibly accelerating due to the instant reduction in torque.

The helicopter will also sink due to the reduction in Rotor RPM. Therefore, the pilot should raise enough collective to prevent the sink, which will increase torque and further decrease RPM.

When torque and tail rotor thrust are equal, the turn will stop, and the pilot should allow the helicopter to sink to the ground.

Once on the ground, lower the collective fully.

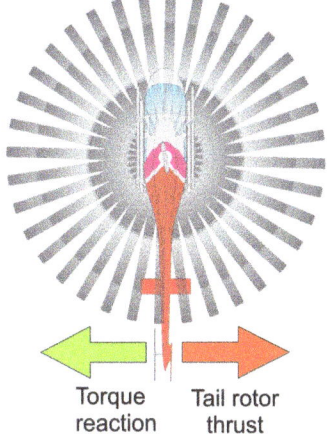

Recovery The recovery technique for a jammed left pedal may require that the pilot put the helicopter in a deliberate over-pitch situation.

If any wind blows, the helicopter will most likely stop its turn 90° to the wind due to the extra drag on the tail. Conversely, the turn may accelerate if the helicopter is still rotating as the tail passes the 180° downwind.

Right pedal forward

When the pedals jam with the right pedal forward of the hover position, tail rotor thrust will be less than torque, and the helicopter will turn to the right.

Right pedal forward position

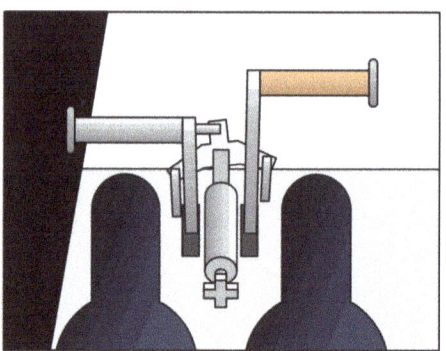

Less tail rotor thrust

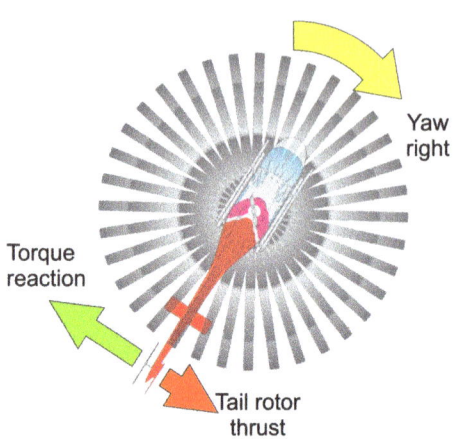

Crack technique

When the pedals jam with the right pedal forward of the hover position, use the "**crack**" technique.

If the pilot rapidly rolls off (cracks) some throttle, then torque will instantly reduce. It is the speed of the roll-off (crack) that is important.

Because Rotor RPM is slower to decay, the turn should stop for a moment. As the helicopter sinks towards the ground, Rotor RPM will reduce, and torque may again exceed tail rotor thrust, at which point the turn to the right may begin again, so another crack of the throttle will be required.

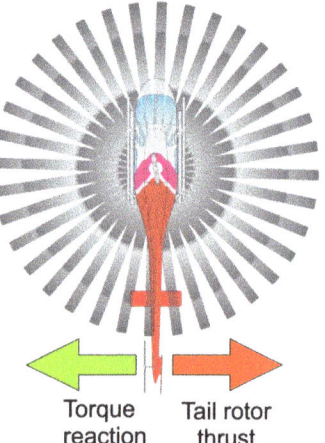

How much throttle

How much the throttle is cracked off depends on the rate of the right turn.

If the turn is:

- **rapid**, then roll the throttle off all the way to the IDLE position immediately.
- **slow**, then crack off a small amount of throttle (approximately one (1) inch) until the turn stops. This may have to be repeated several times until the helicopter skids touch the ground.

Once on the ground, lower the collective fully.

The recovery technique for a jammed right pedal may require that the pilot put the helicopter in a deliberate over-pitch situation. If any wind blows, the helicopter will most likely stop its turn 90° to the wind due to the extra drag on the tail. Conversely, the turn may accelerate if the helicopter is still rotating as the tail passes the 180° downwind.

Jammed Controls while in Forward Flight

Jammed Collective from Forward Flight

Introduction

When the collective jams while in forward flight, the pilot cannot move the collective up or down. The pilot can no longer adjust the power with the collective to manage the altitude.

The helicopter will continue to fly based on the current power setting when the collective jam occurs, so there is time to troubleshoot and formulate a plan.

There are three basic scenarios. A collective jam at:

1. a **high-power setting** above what is required to hover
2. at the **hover power** setting for the day, or
3. a **low power setting** less than what is required to hover.

In general, the:

- **higher** the power setting, the **harder to lose altitude**, but the **gentler and slower** the landing.
- **lower** the power setting, the **easier to lose altitude**, but the landing may be **harder** with **more run-on speed**.

The power curve

Understanding the power curve will assist the crew in determining the best course of action to affect an approach and landing. The power required from the engine to overcome the drag on the main rotor blades and the drag of the fuselage will vary depending on the helicopter's speed.

High power settings are required to hover with increasing power until the helicopter passes through Effective Translational Lift (ETL). On passing ETL, the power requirements drop off until the drag of the fuselage starts to become significant. At this point, the pilot needs more power to keep accelerating until a speed is reached where full power is used, and the helicopter cannot go any faster to maintain straight and level flight.

Normal power curve

The diagram below shows a standard power curve. The **curved line** represents the power **required** to fly straight and level at the corresponding speed on the graph.

The **straight line** represents the power **available** from the engine even if it is not being used.

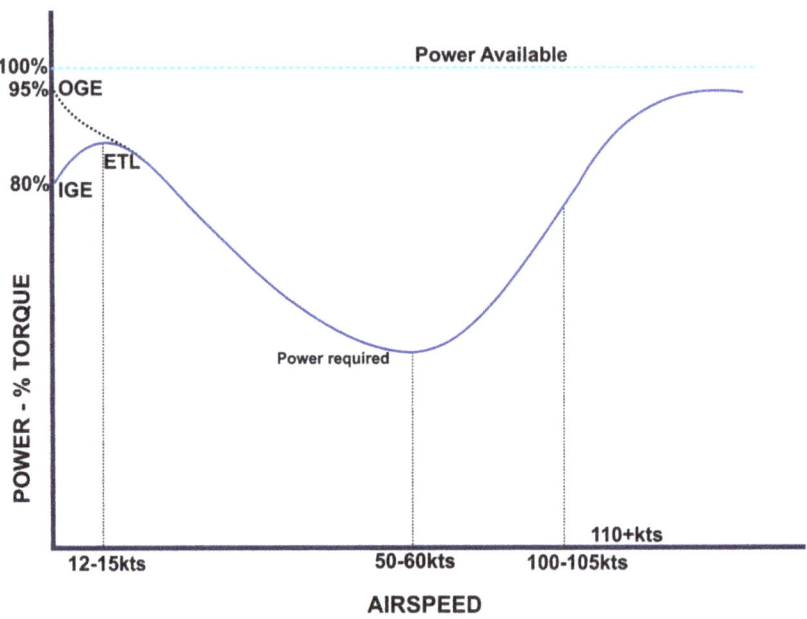

The power curve

The diagram below shows the power curve with the collective jammed at 80% torque, equivalent to roughly 21" MAP in a piston engine helicopter.

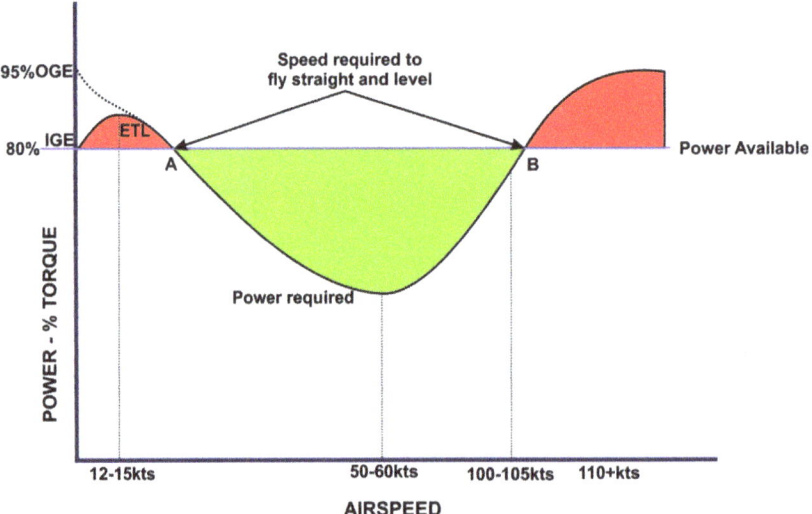

The gap between the two lines represents a power surplus (green shaded area) or a power deficit (red shaded area).

Flying at speeds in the:

- red shaded areas will allow the helicopter to descend.
- green shaded area will allow the helicopter to climb.

Flying at the exact speeds where the power available and power required lines intersect will allow the helicopter to fly straight and level.

The graph above shows two red areas where a descent can be commenced and a large green area where the helicopter can climb.

There are also two speeds the pilot can maintain straight and level:

- **A** is a low-speed straight and level, and
- **B** is a high-speed straight and level.

Chapter 9 Flight Control Emergencies (FCEs)

High Power example

If the collective jams at a higher power setting (say 85%Tq or 22" MAP), then the power curve may look like the following:

Note that the red-shaded areas are reduced, and the green-shaded area has increased. This limits your options to descend but will allow for a slower and smoother touchdown.

Low Power example

If the collective jams at a low power setting, then the power curve may look like the following:

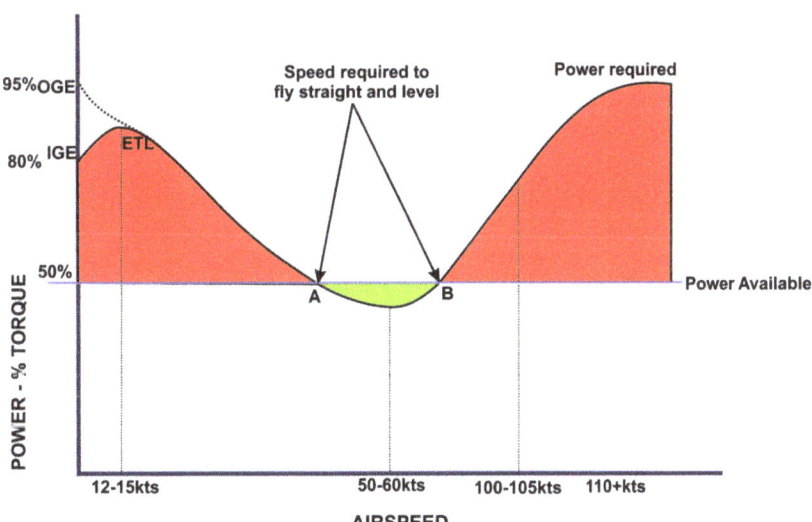

Note that the red-shaded areas have increased, and the green-shaded area has reduced. This increases your options to descend but will require a faster run-on speed and a harder landing.

Pilot options

The pilot now has three options to help control the helicopter's altitude if the collective jams:

- **Vary speed** by adjusting the attitude with cyclic so that the power required changes allowing the pilot to create a climb or descent.
- **Reduce Rotor RPM** within the Engine RPM limits to help decrease rotor thrust by rolling off some throttle. This is only good for a high-power collective jam.
- **Roll the helicopter (steep turn)** so that the vertical component of Total Rotor Thrust is reduced.

Option 1: **Vary Speed**	Vary speed by adjusting the attitude with cyclic so that the power required changes allowing the pilot to create a climb or descent.

Slow descending attitude   **60kt climb attitude**   **High speed descending attitude**

Option 2: Reduce Rotor RPM	Reduce Rotor RPM within the Engine RPM limits to help decrease rotor thrust by rolling off some throttle. This is only good for a high-power collective jam.

Rolling off a small amount of throttle   **RPM gauge**

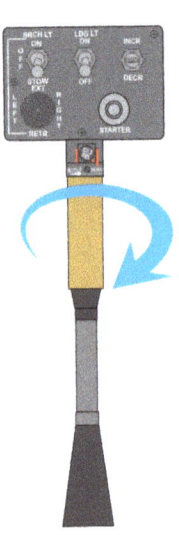 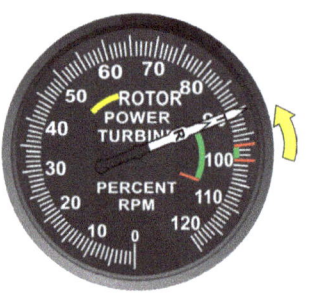

Option 3: **Steep Turn**	Roll the helicopter so that the vertical component of the Total Rotor Thrust is reduced.

Level attitude   **Steep turn allowing a descent**

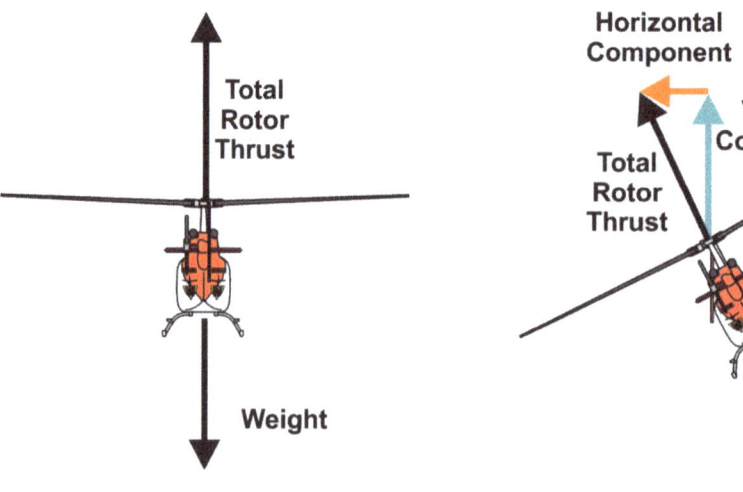

Chapter 9 Flight Control Emergencies (FCEs)

Troubleshoot

It is not necessary to immediately conduct a landing, as in most cases, the helicopter can still fly at a speed that allows the pilot to control the height. This will allow the crew time to troubleshoot the problem and experiment with varying combinations of speed, Rotor RPM and manoeuvring to establish a controlled rate of descent.

It is a delicate balance to adjust speed to control the rate of descent and glideslope to effect a safe landing. The glideslope may vary as the approach is made. The rate of descent may be very slow, so being patient is essential.

In general:

Using	Will...	And...	And...
Aft cyclic	Reduce speed	Increase the rate of descent	Decrease the glideslope
Forward cyclic	Increase speed	Decrease the rate of descent	Increase the glideslope

Glideslope picture

Using the above techniques will allow the glideslope and rate of descent to be adjusted to achieve a chosen landing site. The glideslope picture may look something like the following:

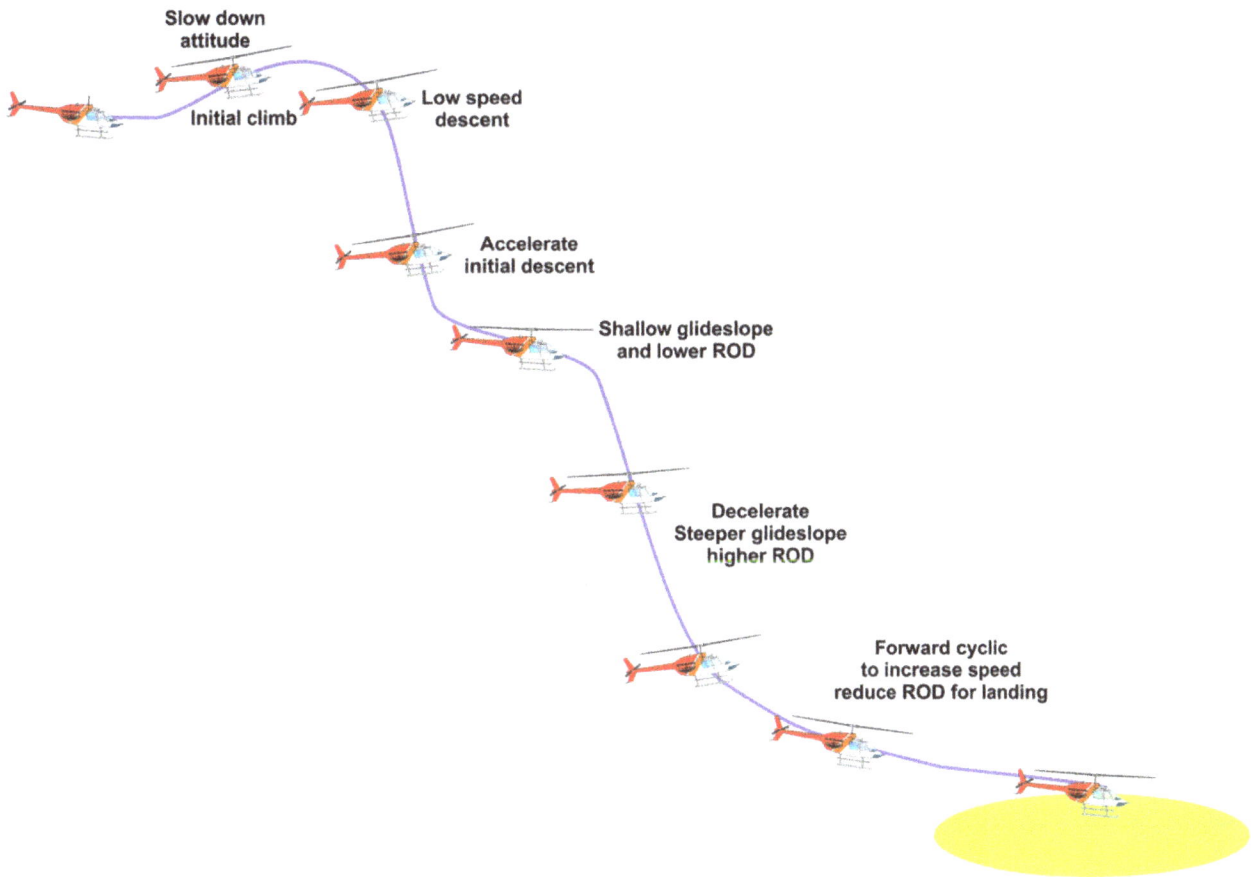

Cautions

When conducting these types of approaches, remember:

- Low speed and high rates of descent can induce Vortex Ring State (VRS). There may be a time when this may be used to create a rate of descent, but the pilot needs to allow sufficient height to affect a recovery. To avoid VRS, increase speed.
- If an emergency requires the pilot to enter autorotation, this will not be possible, so minimise the time in the air by affecting a landing as soon as possible.

Jammed Throttle from Forward Flight

Jammed Throttle

A jammed throttle in forward flight is apparent when the throttle cannot be moved or when the throttle is moved, and nothing happens.

If in forward flight and the Rotor RPM:

- **Is still in the green**, conduct a normal approach to a suitable landing area.
- **Is low:**
  - lower collective to reduce the pitch on the main rotor blades and restore the Rotor RPM, and
  - make a limited power approach as if the collective was jammed at that power setting following the same procedures as stated for a jammed collective.
- **continues to decay**, then continue to lower the collective to the point of entering autorotation and conduct a forced landing procedure with power on.

Troubleshoot

When troubleshooting, initially look for the obvious.

- Is the throttle friction on?
- Is there something binding the throttle?

Try to gently move the throttle but be mindful not to make the situation any worse. It can be made worse by rolling the throttle too far towards the IDLE and not being able to roll it back on or too far towards FULL and not being able to roll it back off. Maintaining RRPM is still important, if the RPM:

- **remains** at a level where the helicopter **can continue flying** under power, then fly to the nearest suitable landing area and make a shallow approach using minimum power. The helicopter will still fly even with low Rotor RPM (in the yellow) but avoid any harsh manoeuvring and be prepared for a run-on landing or a possible overpitch when transitioning back to the hover.
- **continues to decay**, then lower collective to restore the Rotor RPM. The amount of collective lowered may still be sufficient to fly straight and level at a reduced speed to a suitable landing area. If not, the next option is to lower the collective and conduct an autorotation to the ground, knowing that when the collective is raised, some engine power will still provide some cushioning effect.

Once on the ground, follow the same procedure for shutting down as described for a jammed throttle at the hover.

Jammed Cyclic from Forward Flight

Jammed cyclic

With a jammed cyclic, the cyclic cannot be moved in any direction, resulting in the pilot being unable to control the helicopter's attitude (pitch and roll) with the cyclic.

This is usually due to something interfering with the cyclic flight control system.

A jammed cyclic is often a result of a crew seat moving or cockpit objects interfering with the movement, such as a passengers foot.

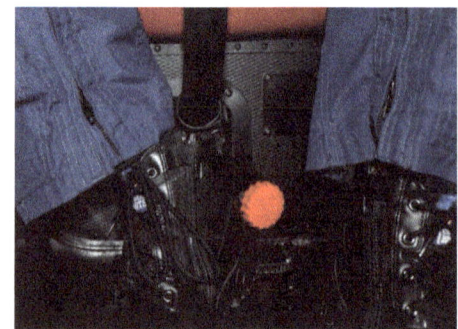

Chapter 9 Flight Control Emergencies (FCEs)

Troubleshoot   A jammed cyclic can become critical if not fixed quickly. Attempt to identify what is causing the jam and fix it. Check the pilot and the co-pilot cyclic, check the seat positions, and communicate with the crew.

Use some force on the cyclic to attempt to move it. Be conservative with the other controls so that you do not make the situation worse.

Jammed Pedals from Forward Flight

Jammed Pedals   When the pedals jam, they cannot be moved left or right, resulting in the pilot not being able to control the **yaw** of the helicopter when making a power change or a speed change while in forward flight.

In forward flight, the pilot will likely notice a pedal jam when a power or speed change is made, such as commencing a climb or a descent from the cruise.

In normal flight, the pilot will automatically move the pedals to maintain balance, and when all of a sudden, the pedals cannot be moved, this will cause the helicopter to be out of balance, and the crew will realise they have a jammed pedal situation. This can be caused by a foreign object caught in the pedals or a control rod in the system binding.

If the pilot returns the collective to the original position, such as the cruise, the helicopter will be back in balance, giving the crew time to troubleshoot the situation and devise a plan.

Troubleshoot   It is important that the pilot does not make the emergency worse. First, return to balanced flight, if possible, then check both sets of pedals to see if there is something obvious jamming the controls.

If something obvious is caught in the pedals, it is essential that when the crew member leans forward to remove the item, they do not inadvertently push the cyclic forward and put the helicopter in a nose pitch-down attitude.

If there is nothing obvious jamming the pedals, try using gentle pressure on them (left and right) to see if they can be moved or repositioned to a new position to help the approach.

If nothing can be done, plan for a jammed pedal approach by flying to a suitable area with a smooth flat surface such as the nearest aerodrome.

Pedal position

Part of the Hover Check is to note the pedal position at the hover. This is done to give the pilot a baseline position if a pedal jam is encountered later in the flight.

Anything greater than...

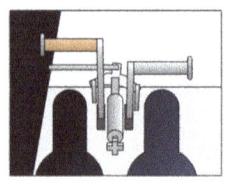

Left Pedal Jam

Base Line Position checked at the hover

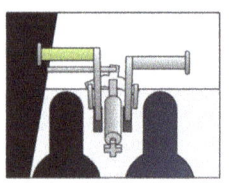

Hover Position for the day

Anything less than...

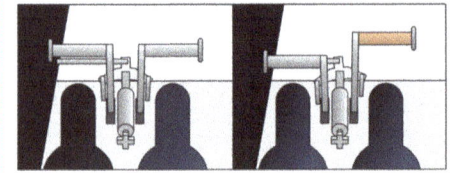

Right Pedal Jam

If the	Then it is a	Result:
left pedal is forward of the hover position for the day	Left Pedal Jam	■ Difficult to descend ■ Landing can be slow, even terminating at the hover
right pedal is forward of the hover position for the day	Right Pedal Jam	■ Easy to descend ■ Landing will be a faster run-on requiring a flat smooth surface

Yaw

When there is a pedal jam in forward flight and the helicopter is no longer in balance, it will yaw either left or right, and this cannot be controlled with the pedals in the usual manner.

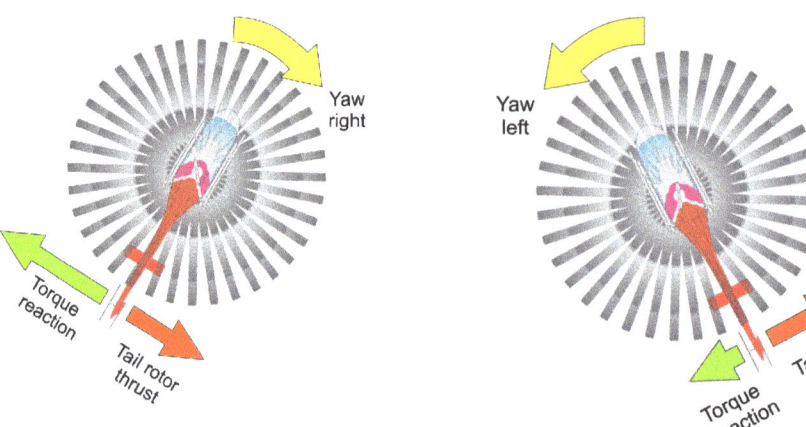

Producing a roll

This yaw will produce a roll in the opposite direction. The helicopter will turn in that direction unless the pilot corrects this using the cyclic.

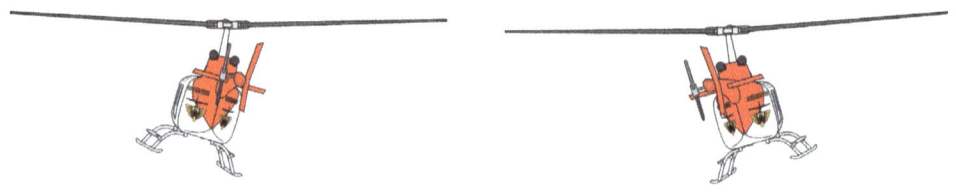

This is undesirable as the pilot will still want to control which direction the helicopter travels and may not want to follow the natural turn of the helicopter.

Chapter 9 Flight Control Emergencies (FCEs)

Control ground track with cyclic: left jam

The pilot will use the opposite cyclic to the yaw to counter the unwanted roll to maintain a desired ground track. With a jammed left pedal and left yaw, the helicopter will want to roll right and turn right. The pilot will use left cyclic to prevent this and maintain the desired ground track. It will feel very uncomfortable.

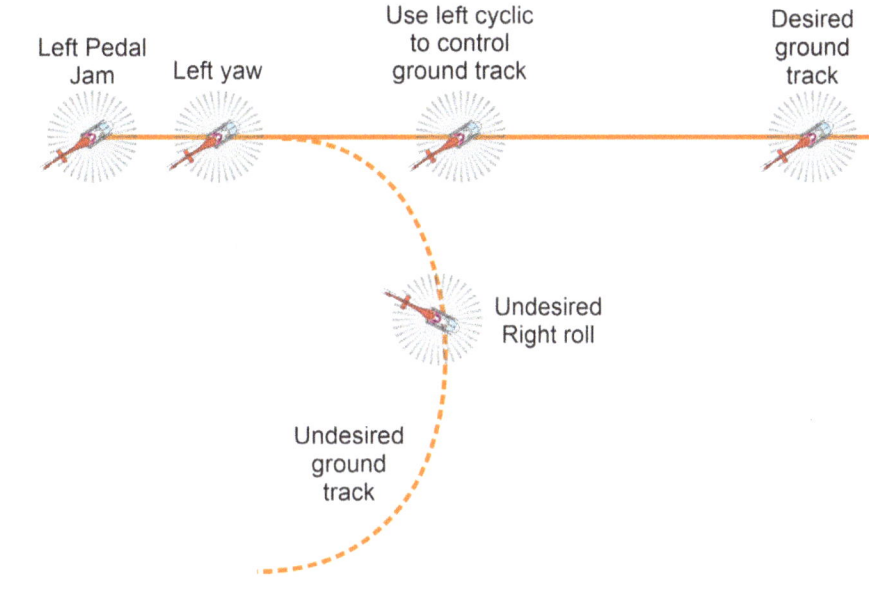

Control ground track with cyclic: right jam

With a jammed right pedal and right yaw, the helicopter will want to roll left and turn left. The pilot will use right cyclic to prevent this and maintain the desired ground track. It will feel very uncomfortable.

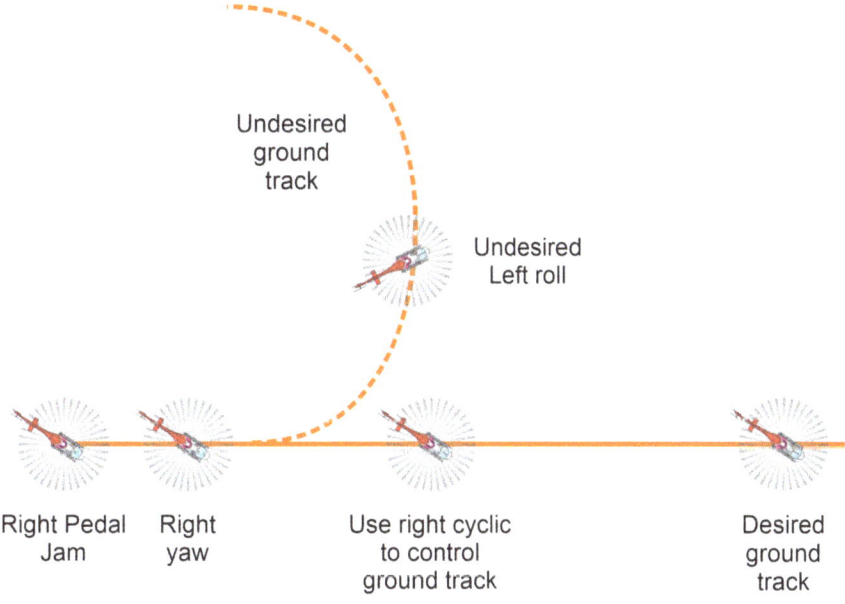

Left Pedal Jam in Forward Flight

Left pedal jam A left pedal jam indicates a higher power setting and tail rotor thrust greater than the torque reaction.

Lowering the collective to descend will **make the situation worse** by putting the helicopter further out of balance. This can be very uncomfortable and will put extra stress on the airframe as the wind, due to the forward movement, causes the helicopter to roll right while the pilot tries to maintain the desired ground track to a landing area.

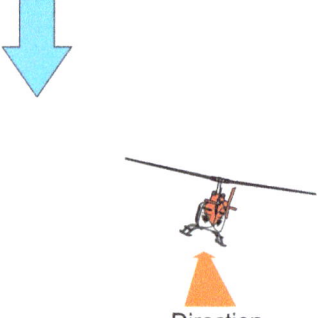

Options To minimise the effects, the pilot has several options to help with the descent and maintain control. In practice, the pilot will use all these techniques to control a descent with a left pedal jam.

Option 1 Roll off some throttle (within the Rotor RPM limits as required) to decrease tail rotor thrust, which will allow the pilot to lower some collective, or it may by itself be enough to reduce enough rotor thrust to allow the helicopter to commence a descent without having to use any collective.

Option 2 Do not lower the collective as much as normal; instead, extend downwind so that the circuit is larger and the approach longer and shallower.

Option 3 Conduct left-hand turns so that the turn helps counter the right roll caused by being out of balance.

Option 4 Maintain speed above minimum drag speed. The extra drag helps weathercock the fuselage and counter the left pedal jam.

Remember, the airspeed indicator may not be reading correctly as the pitot tube will not be receiving dynamic pressure from forward flight, so maintaining speed will be judged by the pilot looking outside.

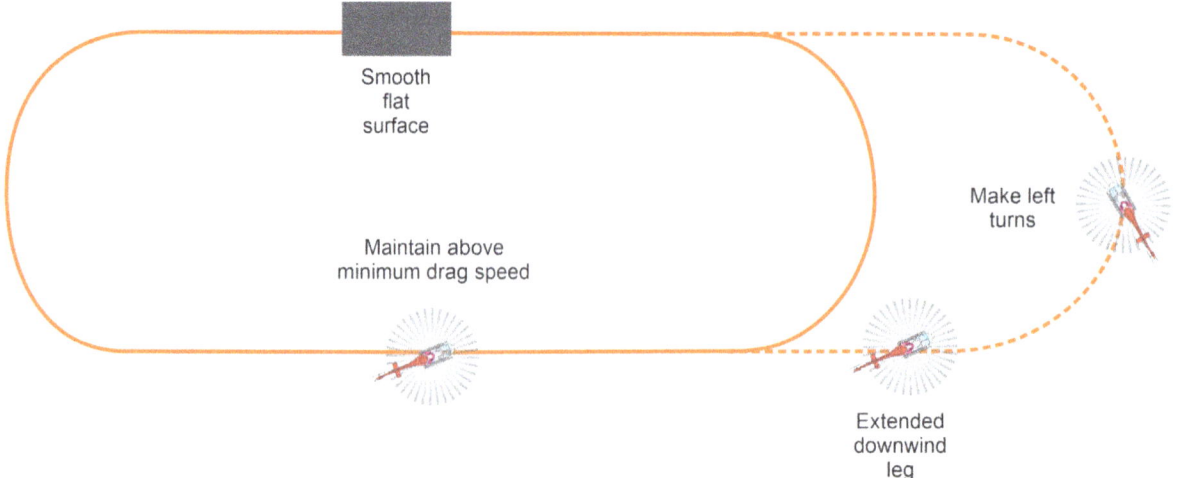

Right Pedal Jam in forward flight

Right pedal jam approach and termination

A right pedal jam indicates a lower power setting and tail rotor thrust less than torque reaction. Lowering the collective to descend will improve the situation by putting the helicopter back in balance.

This allows for a standard approach, but the termination will be faster and require some run-on as there will not be enough pedal available to manage the increased power requirements as the collective is raised when transitioning back to the hover.

Options

To minimise the effects, the pilot has several options to help with the descent and maintain control: In practice, the pilot will use all these techniques together to control a descent with a right pedal jam.

Option 1

Set throttle to maintain a high Rotor RPM (within the Rotor and Engine RPM limits) to maximise the tail rotor thrust, allowing the pilot to lower some collective.

Option 2

Lower the collective as usual and fly a standard approach but maintain the speed above minimum drag speed until confident of making the termination area.

Option 3

Conduct left or right-hand turns as required; it will not make any difference with most right pedal jams.

Option 4

Maintain speed above minimum drag speed for that helicopter type. The extra drag helps weathercock the fuselage and counter the right pedal jam. Remember, if in balance, the airspeed indicator will be reading correctly, but if not in balance, the airspeed indicator will not be reading correctly as the pitot tube will not be receiving the full dynamic pressure from forward flight, so keeping speed up is going to be judged by the pilot by looking outside.

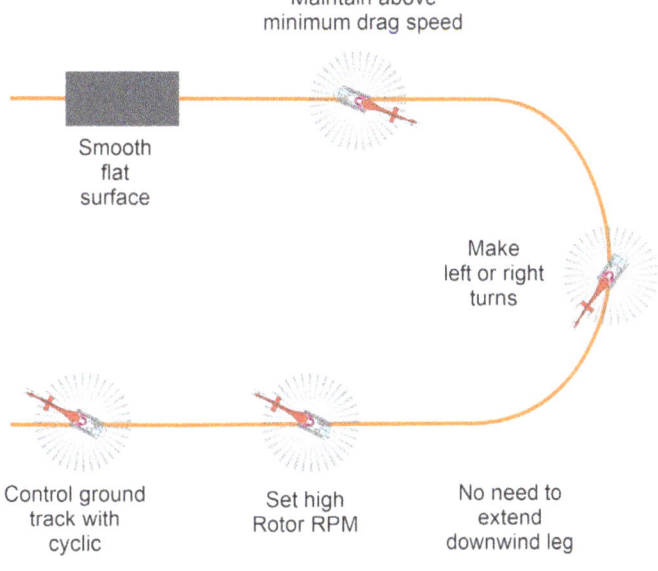

Yaw Control Emergencies (YCE)

What is a YCE A yaw control emergency is any emergency where the tail rotor is not providing thrust to counter the torque reaction of the main rotor blades when the engine is going.

YCE is the generic term given to:

- jammed pedals (left or right) both at the hover and in forward flight
- tail rotor drive shaft failure
- Loss of Tail Rotor Effectiveness (LTE), and
- loss of tail rotor components.

Each of these emergencies will be discussed separately, but it is important to note that they come under the definition of a Yaw Control Emergency (YCE).

Tail Rotor Driveshaft Failure

Introduction The first symptom of a tail rotor driveshaft failure is usually a "bang" or noise from the back, followed by the helicopter yawing or rapidly turning to the right when at the hover. Vibrations and a change in noise levels can follow this. Moving the pedals will have no effect, as the tail rotor blades have stopped turning. This is to be treated in the same manner as a full right pedal jam.

At the hover If at the hover, immediately roll the throttle to the IDLE position to reduce the torque reaction. Aggressive cyclic use is required to maintain ground position and keep the skids level with the ground before landing.

In forward flight In forward flight, there may be the option to maintain flight and forward speed to control the yaw allowing flight to a suitable landing area. Everything will depend on the damage to the aircraft and the crew's assessment of the best course of action. If in doubt, roll the throttle to IDLE and conduct an autorotation and forced landing.

On ground contact On ground contact, immediately lower the collective fully. You are, in effect, controlling a crash. Accept that the helicopter will touchdown while still turning to the right. In most cases, this is unavoidable.

Tail Rotor Component Failure

Introduction A tail rotor component failure will result in severe vibration, noise, and a large centre of gravity shift followed by a nose pitch down and full aft cyclic.

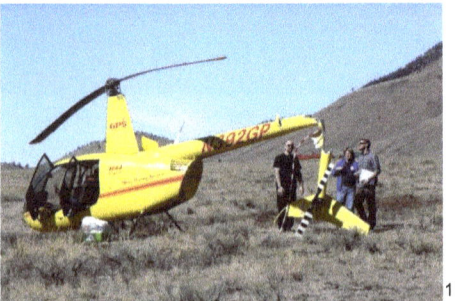

The picture below happened on the ground. If this had happened in flight, it would have been unrecoverable.

---

[1] Reference: http://bloximages.chicago2.vip.townnews.com/taosnews.com/content/tncms/assets/v3/editorial/a/d6/ad60b68e-8cfa-11e2-b798-001a4bcf887a/5142566d8f1c8.image.jpg

Chapter 9 Flight Control Emergencies (FCEs)

Cause

A component failure is usually caused by the tail striking a solid object such as a tree, building, fence post or similar or by poor handling by the pilot where they inadvertently strike the ground hard with the tail, and it breaks off. Whatever caused the failure, the safest place to be is on the ground.

The emergency's severity will depend on how much of the tail (weight) is lost. In most cases, simply losing the tail rotor blades or even the gearbox still allows the helicopter to be managed within its C of G envelope. Anything more (such as above) will result in operations outside the C of G envelope, and the pilot will lose control.

LTE at the Hover and Slow Forward Flight

Introduction

Loss of Tail Rotor Effectiveness (LTE) is evidenced by a rapid turn to the right and is covered in the Aerodynamics theory course. It will only happen if the helicopter is in the hover (IGE or OGE) or very slow flight (less than 35 kts) with high power settings and/or the wind direction is coming at an angle that disrupts the airflow over the tail rotor system. It will not happen in normal forward cruise flight.

Causes

When experiencing LTE, it is important to realise that there is nothing wrong with the tail rotor system; instead, the pilot has either:

1. Exceeded the limits of the tail rotor to counter the torque reaction under a high power setting, low-speed scenario, which is common on take-off, approach or conducting a low-level operation; or
2. Allowed the wind to be coming from a critical area that disrupts the aerodynamics on the tail rotor blades to the point that there is no longer sufficient tail rotor thrust to overcome the torque reaction.

Wind avoid areas

Different helicopters may have different tail rotor effectiveness limits, which will be published in the RFM. For example, here is an example of a Critical Wind Azimuth Chart from an RFM:

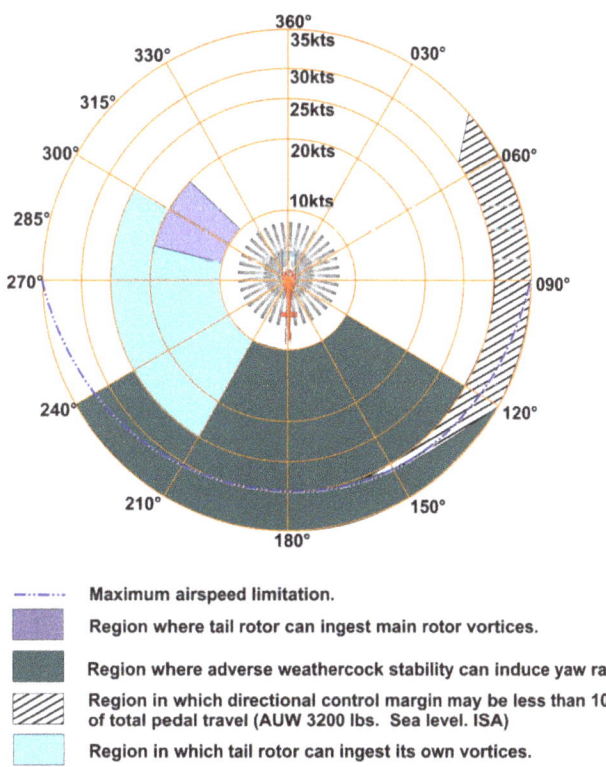

Avoidable	LTE has more to do with the pilot not maintaining situational awareness and allowing the helicopter to be flown or operated close to the helicopter's power limits with the wind coming from a critical angle.
	Good pre-flight planning, an appreciation of the aircraft's performance, and best use of wind on the nose will help avoid LTE.

YCE Summary

YCE summary	Yaw control emergencies can be summarised as follows:

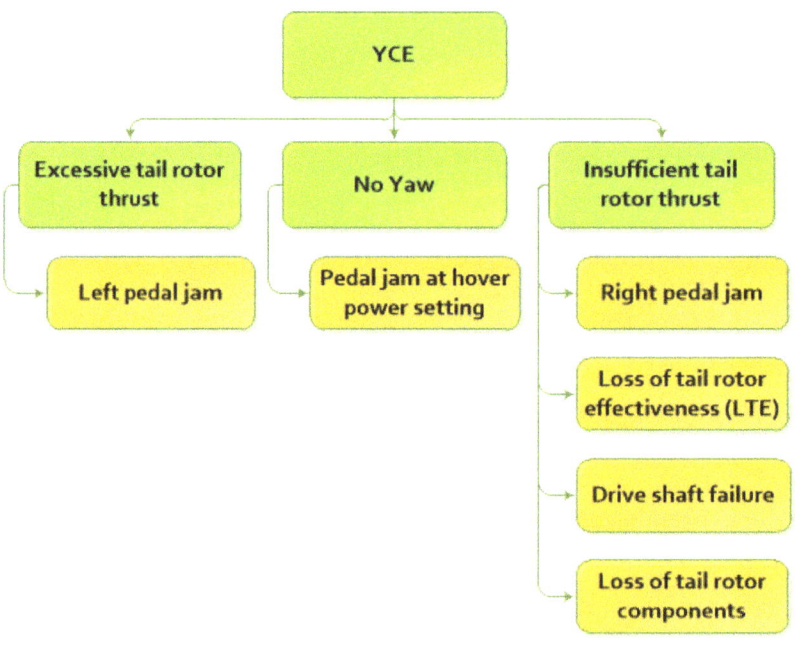

Differences	Difference between a jammed control and a YCE:
	1. Jammed control means that you cannot physically move the affected control.
	2. YCE relates to the loss of tail rotor thrust for any reason.

YCE	Identified by
Jammed pedal	■ Pedals cannot move. ■ Helicopter rotates left or right.
LTE	■ Pedals can still move. ■ Helicopter rotates violently right, even with full left pedal.
Tail rotor drive failure	■ Pedals can still move. ■ Helicopter rotates violently right, even with full left pedal. ■ May be a change in noise and possible vibrations.
Component failure or damage	■ Pedals may or may not be able to move. ■ Helicopter rotates violently right, even with full left pedal. ■ May be a change in noise and possible vibrations. ■ The nose may pitch down excessively.

Hydraulics Failure

Introduction Some aircraft have a hydraulic boost system that will assist in moving one or more of the flight controls (cyclic, collective and pedals).

A driveshaft from the main transmission drives the hydraulic pump. This pump delivers hydraulic fluid from a reservoir via hoses to the hydraulic servos that will be driving the relevant control. As a safety system that can be used for emergencies and training, most hydraulic systems incorporate an electrical isolation switch so the pilot can isolate the hydraulic system from the flight controls.

Symptoms If the hydraulic assistance is lost, the pilot will experience stiff controls that are harder to move and give feedback. The flight controls can still be moved without hydraulics, but it will be much harder and tiring. The controls will also have a lot of "slop" as the pilot has to physically move the hydraulic servo against the pilot valve before any corresponding movement of the swashplate. There will be a large "dead zone" where the cyclic and collective may move, but nothing happens. This can take a bit of practice to get used to.

Points of failure There are several different points of failure that the pilot may encounter with a hydraulics failure, they are:

1. Hydraulic pump or driveshaft failure
2. Hydraulic hose failure where the fluid is lost
3. Individual hydraulic servo failure where the servo blows a seal or the pilot valve becomes stuck
4. Electrical switch failure, and
5. Knowing this information will help the crew troubleshoot the problem and conduct a recovery procedure.

Intermixing bell crank Most helicopter designs have an intermixing bell crank, which means that the collective and the cyclic flight control systems have common components, and moving one control will cause unwanted movement on the other. When the hydraulics are working properly, the pilot will never notice or feel this movement; however, when the hydraulics fail or are turned OFF, there will be a noticeable intermixing of the controls.

If the pilot moves the	Then the
collective UP	Cyclic moves FORWARD
collective DOWN	Cyclic move BACK
cyclic BACK	collective moves UP
cyclic FORWARD	collective moves DOWN

The pilot needs to counter this unwanted movement while the hydraulics are OFF or failed.

Electrical solenoid

Built into the hydraulic system is an electrical solenoid connected to a switch and a circuit breaker in the cockpit. With the hydraulic switch in the ON position, the solenoid will not receive any electrical current and will automatically sit in the OPEN position for normal flight, allowing hydraulic pressure to power the servos. The pilot must select the HYD Switch to the OFF position to close the solenoid. This will allow electrical current to flow to the solenoid to CLOSE it and prevent any hydraulic pressure from reaching the servos.

If, for some reason, the solenoid fails and CLOSES by itself, then the pilot may be fooled into thinking the hydraulics have failed, whereas it is an electrical problem in the hydraulic system.

To check, the pilot can pull the hydraulic circuit breaker, and any power that has inadvertently been provided to the solenoid will be removed, and the hydraulics will be restored to the flight controls.

Troubleshoot

If any of the flight controls become stiff and hard to move simultaneously, then the crew are experiencing a hydraulic failure of some sort.

If only one control has an issue, then the crew are experiencing a servo malfunction or failure.

In both cases, the crew need to check the hydraulic electrical system first in case it is not an easy fix before moving onto the full hydraulics off approach and landing.

It is easy to accidentally turn off the hydraulic system, so check the HYD SWITCH before checking the HYD CIRCUIT breaker. If it is not an electrical problem, then the hydraulics has failed for some reason.

Cyclic Hard-over

During the run-up, the crew will test the hydraulic system, which will require it to be turned OFF.

If the cyclic has an uncommanded movement over to one side that cannot be controlled, this is known as a hard-over. A hard over indicates a problem inside one of the servos and can be overcome by turning the hydraulics immediately back ON.

Once the hydraulic pressure is restored, the hard-over will go away, and normal use of the cyclic will be available. The pilot should shutdown and seek engineering assistance before going flying.

A cyclic hard-over can only occur in flight when practising hydraulic failures and having turned the hydraulic system OFF. Again, if this happens, immediately restore the hydraulics and return to base for engineering assistance.

Air Exercises: Flight Control Emergency (FCE)

Introduction This section describes the air exercises and demonstrations for flight control emergencies.

Airmanship
- Memorise the recovery techniques.
- Be able to make a MAYDAY call from memory.
- Be able to communicate with the crew.
- Maintain situational awareness and a good lookout.
- Always fly the helicopter first.
- Know the secondary effects of each flight control.

Common faults Common faults to guard against during the air exercises:

- No lookout.
- Thinking all the controls have jammed when only one of the four (4) primary controls will ever be jammed at one time during practice.
- Not flying at normal circuit heights or not maintaining a relatively normal circuit shape.
- Over-controlling or overcompensating with the other controls.
- Not managing airspeed.

Air Exercise 9-1: Jammed Controls at the Hover

Jammed Collective at the Hover

Introduction	The instructor will demonstrate a jammed collective, throttle, left and right pedal at the hover and the associated recovery actions.
	After the demonstration, the instructor will give the student all of the controls and position the helicopter at 5ft AGL hover in a suitable area before allowing the student to practice.
Limits	■ Commence from a 5ft stable hover.
	■ Only allow ground contact once the helicopter
	▫ is not turning about the mast, and
	▫ is not drifting backwards or sideways.
Procedure	When the collective jams at the hover, conduct the following steps:

Announce	Announce *"Collective jam"*
Aviate	Maintain the hover position by: ■ Controlling the ground position with cyclic in the normal manner ■ Controlling the heading with pedals in the normal manner
Navigate	Land as soon as possible.
Communicate	Make appropriate radio calls once safely on the ground.
Activate	**From a stable hover IGE** ■ Check for obstacles. ■ Use throttle to control the rate of descent to land. ■ When the skids touch the ground, slowly set throttle to IDLE. **From the hover but the helicopter is descending** ■ Allow it to descend. ■ If necessary, forward cyclic to conduct a run-on landing. ■ On ground contact, slowly set throttle to IDLE. **From the hover but the helicopter is climbing OGE** ■ Conduct the steps as for a stable hover; or ■ Forward cyclic and transition into forward flight and then conduct a jammed collective from forward flight emergency. **When on the ground** ■ Conduct an emergency shutdown ■ Conduct any follow-on actions, and ■ Be aware of blade sailing, as the blades will have positive pitch due to the jammed collective.

Jammed Throttle at the Hover

Procedure: When the throttle jams or fails to respond at the hover, conduct the following steps:

Announce	Announce *"Throttle failure"*
Aviate	Maintain control of the helicopter by - If on the ground, stay on the ground. - If RRPM remains within limits, then lower collective and land. - If RRPM increases above a limit, take steps to reduce engine power output (carb heat, magneto, hover downwind) and land. - If RRPM still increases, conduct an engine failure at the hover by turning the engine OFF. - If RRPM decays, accept this and use collective to cushion on in a deliberate overpitch.
Navigate	Land as soon as possible.
Communicate	Make appropriate radio calls.

Activate	**When on the ground:**
	- Collective full down BUT be aware of a possible overspeed, so do this slowly.
	- Throttle OFF. If there is no effect, then
	    - In a piston engine, MIXTURE LEAN (discussion only)
	    - In a turbine Fuel Valve OFF (discussion only)
	- The battery remains ON until the engine has stopped and the blades have stopped turning.
	- It may take some time for the engine to stop based on the design. A piston engine is almost instant when leaned. A turbine engine can take several minutes as the fuel is used up in the fuel lines.
	- Conduct any follow-on actions.

Discussion Only: Jammed Cyclic at the Hover

Jammed cyclic: Procedure	**Discussion only**
	When the cyclic jams at the hover, conduct the following steps:

Announce	Announce "Cyclic jam"
Aviate	Lower collective. Use pedals to follow any roll to keep the skids level with the ground. Use abrupt up/down collective movements to control large pitch changes if able. **DO NOT FLY AWAY. COMMIT TO THE GROUND.**
Navigate	Land as soon as possible.
Communicate	Make appropriate radio calls once on the ground.
Activate	**When on the ground:** - Collective full down - Conduct an emergency shutdown, and - Conduct any follow on actions.

Pedals Jam in the Hover Position at the Hover

Procedure	When the pedals jam in the hover position at the hover and the helicopter is not turning left or right, then conduct the following steps:

Announce	Announce *"Pedal jam"*
Aviate	- Maintain ground position with cyclic - Control height with collective
Navigate	Land as soon as possible.
Communicate	Make appropriate radio calls once on the ground.
Activate	**Landing:** - Gently lower some collective to land: - Use throttle to control any turn or accept landing with some turning motion if safe to do so. - Maintain ground position with cyclic. **When on the ground:** - Collective full down - Conduct an emergency shutdown, and - Conduct any follow on actions.

Chapter 9 Flight Control Emergencies (FCEs)

Left Forward Pedal Jam at the Hover

Procedure	When the pedals jam in the left forward position at the hover, conduct the following steps:

Announce	Announce *"Left pedal jam"*.
Aviate	- Maintain ground position with cyclic. - Control height with collective.
Navigate	Land as soon as possible.
Communicate	Make appropriate radio calls once on the ground.
Activate	**Bleed technique** - "Bleed" off some throttle slowly. The rate of turn may initially increase as torque decreases but then decrease as Rotor RPM decreases. - Maintain ground position with cyclic. - Control height with collective. - Once the turn stops, lower some collective to land. **When on the ground** - Collective full down - Conduct an emergency shutdown, and - Conduct any follow on actions.

Right Forward Pedal Jam at the Hover

Procedure	When the pedals jam in the right forward position at the hover, conduct the following steps:

Announce	Announce *"Right pedal jam"*.
Aviate	- Maintain ground position with cyclic. - Control height with collective.
Navigate	Land as soon as possible.
Communicate	Make appropriate radio calls once on the ground.
Activate	**Crack technique:** - "Crack" off some throttle quickly. - Maintain ground position with cyclic. - Control height with collective. - Once the turn stops, lower some collective to land. **When on the ground:** - Collective full down - Conduct an emergency shutdown, and - Conduct any follow-on actions.

Air Exercise 9-2: Jammed Controls in Forward Flight

Introduction	The instructor will demonstrate a jammed collective, throttle, left and right pedal while in forward flight and the associated recovery actions.
	After the demonstration, the instructor will give the student all of the controls and position the helicopter downwind in the circuit before allowing the student to practice.
Limits	■ Commence from downwind at either 500 or 1000 ft in the circuit while cruising at 60-100 kts.
	■ Use a power setting that will allow the RRPM to remain in the green range.
	■ Land and discuss the shutdown; do not actually conduct the shutdown by turning off the fuel valve or leaning the mixture.

Collective Jams in Forward Flight

Procedure	When the collective jams in forward flight, conduct the following steps:

Announce	Announce *"Collective jam."*
Aviate	Maintain straight and level flight, if possible, fly to a suitable landing area:
	■ This may be at a reduced speed for low-power collective jams.
	■ This may be at a higher speed for high-power collective jams.
Navigate	Land as soon as possible.
Communicate	Make appropriate radio calls.
	It is a mandatory MAYDAY call for a jammed control while in flight.
Activate	**Troubleshoot**
	■ Check there is nothing obvious causing the jam that the crew can fix immediately.
	■ Check both collectives.
	■ Put a small to medium amount of pressure up and down on the collective to see if the jam can be freed or if it is just a binding. Be careful not to make the situation worse.
	■ Experiment with speed, RRPM and maneuvering to get a feel for the helicopter and what works best to control the rate of descent.
	■ Make a plan.
	■ Brief the crew and passengers.

Chapter 9 Flight Control Emergencies (FCEs)

With a high power setting (at or above hover power)

- Roll off some throttle to reduce the RRPM as required but no less than the bottom of the engine green range; this will also decrease the total rotor thrust to help commence a descent.
- Use some aft cyclic to reduce the speed until the helicopter starts a small rate of descent. Initially, the helicopter will climb as the speed washes off, but it will eventually commence a descent.
- Control the rate of descent by varying airspeed as required.
- Between 50 and 25 ft AGL (depending on the ROD), use some forward cyclic to slightly increase speed to reduce the rate of descent in anticipation of landing.
- If required, increase throttle to increase total rotor thrust to assist in controlling the rate of descent before touchdown.
- As the skids touch the ground, roll the throttle smoothly to the IDLE position to affect a landing.

With a low power setting (below hover power)

- Set the throttle to give maximum available Rotor RPM (Top of the Green).
- Use some aft cyclic to reduce the speed until the helicopter starts a small rate of descent. Initially, the helicopter will climb as the speed washes off, but it will eventually commence a descent.
- Control the rate of descent by varying airspeed as required.
- Between 50 and 25 ft AGL (depending on ROD), use some forward cyclic to increase speed to reduce the rate of descent in anticipation of landing.
- Expect a run-on landing with a speed based on the power available.
- As the skids touch the ground, roll the throttle smoothly to the IDLE position to affect a landing.

When on the ground

- Conduct an emergency shutdown
- Conduct any follow on actions, and
- Be aware of blade sailing, as the blades will have positive pitch due to the jammed collective.

Throttle Jams in Forward Flight

Procedure	When the throttle jams or fails to respond in flight, conduct the following steps:

Announce	Announce *"Throttle failure."*
Aviate	If still able to maintain RRPM, then fly to a suitable landing area If RRPM decays, then lower collective to restore RRPM or enter autorotation.
Navigate	Land as soon as possible.
Communicate	Make appropriate radio calls. It is a mandatory MAYDAY call for a jammed control while in flight.

Activate	**Troubleshoot**
	
	- Check there is nothing obvious causing the jam that the crew can fix immediately.
	- Check both throttles by rolling towards FULL only. Be careful not to make the situation worse.
	- Experiment with speed, Rotor RPM and manoeuvring to get a feel for the helicopter and what works best to control the rate of descent.
	- Make a plan.
	- Brief the crew and passengers.
	
	If able to maintain RRPM
	
	- Make a normal approach, or
	- Approach as if the collective was jammed, or
	- **If unable to maintain Rotor RPM,** enter autorotation and conduct a forced landing.
	
	When on the ground
	
	- Throttle OFF. If there is no effect, then:
	    - Mixture lean or Fuel valve OFF
	    - Conduct an emergency shutdown except for the battery, leave **Battery ON.**
	    - In turbines, it may take up to 60 seconds for the engine to stop as the fuel is consumed from the fuel valve to the combustion chamber. Some engine surging or stuttering may be experienced; this is normal.
	    - Once stopped (and turbine TOT is stable), **Battery OFF.**
	- Conduct any follow-on actions.

Chapter 9 Flight Control Emergencies (FCEs)

Discussion Only: Jammed Cyclic from Forward Flight

Procedure: jammed cyclic	**Discussion only**
	Conduct the following steps when the cyclic jams in flight (Discussion only).

Announce	Announce "Cyclic jam."
Aviate	■ Maintain collective in its current position and use conservatively. ■ If required, use abrupt up/down collective movements to control large pitch changes. ■ Use pedals to follow any roll to keep the skids level with the ground.
Navigate	Attempt landing as soon as possible.
Communicate	Make appropriate radio calls. It is a mandatory MAYDAY call for a jammed control while in flight.
Activate	When on the ground: ■ Collective full down ■ Conduct an emergency shutdown, and ■ Conduct any follow-on actions.

Pedals Jam in the Left Position in Forward Flight

Procedure	When the pedals jam in the LEFT forward position in flight:

Announce	Announce *"Pedal jam."*
Aviate	■ Based on the hover pedal position, confirm a LEFT Pedal Forward Jam. ■ Select a power setting that allows controlled flight.
Navigate	Attempt landing as soon as possible in a suitable area.
Communicate	Make appropriate radio calls. It is a mandatory MAYDAY call for a jammed control while in flight.
Activate	**Establish downwind in the circuit, then** ■ Extend downwind ■ Reduce some tail rotor thrust with throttle if able ■ Make left-hand turns ■ Control ground track with cyclic ■ Make a shallow approach ■ Use collective steering for the approach ■ Use throttle steering for the termination **When on the ground** ■ Collective full down ■ Conduct normal shutdown, and ■ Conduct any follow-on actions.

Approach and termination technique	Before conducting the approach, experiment with various speeds and power settings to get a "feel" for the best combinations of speed and power that will allow the pilot to control the helicopter. Trying to establish a normal sequence of events for the crew is very important to control the helicopter in an abnormal situation.
Left pedal jam approach and termination	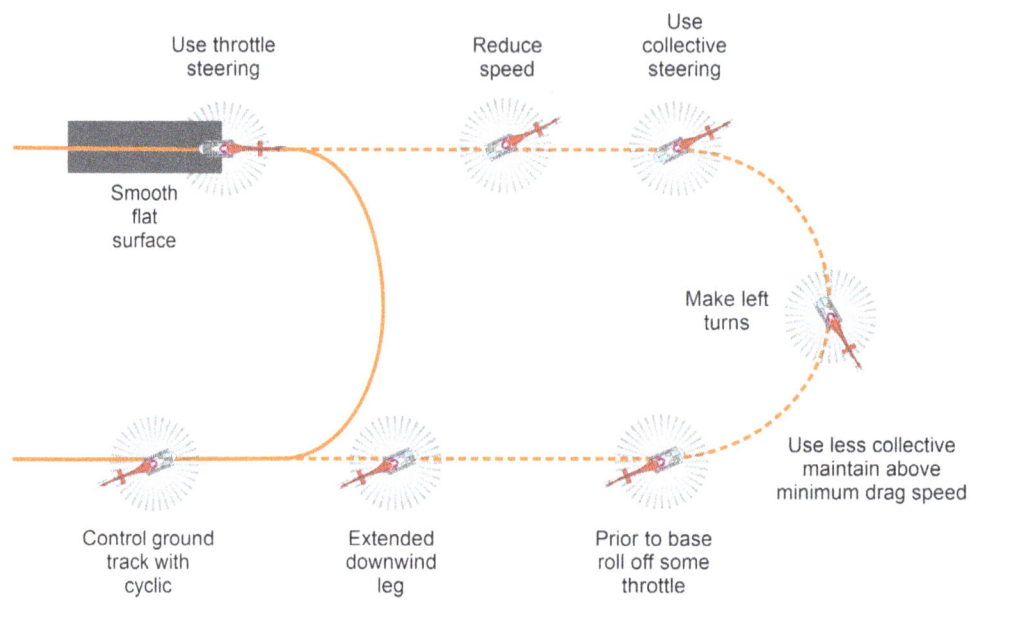

Left pedal jam approach and termination	With a left pedal jam, fly the helicopter to the standard 1000 ft downwind position in the circuit, then conduct the following:

1. Extend the downwind leg.
2. Before base, roll the throttle to the bottom of the RRPM limit and extend the index finger so that when you roll the throttle, you have a reminder of which way the nose of the helicopter will go. This will be very important for the touchdown when throttle steering.
3. Commence a left-hand turn maintaining the desired track across the ground with cyclic.
4. If the yaw is too excessive, use collective steering to help control it.
    - Raising some collective will yaw the nose of the helicopter right.
    - Lowering some collective will yaw the nose of the helicopter left.
5. Once established on a long final, continue to slow the helicopter with some aft cyclic progressively but maintain at least minimum drag speed for that helicopter type until confident of making the selected termination area.
6. As the termination area approaches, continue to reduce speed and raise some collective to control the rate of descent. As the collective is raised, the nose will come right.
7. The moment the nose of the helicopter has come straight, the pilot can no longer use collective steering to control yaw. Instead, the pilot will now revert to throttle steering to control the yaw (in the same manner as for a jammed pedal at the hover) until touchdown.

Pedals Jam in the Right Position in Forward Flight

Procedure	When the pedals jam in the RIGHT forward position in flight:
Announce	Announce *"Pedal jam."*
Aviate	Based on the hover pedal position, confirm a RIGHT Pedal Forward Jam. Select a power setting that allows controlled flight.
Navigate	Attempt landing as soon as possible in a suitable area.
Communicate	Make appropriate radio calls. It is a mandatory MAYDAY call for a jammed control while in flight.
Activate	**Establish downwind in the circuit, then**<ul><li>Make a normal approach</li><li>Maximise tail rotor thrust setting the RPM to Top of the Green</li><li>Make left or right hand turns</li><li>Control ground track with cyclic</li><li>Make a normal approach keeping airspeed above the minimum drag speed until confident of making the termination area</li><li>Use collective steering for the approach</li><li>Use throttle steering for the termination</li></ul>**When on the ground**<ul><li>Collective full down</li><li>Conduct an emergency shutdown, and</li><li>Conduct any follow-on actions.</li></ul>

Approach and termination technique	Before conducting the approach, experiment with various speeds and power settings to get a "feel" for the best combinations of speed and power that will allow the pilot to control the helicopter. Establishing a normal sequence of events for the crew is important to control the helicopter in an abnormal situation.

Right pedal jam approach and termination

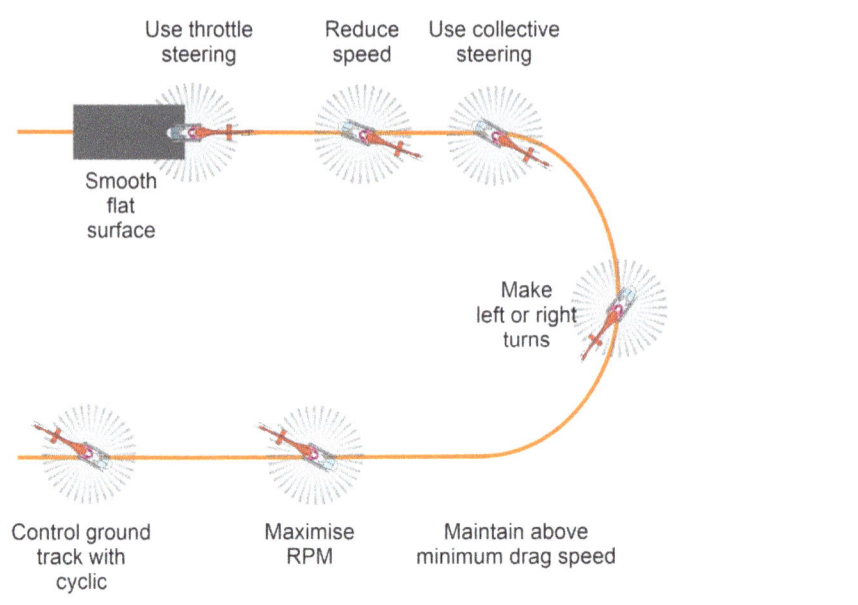

Right Pedal Jam: Approach and termination technique

With a right pedal jam, fly the helicopter to the standard 1000ft downwind position in the circuit, then conduct the following:

1. Fly a closer circuit as the helicopter will be at a lower power setting, so it will be descending quicker. Extending the downwind leg is unnecessary unless the pilot thinks otherwise. Before base, roll the throttle to maximise the RPM. Extend the index finger to remind you which way the nose will go. This is important for the touchdown when throttle steering.
2. Commence a left or right-hand turn maintaining the desired track across the ground with cyclic.
3. If the yaw is too excessive, use collective steering to help control it.
    - Raising some collective will yaw the nose of the helicopter right.
    - Lowering some collective will yaw the nose of the helicopter left.
4. Once established on final, progressively slow the helicopter with some aft cyclic but maintain at least minimum drag speed until confident of making the selected termination area.
5. As the termination area approaches, use aft cyclic to slow down, this may require lowering collective to prevent the helicopter from ballooning up.
6. As the helicopter gets close to the ground and is straight, accept any run-on speed and raise some collective to cushion on.
7. As collective is raised, the nose will come right.
8. The moment the nose of the helicopter has come straight, you can no longer use collective steering to control yaw. Instead, revert to throttle steering to control the yaw (the same as for a jammed pedal at the hover) until touchdown.

Note: The termination for a right pedal jam can happen faster than a left pedal jam.

Air Exercise 9-3: Yaw Control Emergencies

Discussion Only: Tail Rotor Driveshaft Failure at the Hover

Procedure	Discussion only
	When the tail rotor drive shaft fails at the hover, conduct the following steps:

Announce	Announce **"Tail Rotor Drive failure."**
Aviate	<ul><li>Immediately roll off the throttle to the IDLE position.</li><li>Maintain ground position with cyclic.</li><li>Allow the helicopter to descend.</li><li>Cushion the landing with collective.</li></ul>
Navigate	No choices here!
Communicate	Make appropriate radio calls .
Activate	**When on the ground:**<ul><li>Collective full down</li><li>Conduct an emergency shutdown of the engine, and</li><li>Conduct any follow-on actions that may be required.</li></ul>

Discussion Only: Tail Rotor Driveshaft Failure in Forward Flight

Procedure	Discussion only
	When the tail rotor drive shaft fails in forward flight, conduct the following steps:

Announce	Announce **"Tail Rotor Drive failure."**
Aviate	Fly the aircraft and maintain control.<ul><li>If the helicopter is still responding to cyclic and collective control inputs and the fuselage can be kept semi-straight at speed, then maintain forward flight for controllability and fly to a suitable area.</li><li>If the helicopter is not responding enough to maintain control, and there are severe noises and vibrations, then roll the throttle to IDLE and conduct an autorotation.</li></ul>
Navigate	Land as soon as possible. There may be a decision to make:<ul><li>Can I fly to a suitable landing site, or</li><li>Do I need to enter autorotation and accept the terrain below.</li></ul>
Communicate	Make appropriate radio calls.
Activate	**When on the ground:**<ul><li>Collective full down</li><li>Conduct an emergency shutdown of the engine, and</li><li>Conduct any follow-on actions that may be required.</li></ul>

Mike Becker's Helicopter Handbook

Discussion Only: Tail Rotor Component Failure at the Hover

Procedure   **Discussion only**

When there is a loss of a section of the tail at the hover, conduct the following steps:

Announce	Announce **"Tail Rotor failure."**
Aviate	Immediately roll the throttle to IDLE and manage the landing as best as you can.
Navigate	Land as soon as possible.
Communicate	Make appropriate radio calls.
Activate	**When on the ground:** ■ Collective full down ■ Conduct an emergency shutdown of the engine, and ■ Conduct any follow-on actions that may be required.

Chapter 9 Flight Control Emergencies (FCEs)

Discussion Only: Tail Rotor Component Failure in Forward Flight

Procedure	Discussion only
	When there is a loss of a section in forward flight, conduct the following steps:
Announce	Announce **"Tail Rotor failure."**
Aviate	Fly the aircraft and maintain control. - If the helicopter is still responding to cyclic and collective control inputs and the fuselage can be kept semi-straight at speed, then maintain forward flight for controllability and fly to a suitable area and make a long shallow approach keeping speed above the minimum drag speed until you reach the touchdown area (a runway preferably). - If the helicopter is not responding enough to maintain control, and there are severe noises and vibrations, attempt a long shallow high-speed descent until the touchdown area is reached. Then attempt to slow the aircraft based on the surface. This will be messy! Note: Rolling the throttle to IDLE and conducting an autorotation may make the situation worse. Losing weight from the tail means that the pilot will already be using aft cyclic to maintain control, which may already be very close to the limit. Lowering the collective and entering autorotation requires a lot of aft cyclic to prevent the nose pitching down, and there may not be enough aft cyclic left to control this manoeuvre. Now is not the time to find out!
Navigate	Land as soon as possible. There may be a decision to make. - Can I fly to a suitable landing site, or - Do I need to attempt a landing now.
Communicate	Make appropriate radio calls.
Activate	**When on the ground:** - Collective full down - Conduct an emergency shutdown of the engine, and - Conduct any follow-on actions that may be required.

Discussion Only: LTE at the Hover and Slow Forward Flight

Procedure	**Discussion only**
	When there is Loss of Tail rotor Effectiveness (LTE) at the hover (IGE or OGE) or a slow forward speed, conduct the following steps:
Announce	Announce **"LTE."**
Aviate	Regain control of the helicopter by: **If at the hover:** - Full left pedal (opposite to the un-commanded turn) - Immediately roll off the throttle to the IDLE position - Maintain ground position with cyclic - Allow the helicopter to descend - Control the landing with collective **If in slow forward flight:** - Full left pedal (opposite to the un-commanded turn) - Some forward cyclic to increase speed and transition into normal flight (using weather cocking to reduce the rate of turn) - Lower some collective to reduce the torque reaction - Confirm Rotor RPM is in the green If the turn stops, then resume normal flight. If the turn does not stop and there is sufficient height to effect a recovery then: - Immediately roll off the throttle to the IDLE position - Lower collective fully to enter autorotation, and - If the turn stops, then effect a power recovery and resume normal flight.
Navigate	Maintain situational awareness of your position.
Communicate	Make appropriate radio calls if required.
Activate	**Do not go back to the same task in the same configuration.** Instead, consider where the wind is coming from and adjust your technique.

Air Exercise 9-4: Hydraulic Failure

Introduction
The instructor will demonstrate a hydraulic failure at the hover and while in forward flight and the associated recovery actions.

After the demonstration, the instructor will give the student the controls and position the helicopter as follows:

- Downwind in the circuit while in the cruise configuration before allowing the student to practice a hydraulic failure while in forward flight.
- At a 5 ft skid height, hover in a suitable area before allowing the student to practice a hydraulic failure at the hover.

Limits
Hydraulic failures can be specific to the helicopter type.

Many small helicopters do not have hydraulics, so this emergency is irrelevant.

Helicopters with hydraulics may vary in their configuration. Some only have hydraulics on the cyclic, others the cyclic and collective and others may then also have hydraulics on the pedals.

Specific helicopter-type information relative to the emergency must be obtained from the RFM. This book will detail a very generic scenario.

Discussion Only: Cyclic Hard-over

Procedure
Discussion only

When the cyclic encounters a hard-over either on the ground or in flight, conduct the following steps:

Announce	Announce "**Cyclic Hard-over.**"
Aviate	Maintain control of the helicopter by: **If on the ground:** - Turn the hydraulic system back on, and - Shut down and seek assistance. **If in flight:** - Turn the hydraulic system back on, and - Return to base and seek assistance.
Navigate	Maintain situational awareness of your position.
Communicate	Make appropriate radio calls if required.
Activate	**Abort the current task and return to base, leaving the hydraulics ON.**

Hydraulic Failure from Forward Flight

Procedure When the hydraulics fail while in forward flight, conduct the following steps:

Announce	Announce *"Hydraulic failure."*
Aviate	Reduce or increase speed to the Hydraulics OFF speed stated in the RFM. - If in the cruise, reduce speed. - If operating for some reason at a low speed, accelerate to the Hydraulics OFF speed. - Maintain straight and level or an appropriate height to conduct a circuit to a landing area.
Navigate	Land as soon as possible.
Communicate	Make appropriate radio calls.
Activate	**Troubleshoot** - Cycle the hydraulic switch OFF, then ON. If hydraulics are restored, then end of malfunction. If hydraulics are not restored, then: - Hydraulic circuit breaker OUT. If hydraulics are restored, then end of malfunction and return to base. If hydraulics are not restored, then - Hydraulic circuit breaker IN - Hydraulic switch OFF - Divert or make a precautionary landing and maintain control with the Hydraulic system OFF at the power, speed and manoeuvre limits stated in the RFM. **Approach** - Make a shallower than normal approach and plan for a slow run on landing. **When on the ground** - Collective full down - Conduct a normal shutdown, and - Conduct any follow-on actions.

Hydraulic Failure at the Hover

Procedure	When the hydraulics fail at the hover, conduct the following steps:

Announce	Announce *"Hydraulic failure."*
Aviate	Maintain the hover or allow a slow hover taxi forward.
Navigate	Land as soon as possible.
Communicate	Make appropriate radio calls once safely on the ground.
Activate	**From a stable hover IGE or a slow forward hover taxi** ■ Lower some collective to start a descent ■ Maintain ground position or a straight hover taxi ■ Keep straight with pedals, and ■ Land. **When on the ground** ■ Collective full down ■ Conduct a normal shutdown, and ■ Conduct any follow-on actions.

10

Limited Power Operations

Aim
To take-off and land a helicopter with limited power available.

Objectives
On completion of this lesson, the student will be able to:

- locate and use a Performance Chart in the RFM
- state the difference between Power Required versus Power Available
- identify the Power Margin
- list the power categories and what sort of take-off and approach can be utilised for each
- state the meaning and use of a transient limitation
- calculate a Pressure Altitude (PA) and Density Altitude (DA)
- identify the wind direction using the rotors' downwash
- explain the ground cushion and the effect of varying surfaces, and
- list and describe when and how to use four (4) types of take-off and approach profiles.

Motivation
On most flights, the helicopter must operate at maximum power for lift off, take-off and landing. We tend to fill them with fuel, passengers and a payload, making the helicopter heavy. Often, we are coming and going to an area with limited space in conditions that are not consistent. In addition, the weather and atmospheric conditions can change from one place to another, meaning the helicopter's performance may also change on the same flight. For example, imagine taking off at sea level with an outside air temperature (OAT) of 25°C and 15 kts of wind from a runway environment at an airport and then landing on an 8000 ft mountain with an OAT of 15°C, nil wind over a rocky surface surrounded by trees. The Power Available versus the Power Required will change, and the pilot will have to manage this.

Also, don't be fooled into thinking that bigger or turbine helicopters always have more power available than required. Whether helicopters are big or small, we all tend to fill them up with maximum payload and fly in varying and changing conditions. Therefore, Power Available versus Power Required is always a prime consideration that must be considered and planned for by the pilot *continually*. Many of the skills to calculate, determine and decide a power margin need to be done mentally in real time with little to no access to documentation. It's a skill of judgement you need to learn.

Preparation: Limited Power Operations

Introduction

Until this point in flight training, the student will have had adequate power during air exercises. In fact, you would have had excess power. Understanding power and adjusting the piloting technique will be important as you progress onto more advanced operations such as Confined Areas, Low Flying, Navigation, and Mountain Flying.

Pilot technique is going to be influenced by the Power Margin, which is influenced by:

- the Density Altitude (elevation corrected for pressure and temperature)
- the weather conditions, including wind, cloud and visibility
- the helicopter's All Up Weight (AUW)
- the landing and take-off area available, and
- how the helicopter is performing on the day.

Pitch and Power

Rotor Thrust and engine power

The purpose of Main Rotor Thrust is to oppose the helicopter's weight, allowing it to lift off the ground and then move in any direction the pilot chooses.

The purpose of Tail Rotor Thrust is to oppose the torque reaction so the helicopter does not yaw due to the changes in torque reaction as engine power is changed.

The heavier the helicopter, the greater the rotor thrust required from the main rotor to oppose the weight. Increased main rotor thrust requires more power from the engine. An increase in engine power produces a greater torque reaction, which requires more tail rotor thrust to oppose it. This increase in tail rotor thrust requires even more power from the engine.

Therefore, the heavier the helicopter, the more thrust is required from BOTH the main and tail rotor blades and the greater the power required from the engine to drive the main and tail rotor blades at a constant RPM.

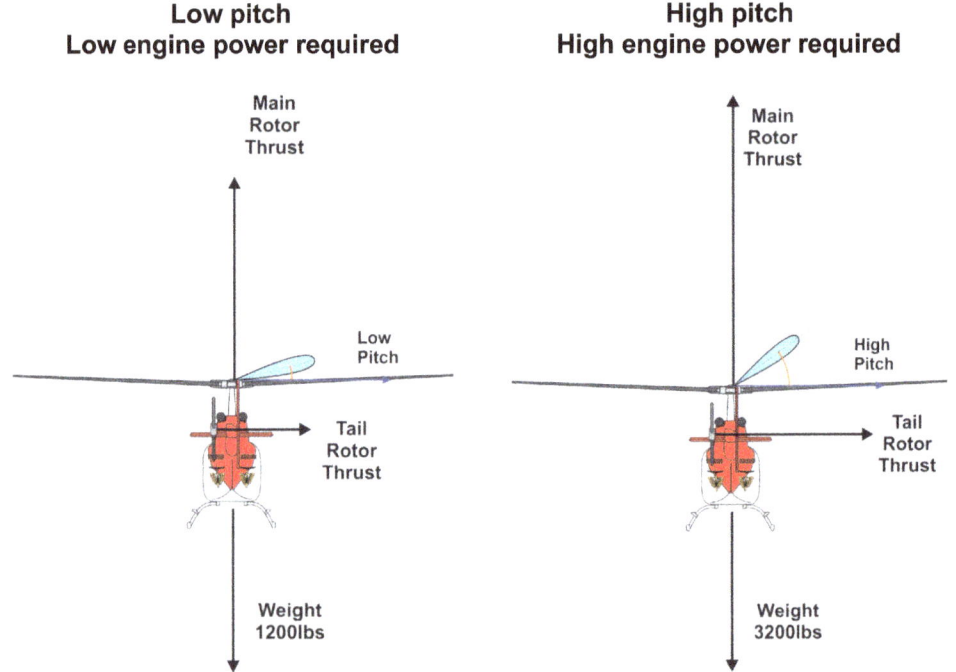

Anything that alters the weight of the helicopter, the performance of the main rotor system or the engine's performance will affect the amount of engine power available to the pilot to lift off, land or fly.

Performance Charts

Performance charts

Performance Charts indicate the maximum gross weight possible for a helicopter to hover In Ground Effect (IGE) or Out of Ground Effect (OGE) at a specified Pressure Altitude (PA) and Outside Air Temperature (OAT).

Some examples of performance charts are provided later within this chapter for explanation purposes, but those specific to your particular helicopter and used for actual calculations are found in the RFM.

Factors considered

Performance Charts consider:

- Pressure Altitude (PA)
- Outside Air Temperature (OAT), and
- All Up Weight (AUW).

But these charts **DO NOT** usually consider:

- wind
- humidity
- the type of surface, or
- pilot technique or experience.

Therefore, the charts provide a baseline indication only and should be used during pre-flight planning or before lift off, if applicable, as a rough guide to assist the crew in making decisions before the flight. They are not in themselves a guarantee that you can do it. In many commercial operations, performance charts are only ever used to prove a helicopter's capability to a client during a contract negotiation or before a job.

In real operations, a pilot will then have to make an operational decision in-flight based on consideration of all the prevailing factors at the time.

All Performance Charts are created by the *helicopter manufacturer,* assuming the engine produces the minimum specified power according to the *engine manufacturer.* However, consider:

1. The engine manufacturer makes an engine capable of delivering an estimated MAX Brake/Shaft HP at a specified RPM. They usually do not make an engine for a particular machine but make it available for sale to multiple aircraft manufacturers (fixed wing and helicopter).
2. The aircraft manufacturer will take this engine and place it within their airframe, and they will determine the actual limitations of the engine based on airframe requirements.
3. The engine manufacturer, if asked, will refer to the AFM or RFM to determine the actual condition of the engine based on how it is used in that aircraft.

Meeting the specification

As we all know, engines can suffer from wear and deterioration, or there may be a mechanical or technical issue where it is simply not performing to the manufacturer's minimum specification on a particular day.

For this reason, a pilot needs a process to confirm engine performance.

If the engine:

- **meets** the minimum specifications, then the data on the Performance Charts will be accurate and able to be used by the pilot to estimate helicopter performance.
- **does not meet** the minimum specifications, then the Performance Charts will **not be accurate** and cannot be relied upon by the pilot to estimate helicopter performance.

Checking engine performance is different for a piston engine than a turbine engine. In a piston, it is simply referred to as a Performance Check, but in a turbine, there is a formal documented procedure referred to as the Engine Power Assurance Check.

Piston Engine Performance Check

Piston engine performance checks

In a piston engine helicopter, engine performance is checked in a multi-layered process as follows:

- A maintenance check every 100 hours
- After every engine start, the pilot will conduct a performance check during the engine runup procedure before lift off, and
- Conducting a power check during flight.

100 Hour maintenance checks

The 100 hourly maintenance will check the following to confirm engine performance:

- Check the compression of each cylinder and record it so any downward trends can be noticed. The compression check uses a Delta Pressure (Difference comparison) reading. 80 PSI of compressed air is pumped into a cylinder while the piston for that cylinder is at Top Dead Centre (TDC). The resulting Delta (difference) will determine the potential pressure loss occurring either from around the piston rings or from either the inlet or exhaust valves which may not have full contact with their individual valve seat or are being held open by the valve stem sticking in the valve guide, resulting in the air pressure dropping due to leakage. If the compression reading drops to a reading of 60-50 PSI or less and the pressure drop cannot be rectified, the affected cylinder and piston are to be repaired, overhauled or replaced. This is more commonly referred to as a Top Overhaul.
- Each spark plug is cleaned and checked for the quality of the actual spark in a pressurised atmosphere similar to the pressure they would experience within the cylinder. If they fail or are faulty, they are replaced.

Performance check during engine runup

After every engine start, the pilot will conduct a performance check during the engine runup procedure before lift off, which involves:

- Oil pressure increases instantly after the engine start.
- Operating temperatures and pressures are achieved after a short period after the engine start.
- Engine RPM is then increased to approximately 1500 RPM (specified by the manufacturer in the RFM), and each magneto is checked. This will confirm the TIMING and QUALITY of the spark in each cylinder and magneto.
- Engine RPM is then taken up to full operating RPM with the corresponding MAP and Oil Pressure noted. This will be the ultimate performance check and can be affected by the rotor blades' flat pitch setting with the collective full down.
- In this state, the lower the MAP, the more power is available from the engine to the rotor system. Most small piston engines sit around the 12-15" MAP at flat pitch and full operating RPM. If it is noted on a particular day that the flat pitch MAP at full operating RPM is excessively high (say 17-20" MAP) or it is significantly different from the previous start, then there will be a performance issue with the engine, and it should be investigated before flight.

In-flight performance check

A pilot can conduct a performance check in flight. A power margin can be determined by noting the power required to hover and then by pulling full power available. This will be described in more detail later on in the chapter.

Turbine Engine Power Assurance Check

Turbine Engine Turbine engine helicopters have a formal *Engine Power Assurance Check* procedure, and an associated Power Assurance Chart detailed in the RFM. Each helicopter type may have a different method that should be followed, but for this book, we will consider the one used for the Bell206 Jet Ranger.

Below is an example of a Bell206 Power Assurance Chart

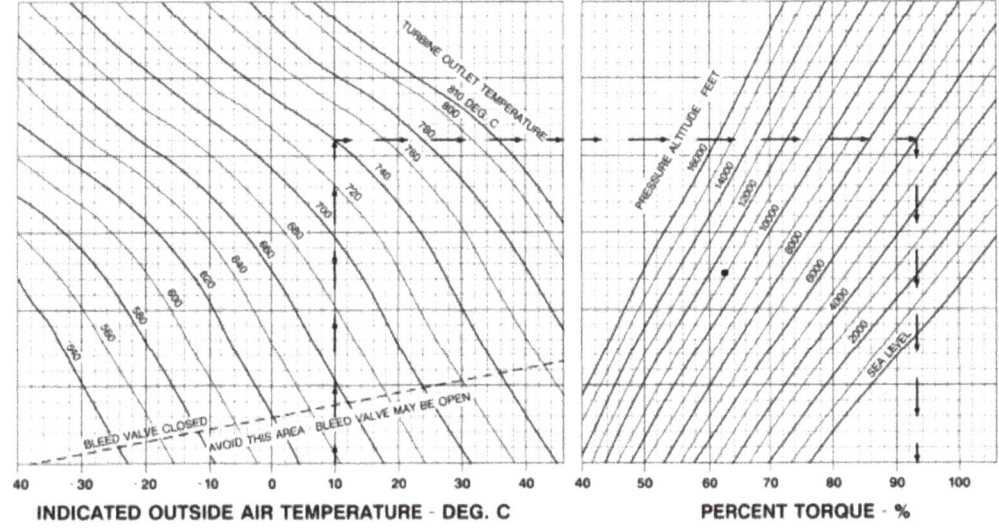

BASIC INLET
MODEL 206B-III POWER CHECK - ALLISON 250-C20B/J ENGINE

Engine Power Assurance Check The *Engine Power Assurance Check* involves:

- The pilot turns off any ancillary equipment not required, such as bleed air for heaters and air conditioners, the generator and other items that may be drawing power from the engine that is not used for powering the drive train.
- Pulling a nominated power setting at a nominated Pressure Altitude, then recording the OAT and engine parameters such as turbine temperature (TOT) and compressor speed (N1) at full operating RPM (N2).
- This information is then plotted on the Power Assurance Chart, and the pilot can determine if the engine is performing according to the airframe manufacturer's specifications.
- If the engine performs better than the minimum, it is called a plus engine.
- If the engine performs less than the minimum, it is called a minus engine.

Chapter 10 Limited Power Operations

Example

Consider a pilot about to conduct a limited power operation in the mountains in a turbine-powered helicopter (a B206 in this case).

They have gone into the Performance Section of the RFM and used the Performance Chart to determine that at maximum AUW with an OAT of +4°C at 7000 ft Elevation, the helicopter can hover IGE.

Before leaving the airport, they may conduct an Engine Power Assurance Check so that they know the information they rely on within the Performance Chart will be applicable given the engine's current state.

Using the B206 Engine Power Assurance Chart, the pilot lifts to the hover and notes:

1. Pressure Altitude of 6000 ft
2. OAT+10°C
3. TOT 740°C, and
4. Tq of 85%.

Using the chart above, the pilot calculates that the engine should use no more than 93% Tq at the hover. Because the amount of power required to hover was actually less than 93%Tq (85% Tq in this case), the engine is producing more power than the minimum specified by the engine manufacturer and the Performance Chart information will be valid. In this case, the pilot has a plus 8% Tq margin engine (commonly called a plus 8 engine).

If the pilot was pulling more than 93% Tq (say 96% Tq), then the engine would not meet the minimum specifications, the helicopter would not be able to meet the Performance Chart information, and the pilot shall not rely on it. In this case, the pilot has a minus 3% Tq margin engine (commonly called a minus 3 engine), and maintenance should be consulted before flying.

Using a Performance Chart

Using a Performance Chart

In the Performance Section of the RFM, there will be multiple charts the pilot can refer to, giving combinations of All Up Weight (AUW) and Pressure Altitudes at different OATs. Entering the data, the pilot can decide whether the helicopter should be able to hover IGE or OGE at a particular location. These graphs vary and need some interpreting to use, so it helps if you get familiar with your particular charts.

Below is an example of an R44 Performance Chart based on the helicopter hovering at a 2 ft skid height IGE hover in zero wind.

To use the chart, enter at the current AUW and move up from the lb axis (or down from the kg axis) to the current OAT. Then move left to determine the maximum IGE hover Pressure Altitude the manufacturer says is possible. Any hovering at or below this Pressure Altitude at this temperature and weight should be possible.

If the location of the helipad is above this Pressure Altitude, then the helicopter will not have the power available to hover.

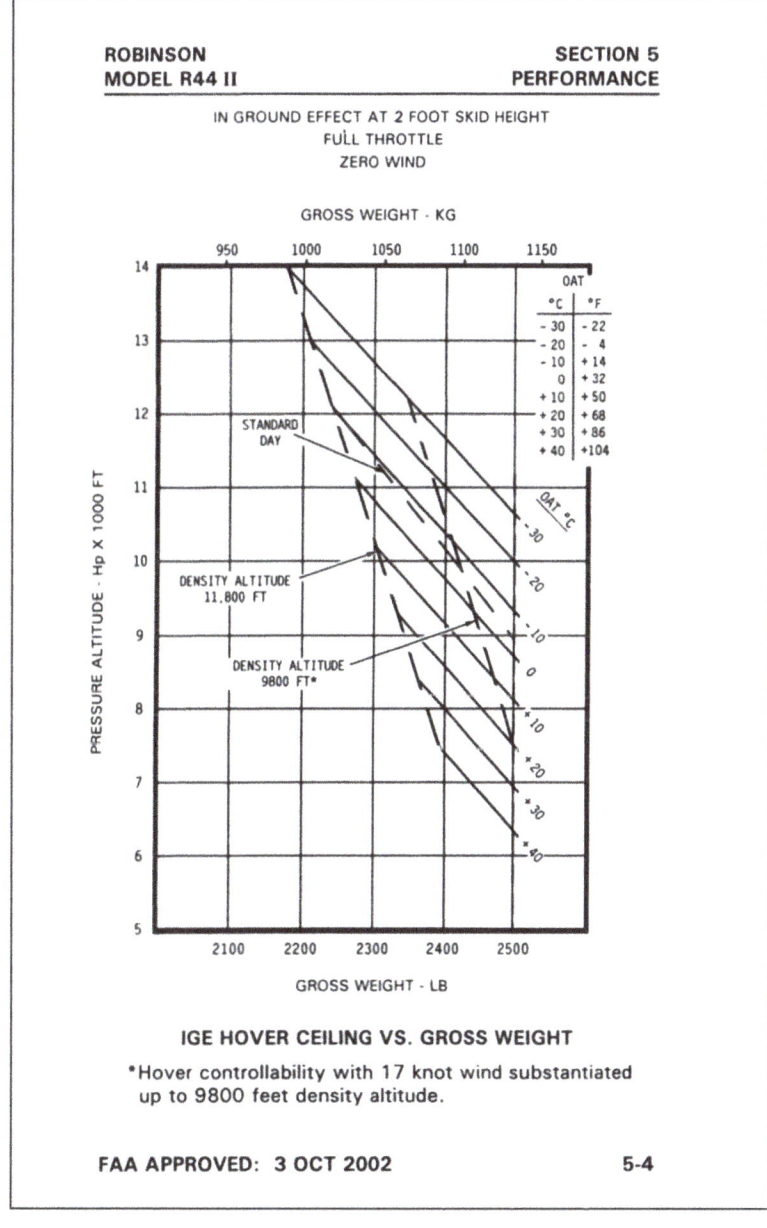

Chapter 10 Limited Power Operations

Example

Method 1:

If a pilot is about to fly to a remote location different from the departure point, it would be good to know if:

1. the helicopter can lift off and hover at the departure location, and
2. more importantly, the helicopter can hover safely at the destination.

For this example, assume the destination has a Pressure Altitude (PA) of 8000 ft and an OAT of +15°C.

Enter the chart at a Pressure Altitude value of 8000 ft (8 on the left axis). Move horizontally across to halfway between the +10 and +20 OAT lines (this represents +15°C).

From this point, either:

- move straight up (to have an allowable AUW in kg), which is approximately 1125 kg, or
- straight down (to note the allowable AUW in lb), which is approximately 2475 lb.

If the arrival AUW of the helicopter is less than the maximum gross weight calculated on the chart, then the helicopter should be able to safely hover (and, therefore, lift off and land) at the destination. In this example, the helicopter is right on its maximum limit, so the pilot should prepare for a cushion creep type take-off and landing.

Method 2:

Using the same graph, the pilot can enter the data differently. If they know the departing AUW is 2400 lb but want to know the maximum Pressure Altitude that can be safely landed at IGE, then the graph can be entered from a different point.

For this example, the departure aerodrome is at sea level and the OAT is +24°C. The landing area is on a 6000 ft mountain, and the anticipated temperature at this location is estimated to be +12°C (based on a lapse rate of 2°C per 1000 ft).

Enter the chart at the known AUW and move up (or down) to the temperature estimated at the top of the mountain (+12°C), then move directly left to read a Pressure Altitude of 9000 ft (approximately).

If the pilot intends to land on a helipad located at 6000 ft, then there should be some confidence that this is achievable according to the performance chart.

Real life

Having done the above preparation, the pilot can depart with confidence. **However,** the Power Available and the Power Required will need to be **confirmed in flight** by the pilot on departure (Hover Check) and on arrival (Full Power Check) as conditions may change. Additionally, the landing area may not be what is expected, so a different type of approach and departure may be required. All this means is the Performance Charts are good at indicating what may be possible, but in real life, the pilot ultimately needs to make the determination based on real-time information.

Power Required and Power Available

Power required and available

The **Power Required** is the amount of power required from the engine to make the helicopter perform as the pilot desires at a particular time. The amount of Power Required to hover can differ from that required to fly straight and level at 60 kts to that required to fly at 80 kts.

The **Power Available** is the total amount of Power Available from the engine, whether the pilot uses it or not. In most cases, the Power Available remains constant until atmospheric factors such as Pressure Altitude, temperature, and humidity affect the engine's performance.

Piston example

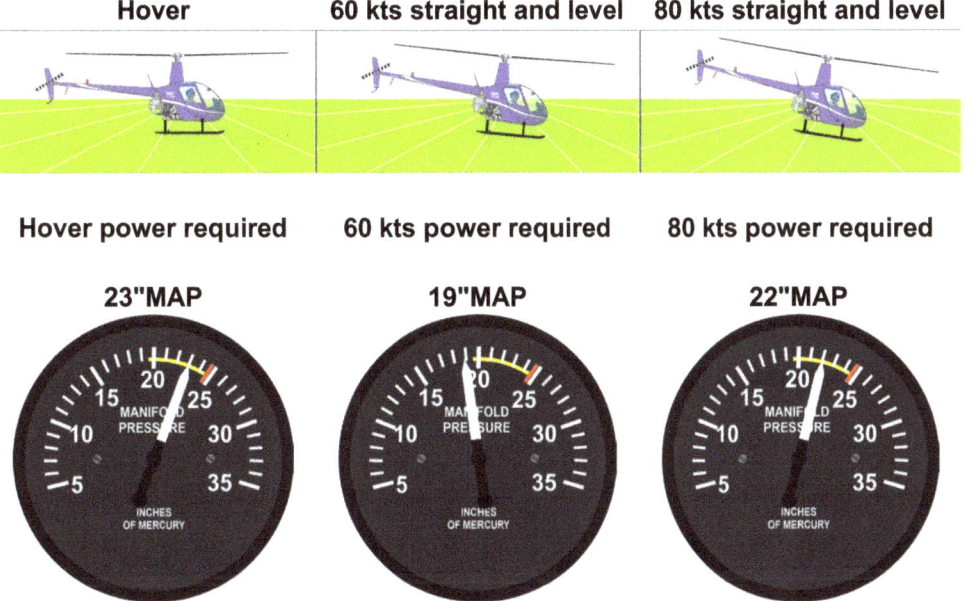

If Maximum Power Available is 25" MAP

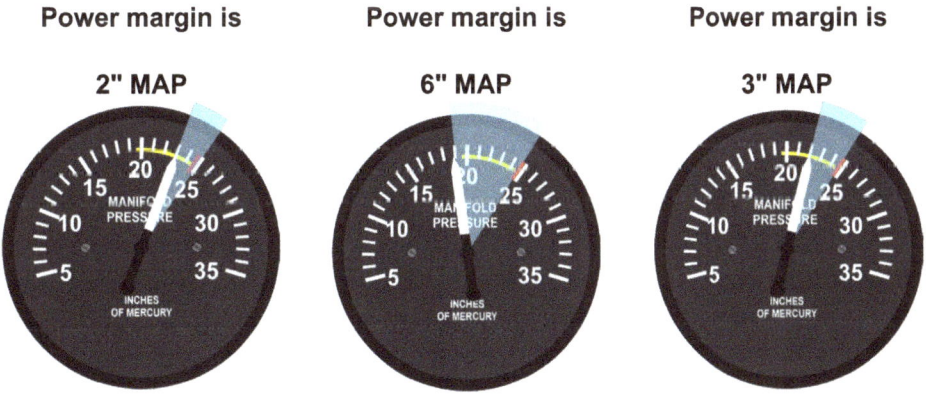

The difference between the Power Required and the Power Available is the

Power Margin

Chapter 10 Limited Power Operations

Turbine example

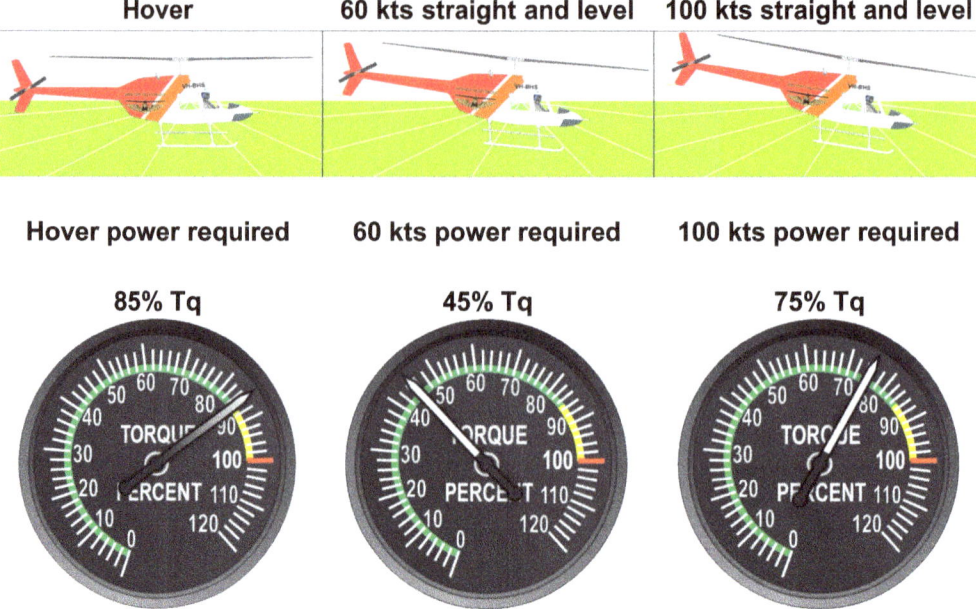

If the maximum Power Available is **100% Tq**

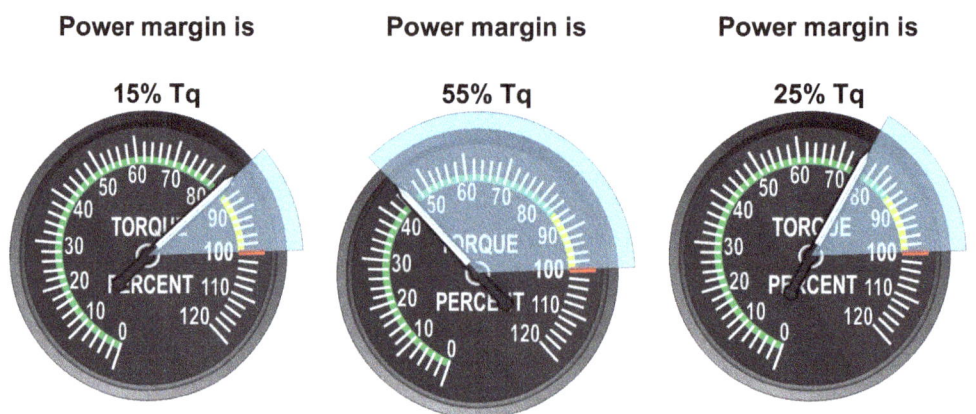

The difference between the Power Required and the Power Available is the

Power Margin

Power Margin and Power Curve

Power Curve

Helicopter power requirements vary with forward speed (and wind).

The Power Curve below is typical of any helicopter at sea level.

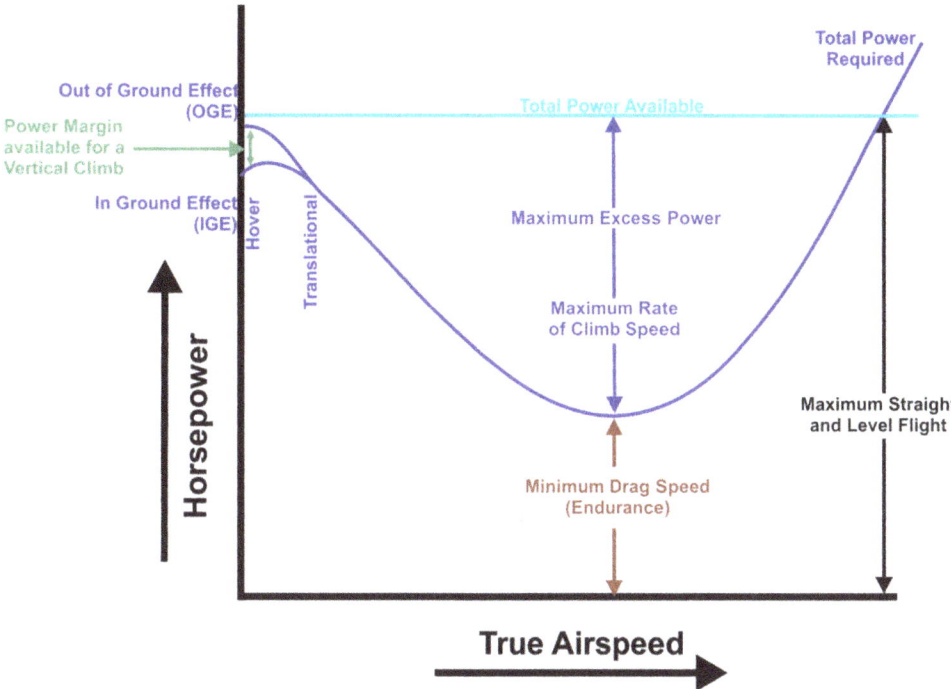

Considering the Power Graph, it can be determined that from the hover In Ground Effect (IGE), as forward speed is increased, more power is required until translational lift is reached. From this point, the power required decreases to reach a minimum at the minimum drag speed.

Minimum drag speed is the speed at which the power margin is the greatest and corresponds with the best rate of climb speed. Above this speed, power demands increase steeply due to the increased parasite drag on the fuselage until maximum power and speed are reached.

Power margin at the hover

Since the **Power Required to hover** is often the most power-demanding manoeuvre, at the most critical time, it is the most commonly used reference point in calculating a baseline power margin.

Piston Example

For example, if a helicopter requires 20" MAP to hover on a particular day and the maximum power available is 24" MAP, then the power margin is 4" MAP.

From experience, the pilot may know that a power margin available of 4" MAP will allow this helicopter to make a vertical departure over an obstacle.

If, however, the helicopter requires 23" MAP to hover on a different day, and the maximum power available is 24" MAP, then the power margin is only 1" MAP.

From experience, the pilot may know that a power margin of 1" MAP will only allow this helicopter to make a cushion creep take-off.

Chapter 10 Limited Power Operations

Turbine example

For example, if a helicopter requires 85% Tq to hover on a particular day and the maximum power available is 100% Tq, then the power margin is 15% Tq. From experience, the pilot may know that a power margin available of 15% Tq will allow this helicopter to make a vertical departure over an obstacle.

If, however, the helicopter requires 95% Tq to hover on a different day, and the maximum power available is 100% Tq, then the power margin is only 5% Tq. From experience, the pilot may know that a power margin of 5% Tq will only allow this helicopter to make a cushion creep take-off.

Power margin while in forward flight

The trick now is determining if there is a sufficient power margin to land in an area you have not been into before. Therefore, if you have not been able to conduct a lift off, hover and take-off in the same area under the same conditions, how do you gauge a power margin before going into an area?

There are multiple techniques often discussed on how to do this. Some are appropriate and work, while others offer no value and a false sense of safety.

There are several knowns when conducting a power check while in flight.

1. The pilot has a baseline from conducting the first lift off and hover at the departure helipad.
2. The pilot can do a full power check enroute or over the landing site before conducting the approach to confirm what power the engine will give by raising the collective to the maximum allowable.

Piston example

Turbine

Summary

Once the power margin has been determined, the pilot can consider additional information such as wind, landing area and surface, obstacles and their own personal limits to formulate a plan and make decisions on the type of approach and departure that may now apply to the landing site.

Power Margin and Power Categories

Power Categories Power margins can be described differently depending on where you have conducted your initial training, so different instructors may have varying ideas on how to describe it. In general, a civil Flight School only talks about a margin as a percentage of Tq or a difference in MAP. A military Flight School will split the calculated margin into a category system.

The chart below describes, in a general manner, the power margins for both piston and turbine helicopters, how they are described and what can be done with them.

Any power margin less than CAT4 is deemed to be a Limited Power operation.				
Power margin in excess of that required to hover IGE				**CAT4** 4" MAP 16-20% Tq **Options**
			CAT3 3" MAP 11-15% Tq **Options**	Vertical
		CAT2 2" MAP 6-10% Tq **Options**	Constant Angle or Towering Steep	Constant Angle or Towering Steep
CAT1 1" MAP 0-5% Tq **Options**		Normal	Normal	Normal
Cushion Creep Shallow Running	Cushion Creep Shallow Running	Cushion Creep Shallow Running	Cushion Creep Shallow Running	Cushion Creep Shallow Running
Limited Power Operation				

Limited Power Each Power Category allows for the take-off, approach and landing techniques for that Category and those in the Categories below it.

CAT4 power allows ALL take-offs and approaches; the type of take-offs and approaches the pilot can choose from are not limited by power.

If the pilot determines there is **CAT3** power available. The options include all options except Vertical, so there is a limitation.

A low-power **CAT2** power available will not allow any take-off, approach and landing techniques that are greater than it, so you cannot do Constant Angle, Towering, Steep or Vertical take-off.

A limited power **CAT1** will not allow any take-off, approach and landing techniques that are greater than it, so you cannot do a Normal, Constant Angle, Towering, Steep or Vertical take-off.

Chapter 10 Limited Power Operations

Considerations for landing Because helicopters can land almost anywhere, among tall trees and buildings, riverbeds, on sand and open runways, the crew must now carefully choose where they can take-off and land, based not only on the terrain, surrounding obstacles and surface but also on the *power margin* they have available to them on the day to be able to get *in* and *out*.

Transients and Power limits

Transients A transient is an extension to a limit for a short period of time to conduct the take-off or approach during a limited power operation. Remember, it is better to safely exceed a normal limit by a small amount for a short period rather than hit the ground or hit an obstacle within limits. A transient should be considered a "Get out of jail free!" card (referring to the game of Monopoly) and not be considered a normal operation.

Piston and turbine helicopters manage transients slightly differently because the manufacturers disguise how much power is available to the pilot:

- for safety reasons
- to make the helicopters appear better on paper and within their performance charts, and
- to reduce wear and stress on some components.

Power Limits Helicopter manufacturers do not make engines. Instead, they go out to the open market and negotiate a deal with an engine manufacturer to supply engines for their new helicopter design. This opens the door for two things:

1. The engine manufacturer will produce an engine capable of different applications in different machines, operating at different RPMs and power outputs depending on the power band required for the machine in which they will be installed.
2. The helicopter manufacturer can manipulate the engine power output displayed based on the markings on the gauges so that the performance suits the design limits of the components and the performance chart numbers can meet criteria in the sales process.

What does this mean These limits imposed on the helicopter regarding power may not have anything to do with what the engine can actually give you if needed. The limitations are a combination of what the engine can deliver and what the drivetrain within the rotor system can absorb. The bottom line is, if you need more, you can typically get it, but the ramification is an exceedance of a limit and the requirement to report it and possibly complete some form of extra maintenance.

Piston example

The R22 has a Lycoming engine that produces 145 Hp. Yet only 131 Hp is available for take-off, and 124 Hp is available for continuous use. Robinson puts a red line at approximately 25" MAP on the MAP gauge.

This effectively limits the power the pilot can use and '*derates*' the engine to 131 Hp. If the pilot needed to use more Hp, pulling the collective up further will increase the MAP past the redline.

Ignoring the redline, at sea-level, the MAP can be pulled all the way up to 28" MAP to get the full 145 Hp, but this will put extra strain and wear on the drive belts and the drive system. If the pilot temporarily went over the 25" MAP redline on take-off, this would be considered a transient over-boost.

If the pilot continued to pull the collective up even more, then the engine would have no more to give, and at that point, the RRPM will decay, and the helicopter could enter an overpitching situation.

Turbine example

The Bell206 Jet Ranger has a Rolls Royce 250C20 turbine engine that produces 450 Hp. Yet only 318 Hp is available to use for take-off and continuous use. Bell does this by putting red lines on the gauges that display engine information. This effectively limits the amount of power the pilot can use and 'derates' the engine.

The engine limits are affected by changes in Density Altitude, engine condition and configuration. Exceeding any one of the redlines will get more power, but this will put extra strain and wear on the components. This would be considered a transient if the pilot temporarily pulled over the red line on take-off.

If the pilot continued to pull the collective up even more, then the engine would have no more to give, and at that point, the RRPM will decay, and the helicopter could enter an overpitching situation.

Example

Below are the transients for the Bell 206BIII Jet Ranger:

Item	Normal limit	Transient	Description
Torque	100%	110% for 5 seconds	Allows the pilot to exceed 100% Torque by 10% for 5 seconds (unintentional use only).

Chapter 10 Limited Power Operations

Item	Normal limit	Transient	Description
TOT	810°	810-843 for 6 seconds	Allows the pilot to exceed 810° TOT by 33° for 6 seconds (unintentional use only).
N1 Gas Producer	105%	106% for 15 seconds	If the pilot uses one of the above transients (Tq and/or TOT) the N1 RPM will also be higher. N1 in excess of 106% may result in N1 topping and a subsequent compressor stall or surge.
RRPM	97-100%	Rotor droop	At high power settings, the engine may not be able to overcome the added drag on the rotor blades, and the Rotor RPM may droop until the collective is again lowered.

Power, Atmospheric Conditions and Altitude

Introduction As atmospheric conditions vary from day to day, so does helicopter performance. This variation affects helicopters to a greater degree than fixed-wing aircraft mainly because helicopters require maximum power to lift off vertically to the hover compared to a fixed wing that may only require a longer runway to build up more speed. Sometimes a task completed one day or even 5 minutes ago cannot be flown the same again due to a change in performance.

Power and altitude At altitude, the engine's power will be less than sea level, so the Power Available and Power Required curves become closer together. The power margin will be less, and aircraft performance will be reduced. Also, the aerodynamic efficiency of the helicopter will be reduced as altitude increases, which means more pitch must be applied to produce the necessary rotor thrust to balance weight. This, in turn, means more power is required, and the power required curve is displaced upwards.

ISA

The International Civil Aviation Organisation has adopted a standard, average atmosphere called the **I**nternational **S**tandard **A**tmosphere (**ISA**). This standard allows comparison between aircraft performance over a wide range of atmospheric conditions. The International Standard Atmosphere is specified in the table below.

ISA	Standard	Fixed Lapse Rates
Sea Level **Pressure**	1013.2 Hp or 29.92" Hg	30 ft for each Hp/Hg
Sea Level **Temperature**	15°C	120 ft for every 1°C

Hectopascal (Hp) or Millibar (Mb)

The terms Hectopascal (Hp) and Millibar (Mb) refer to the same measurement and are interchangeable. Different name, same meaning!

The difference between the numbers 1013.2 Mb and 29.92 Hg, although also the exact same measurement, is simply due to a metric or imperial system of measurement.

The Hp or Mb is common in Countries utilising the metric system.

The Hg or mercury system is used in Countries still utilising the imperial system (USA).

Accepted rates of variance

The accepted rate of decrease of temperature with height (referred to as a lapse rate) under normal conditions is 2° per 1000 ft.

The accepted rate of decrease of pressure with height under normal conditions is 1 Mb per 30 ft.

The accepted variance of height for every degree different from the ISA at any given pressure height is 120 ft.

These rates of variance are used to calculate Pressure Altitude and Density Altitude.

Q-codes

During World War II, the British developed a Q-code system to communicate information to pilots over the radio. The two most common Q codes still in use today are:

QNH   Question Nil Height and

QFE   Question Field Elevation

Chapter 10 Limited Power Operations

QNH

QNH = Pressure at Sea Level

QNH is obtained by measuring the existing surface pressure and converting it to a pressure that would theoretically exist at sea level at that point (assuming ISA conditions).

With QNH set on the altimeter subscale, the Altimeter's main scale will show the approximate true altitude above mean sea level (i.e. the elevation AMSL).

If the QNH is:

- 1013.2 Hp, then the elevation and pressure altitude are the same value
- Is less than 1013.2 Hp, then the pressure altitude is higher than the elevation, and
- Is greater than 1013.2 Hp, then the pressure altitude is lower than the elevation.

An aircraft parked on the ground at the aerodrome reference point will have the station elevation indicated on the Altimeter. If you are at a remote helipad and do not know the QNH but know the helipad's elevation, then setting the elevation on the altimeters main scale will automatically set the correct QNH on the subscale.

QFE

QFE = Pressure at the location

QFE is the actual surface pressure for a particular position and is not corrected to sea level pressure. With QFE set on the altimeter subscale, the Altimeter will read zero when on the ground at that position, regardless of the actual elevation.

Definitions

The following are important definitions to understand when using performance charts.

Elevation is the true distance of the ground above mean sea level.

Height is the distance of an object above ground level.

Altitude is the distance of an object above mean sea level.

The diagram below helps illustrate the definitions.

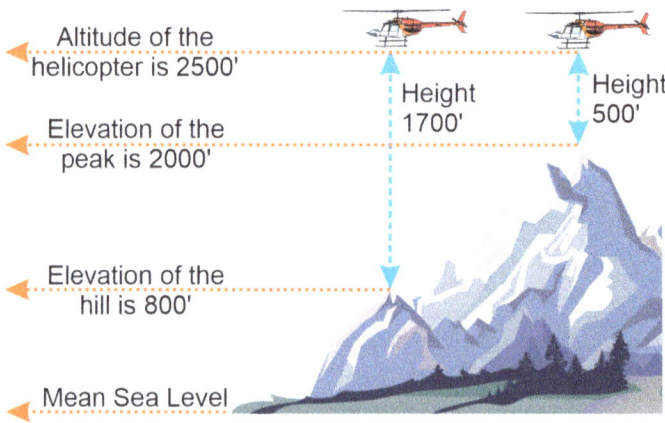

Pressure altitude

Pressure Altitude is the altitude in the ISA where the prevailing pressure exists. In other words, it is that altitude in the atmosphere where the 1013 Hp (29.92 Hg) pressure actually exists on the day. Pressure Altitude is obtained using the following formula:

Pressure Altitude = (1013 − QNH) x 30 ft
= the pressure difference +/− the Elevation

Calculating pressure altitude

To calculate Pressure Altitude, list the relevant information and then calculate it out.

For example:

Given an elevation of 1900 ft and a QNH of 1020 Hp, what is the Pressure Altitude?

Elevation	=	1900 ft
QNH	=	1013-1020 = minus 7 Hp
		Minus 7 Hp x 30 ft = minus 210 ft
		1900 ft minus 210 ft = 1690 ft
Pressure Altitude =		**1690 ft**

In flight

To quickly calculate the Pressure Altitude in the helicopter without having to conduct mental arithmetic, set 1013.2 Hp (29.92 Hg) on the Altimeter's subscale, then note the reading on the Altimeter's main scale. This is Pressure Altitude.

Density altitude

Density Altitude is Pressure Altitude corrected for temperature variances away from ISA at the calculated Pressure Altitude.

If the OAT at the Pressure Altitude is:

- hotter than the expected ISA, then Density Altitude is higher than Pressure Altitude
- colder than the expected ISA, then Density Altitude is lower than Pressure Altitude.

Each degree in temperature too hot or too cold equals a change in the Density Altitude of 120 ft. Density altitude is obtained using the following formula:

Density Altitude = Pressure Altitude +/- (Temperature Variance x 120 ft)

Chapter 10 Limited Power Operations

Calculating Density Altitude

To calculate Density Altitude, list the relevant information and then calculate it out.

For Example:

Given an elevation of 1900 ft, a QNH of 1020 Hp and an OAT of +21°C, what is the Density altitude?

Elevation	=	1900'
QNH	=	1013-1020 = minus 7 Hp
		Minus 7 Hp x 30 ft = -minus 210 ft
		1900 ft minus 210 ft = 1690 ft
Pressure Altitude =		**1690 ft**
OAT	=	+21°C
		At 1690 the ISA temperature should be +12°C but it is +21°C which is +9°C hotter than expected. This is referred to as ISA +9
		9 x 120 ft = 1080 ft higher.
		1690 + 1080 = 2770
Density Altitude =		**2770 ft**

What does all this mean?

The question to ask is:

What does all this have to do with flying a helicopter and working out a power margin?

The point is the helicopter will perform to the Density Altitude even though the pilot may only be considering it performing at the displayed Pressure Altitude or Elevation.

The management of Power Available vs Power Required and the cognitive recognition of the difference between the two may become critical in managing the helicopter operation.

For example:

Going back to the previous calculation, where we said the elevation was 1900 ft and the calculated Density Altitude was 2770 ft. The pilot may think they are only operating at 1900 ft, but the helicopter will perform as if it is at 2770 ft.

As a pilot, you need to be aware of Density Altitude and its effect on helicopter performance, regardless of the elevation.

Effect on power available	As Density Altitude increases, Power Available decreases. **As a general rule, in a normally aspirated piston engine helicopter, the engine will lose 1" MAP per 1000 ft.** So at sea level, the piston engine on a standard day should always have 28" MAP available (even if you cannot use it). At 1000 ft, there will only be 27" MAP available, and so on. The table below summarises this: | Density Altitude	Loss of MAP	MAP Guage reading
---	---	---
Sea level	0	28
1000	1	27
2000	2	26
3000	3	25
4000	4	24 (red line for the R22)
5000	5	23
6000	6	22
7000	7	21
8000	8	20
9000	9	19
10000	10	18	
Piston Example	For example, if lifting off at sea level on a standard day takes requires 24" MAP, then there is no way a landing at 5000 ft is even possible in the current configuration. You don't actually need a Performance Chart to tell you that!
Turbine	The turbine helicopter is not affected by Density Altitude to the same degree as a piston. A pilot can always pull the required Tq **BUT** turbine temperature (TOT) and compressor efficiency (N1) will be affected, and these can then become the limiting factors which will then cause a limit on the Tq. This is not usually a consideration until going above Density Altitudes in excess of 6000 ft. Power management in a turbine is a bit more complicated and will typically be covered during your first turbine rating rather than going into detail in this book.

Effect on the power curve

When related to the power curve, this means the Power Available line moves down the X axis, and the Power Required line moves up.

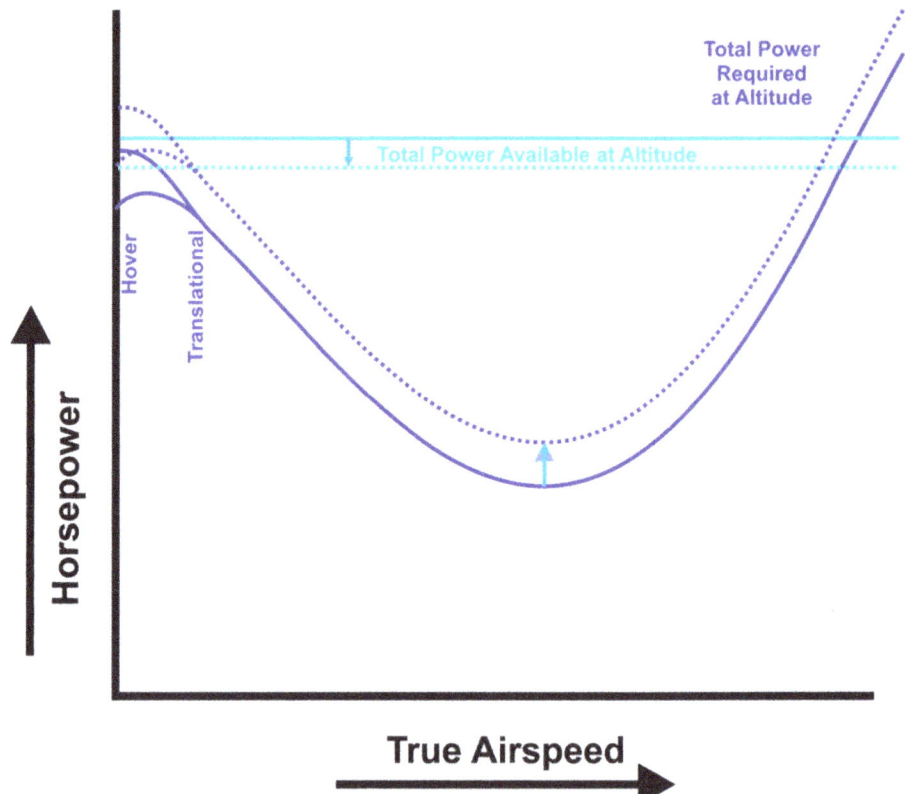

Considering this graph, the helicopter would have sufficient power to hover but could not maintain height during the transition to forward flight. Any further increase in Density Altitude or weight would mean the helicopter could not even hover.

Environmental factors

Atmospheric conditions and altitude will affect helicopter performance and the helicopter's power margin. Some flight manuals contain performance charts related to humidity factors, and some do not.

When planning a Limited Power Operation, the pilot must consider the:

- elevation of the landing site (Altitude)
- atmospheric pressure on the day (QNH for calculating Pressure Altitude)
- outside air temperature at the landing site (OAT for calculating Density Altitude)
- humidity, and
- wind.

The following table summarises the effects:

Item	Effect on performance	Effect on power margin
High AUW	Performance ↓	Power margin ↓
Low AUW	Performance ↑	Power margin ↑
Increasing ALTITUDE	Performance ↓	Power margin ↓
Decreasing ALTITUDE	Performance ↑	Power margin ↑
Increasing PRESSURE	Performance ↑	Power margin ↑
Decreasing PRESSURE	Performance ↓	Power margin ↓
Increasing TEMPERATURE	Performance ↓	Power margin ↓
Decreasing TEMPERATURE	Performance ↑	Power margin ↑
Increasing HUMIDITY	Performance ↓	Power margin ↓
Decreasing HUMIDITY	Performance ↑	Power margin ↑
Strong Wing WIND (> 25 kt)*	Performance ↓	Power margin ↓
Increasing WIND (5-25 kt)	Performance ↑	Power margin ↑
Decreasing WIND (< 5 kt)	Performance ↓	Power margin ↓
Smooth flat hard SURFACE	Performance ↑	Power margin ↑
Rough, uneven soft SURFACE	Performance ↓	Power margin ↓

Wind

Wind

Wind is the most significant environmental factor that can be utilised as an advantage when conducting limited power operations; therefore, it is essential to know the winds:

- direction
- strength, and
- consistency.

Wind will affect the rotor system in the same manner as airspeed. When viewing a standard power curve, wind is not normally considered, but we know that there is always some wind blowing on most days. As wind increases, it is as if the helicopter is already moving forward, and the power required will be affected.

10 kts headwind

If the wind is blowing at 10 kts and the helicopter is about to take-off into the wind, then the power required to hover will relate to that part of the power curve as if the helicopter was already moving forward at 10 kts.

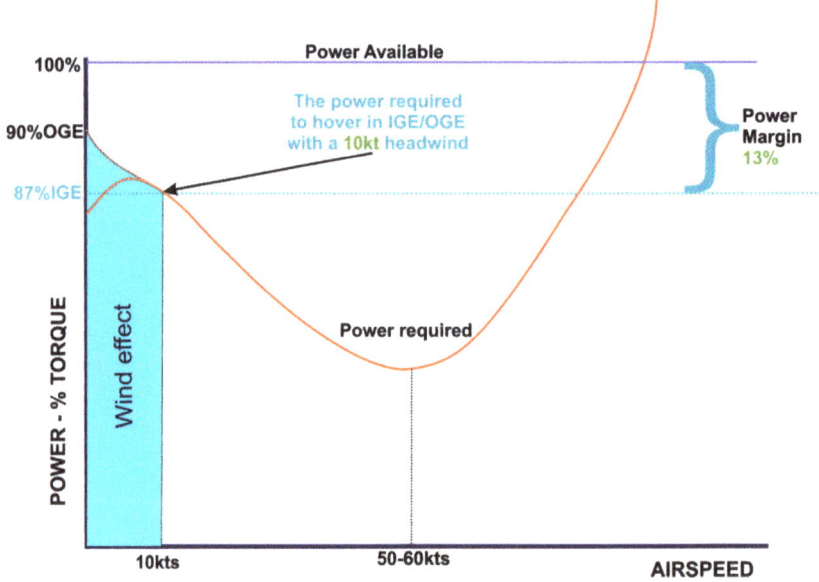

15 kts headwind

If the wind is blowing at 15 kts and the helicopter is about to take-off into the wind, then the power required to hover will relate to that part of the power curve as if the helicopter was already moving forward at 15 kts.

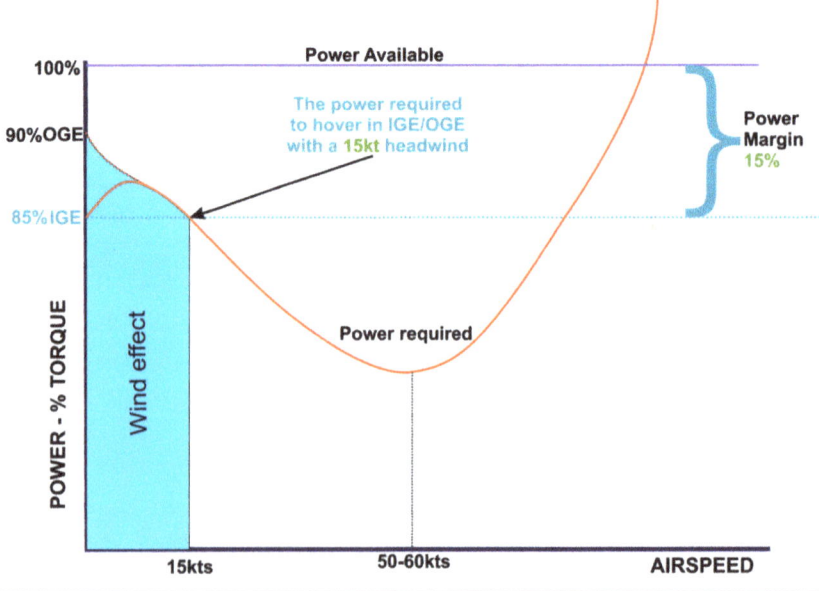

20 kts headwind

If the wind is blowing at 20 kts and the helicopter is about to take-off into the wind, then the power required to hover will relate to that part of the power curve as if the helicopter was already moving forward at 20 kts.

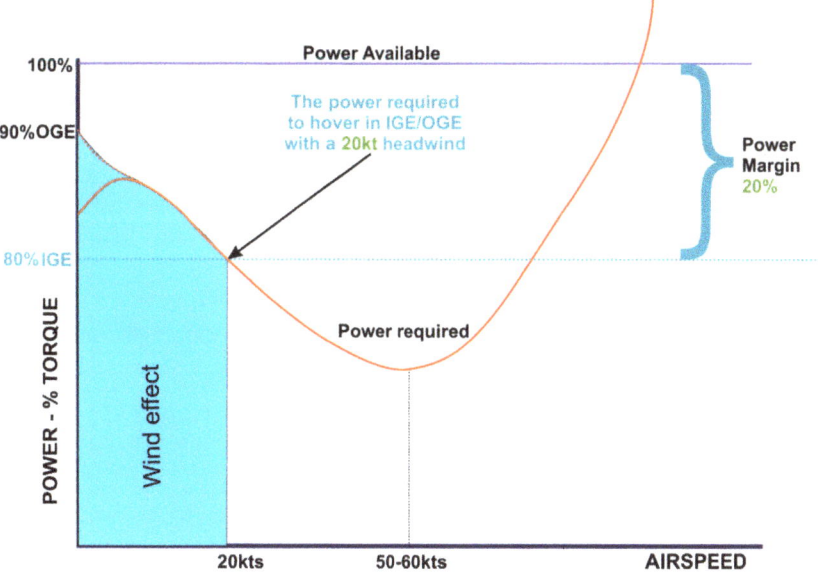

Translational lift and headwinds

When the wind is:

- **greater than 12-15 kts**, the helicopter has already passed through translational lift, and unlikely to experience any sink on take-off.
- **less than 12kts**, the helicopter will likely experience the sink before entering translational lift.

Wind direction and strength

Understanding basic wind effect will allow the pilot to determine how strong the wind is and where the helicopter will reach translational lift on the take-off while still at the hover.

In nil wind conditions at the hover, the downwash from the rotor blades will be equal around the helicopter.

As the wind increases, the downwash will be blown backwards until the helicopter has entered translational lift at approximately 12-15 kts.

Knowing this, the crew are now able to look outside and make a determination on the wind direction and strength.

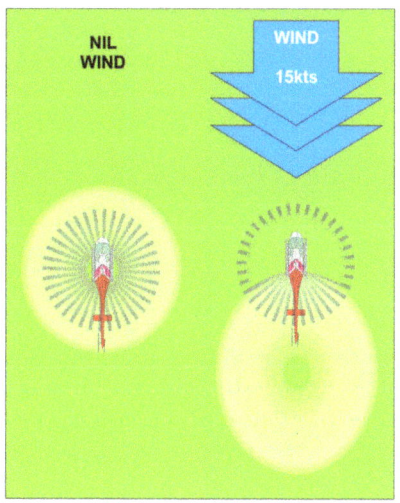

Chapter 10 Limited Power Operations

Direction By conducting a 360° pedal turn at the hover, the crew can see where the downwash is closest to the nose of the helicopter. This indicates the wind direction.

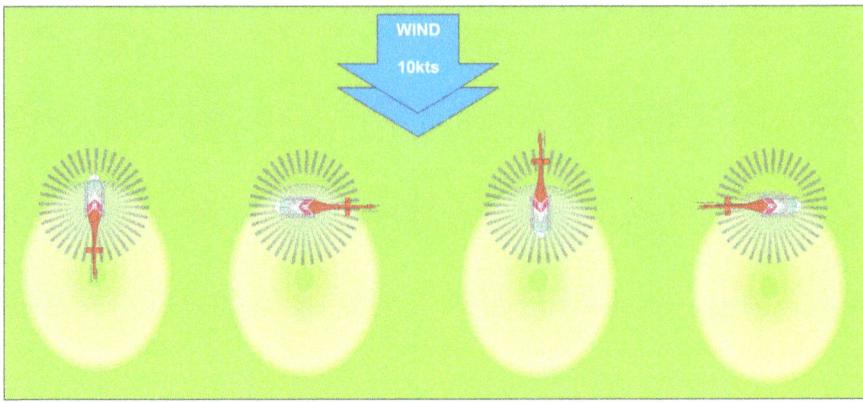

Strength To determine the wind strength, the following can be applied:

If the wind is on the nose, then at:

- 0 kts, the downwash will be equal distance from the rotor tips throughout a full 360-degree turn
- approximately 5 kts, the downwash will be at the tips of the rotor blades.
- 10 kts, the downwash will be at the pilot's feet.
- 15 kts, the downwash will be in-line with the main rotor mast.
- greater than 15 kts, the downwash may be blown further back.

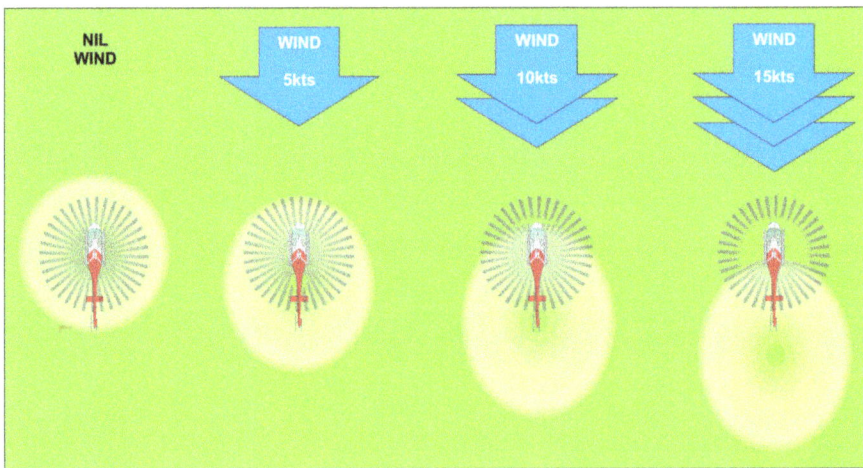

Translational lift Knowledge about the wind allows the pilot to determine where the helicopter will pass through translational lift before the take-off. The dividing line where the downwash is affecting the ground and where it is not affecting the ground will best indicate where the helicopter will enter translational lift and start to fly on the take-off.

Although this cannot be used in all circumstances, it does allow the pilot to make some decisions prior to the take-off, particularly regarding the surface and obstacles on the take-off path.

Ground Cushion and Surfaces

Ground cushion

Hovering In Ground Effect (IGE) in nil wind requires less power than hovering Out of Ground Effect (OGE). The closer the helicopter is to the ground, the more effect the ground cushion has as it intensifies and the less power required.

Most lift offs to the hover have a power margin that allows a standard 5 ft skid height hover. This provides a compromise between power and safety so that the helicopter does not accidentally come into contact with an obstacle on the ground, and the rotor wash does not stir up debris that can be recirculated through the rotor disk.

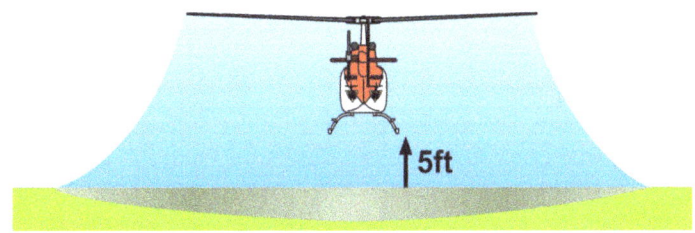

5 ft skid height hover

If in a situation where power is severely limited, it is a good technique to hover lower to the ground to increase the power margin, but this requires more accurate flying and vigilance by the pilot and crew.

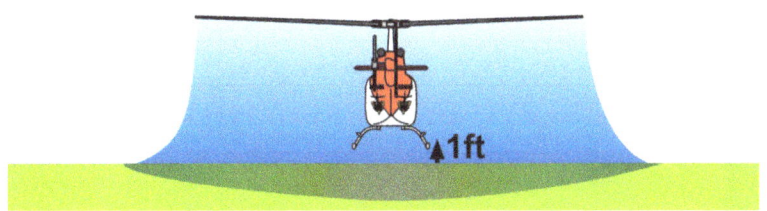

1ft skid height hover

General rule

As a general rule: For every
1 ft closer to the ground, the power margin will increase by ½" MAP or 1% Tq.

Hover height	Power required	Margin	Effect on the Power Category
5 ft AGL	25" MAP or 90% Tq	Nil and 10% Tq	**CAT2** power
4 ft AGL	24½" MAP or 89% Tq	½" MAP and 11% Tq	**CAT3** power
3 ft AGL	24" MAP or 88% Tq	1" MAP and 12% Tq	**CAT3** power
2 ft AGL	23½" MAP or 87% Tq	1½" MAP and 13% Tq	**CAT3** power
1 ft AGL	23" MAP or 86% Tq	2" MAP and 14% Tq	**CAT3** power

Surface The type of surface the helicopter hovers over can affect the helicopter's performance and the available power margin. Generally, any uneven surface absorbs energy and will decrease helicopter performance. Any surface that is smooth, flat, hard and even is best and will increase helicopter performance. Following are some typical surfaces and their effect on performance:

Surface	Example	Power margin
Concrete pad Smooth flat hard surface is best.		Increases ↑
Long grass Requires a higher hover and the grass absorbs the energy of the rotor system.		Decreases ↓
Short grass Smooth, flat and reasonably hard.		Increases ↑
Water Water can be the worst surface to hover over as it absorbs energy from the downwash forcing the water away from the helicopter. Hovering over water can be dangerous unless excess power is available.		Severely decreases ↓↓
Slope An uneven slope means an uneven ground cushion.		Decreases ↓

---

[1] Hovering over water: Reference : NY ANG [Public domain]

[2] Hovering over rocky terrain. Reference: http://www.news.com.au/national/helicopter-pilot-insists-he-aint-a-hero/story-e6frfkp9-1226409080608

Surface	Example	Power margin
Rocky terrain Requires a higher hover and the ground cushion is uneven.		Decreases
Sand, snow, dust The downwash will stir up the small particles and recirculate them through the rotor disc. This absorbs energy. Additionally, visibility can be reduced.		Decreases

Surface and take-off path The lower the power margin, the bigger and smoother the take-off and landing area needs to be. In general, the following will apply in nil wind conditions:

Power Category	Take-off and approach area	Surface
CAT1	At least half a football field	Smooth and firm open area
CAT2	At least 25 meters	Open area
CAT3	At least 10 meters	Low obstacles
CAT4	A small or 'confined area'	Allows for higher obstacles

---

[3] Hovering over sand. Reference: Sgt Tammy Hineline [Public domain]

Chapter 10 Limited Power Operations

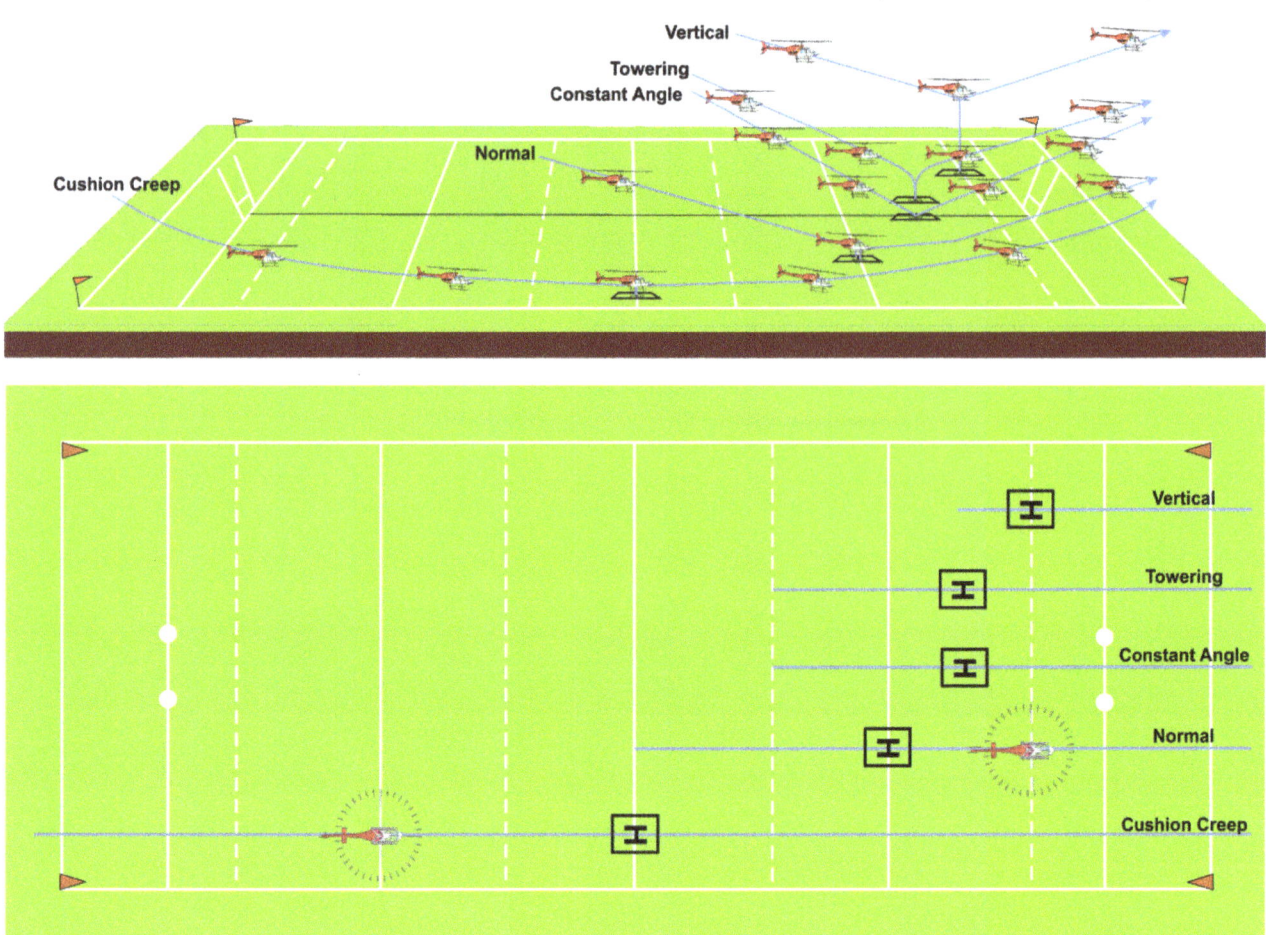

Take-off and Approach Profiles

Introduction Initially, the circuit is conducted at an airfield or large open area with a large power margin. The purpose is for the student to gain confidence and increase their skills to take off and approach in a standard manner. As the student's skills increase, the instructor will introduce the four (4) standard take-off and approach profiles to be practised while conducting circuit training.

Take-off and approach profiles The four (4) standard take-off and approach profiles are:

1. Normal
2. Constant Angle
3. Steep, and
4. Vertical.

Normal Take-off and Approach

Normal A normal take-off and approach is used when the power margin between the Power Available and the Power Required is greater than 2" MAP or 6% Tq (**CAT2** or better).

The normal approach is the most commonly accepted method for the pilot to conduct a safe and standard take-off and approach.

The normal approach is further divided into two (2) sub categories. They are the:

- Standard Airfield take-off, and
- Constant Angle take-off.

Standard Airfield take-off and approach

To conduct a Standard Airfield take-off, establish the helicopter in a standard 5 ft hover, then conduct a normal transition into forward flight as you have already been doing.

The Standard Airfield Approach allows the helicopter to conduct a normal 6-degree glideslope from the 500 ft final point and progressively arrive at the landing site terminating to a 5 ft hover. This is what you have already been doing in circuit training.

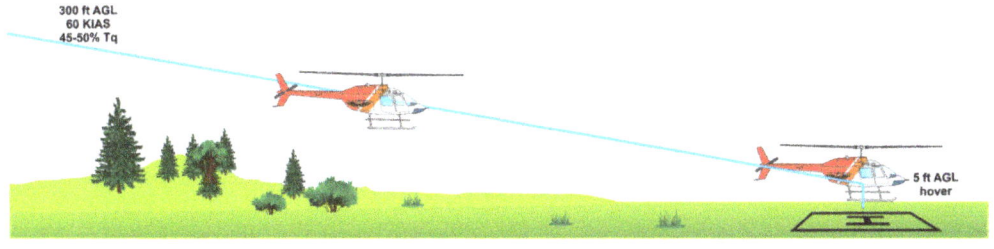

Constant Angle Take-off and Approach

Constant Angle take-off

The constant angle take-off and approach are used when the pilot wants to move away from the ground sooner rather than build up speed. This type of take-off and approach is often used over sandy or dusty areas and at night so that there is always a positive rate of climb on the take-off and a controlled, slightly steeper approach before landing.

To conduct a Constant Angle take-off, from a standard 5 ft hover, start to move forward and up to transition into forward flight at a constant 12-degree angle. Once reaching 55 kts, resume the normal take-off profile.

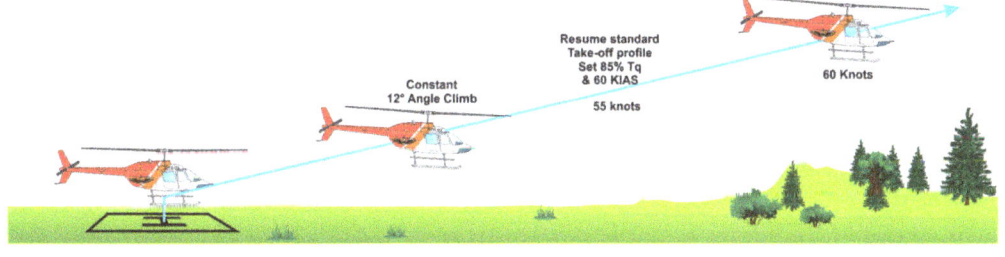

Constant Angle approach

The constant angle approach is set up the same as a normal approach, except the final approach may be slightly steeper, and the angle is held constant all the way to the hover position over the landing site.

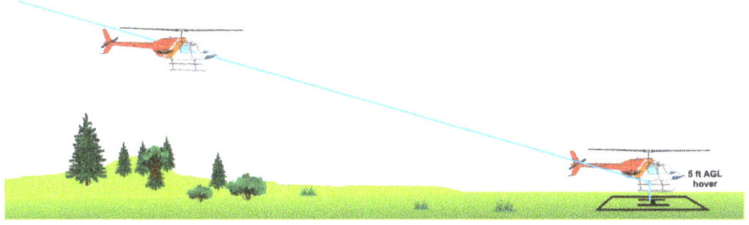

Steep Take-off and Approach

Steep

A steep take-off and approach (often referred to as a towering take-off and a steep approach) are used when the power margin between the Power Available and the Power Required is greater than 3" MAP or 11% Tq (**CAT3** or better).

The Steep take-off and approach are often used to avoid obstacles that the helicopter may have to depart or approach over.

Steep or Towering take-off

To conduct a Steep take-off, from a standard 5 ft hover, raise some collective so the helicopter commences a small vertical climb. Once there is a positive rate of climb (ROC) but no later than ½ to ⅔ of the rotor diameter (in other words, the helicopter is still in ground effect), start to move forward and commence a steep constant angle transition into forward flight. The take-off angles can be quite steep, allowing a 45-degree angle. Once reaching 55 kts, resume the normal take-off profile.

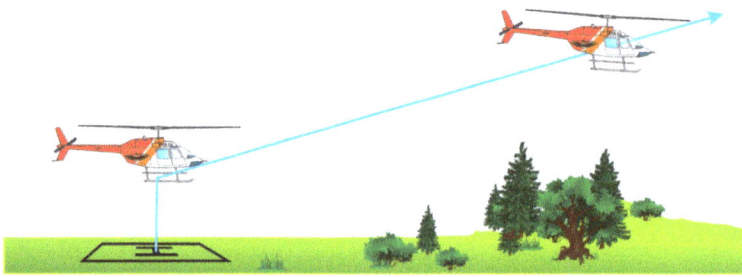

Steep approach

The steep approach is the same as a constant angle approach, except it is often established late on finals and much steeper. The angle is held constant all the way to the hover position over the landing site and can give a glideslope up to 45 degrees.

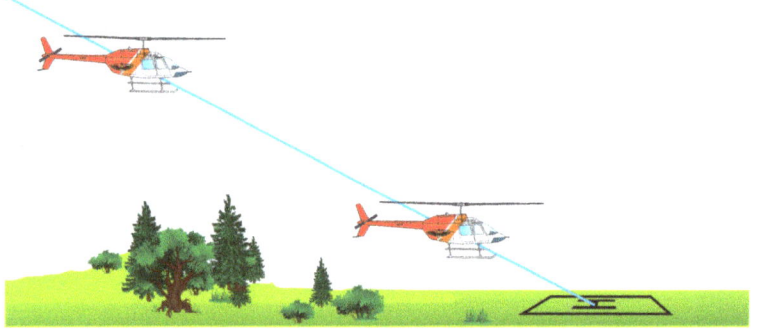

Double angle approach

A **double angle approach** may be used when approaching a confined area, and the glideslope has to be increased for the helicopter to get into the Confined Area.

In the diagram below, the helicopter is making a normal approach, but once over the trees, the glide slope is increased so that the final part of the approach is steep.

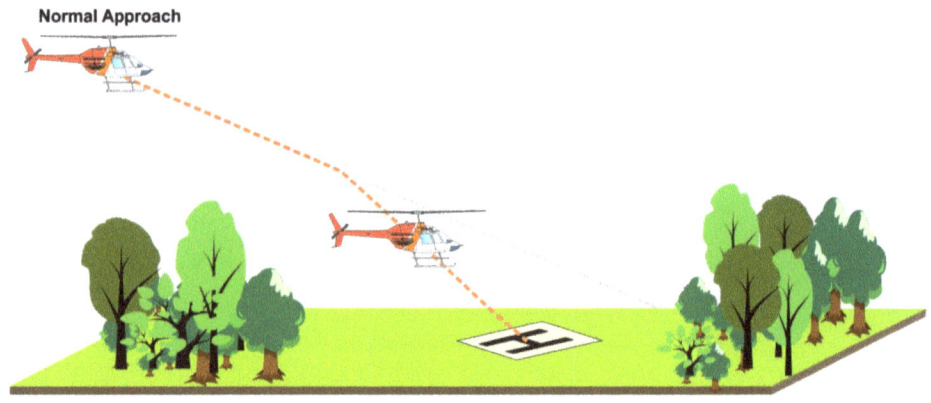

Vertical Take-off and Approach

Vertical Vertical take-off and approach are used where there is a large power margin over 4" MAP or 16% Tq (CAT4), and there is a reason to do it. More often than not, it is to avoid obstacles on take-off or approach or to minimise the effect of the downwash on the ground to avoid sand, dust, snow etc.

Vertical take-off To conduct a vertical take-off, with the helicopter on the ground, increase power so that the helicopter rises vertically. Continue to rise vertically until the obstacles are cleared, then conduct a normal transition into forward flight.

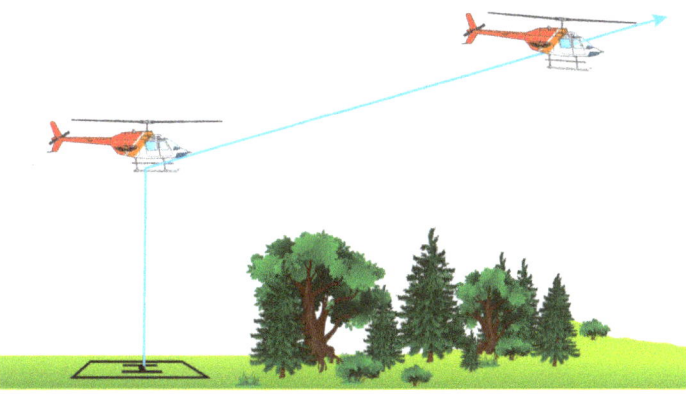

Vertical approach The vertical approach is done in the opposite manner. Conduct a normal approach to overhead the helicopter landing site terminating in an Out-Of-Ground effect (OGE) hover. From the hover, descend vertically to the ground.

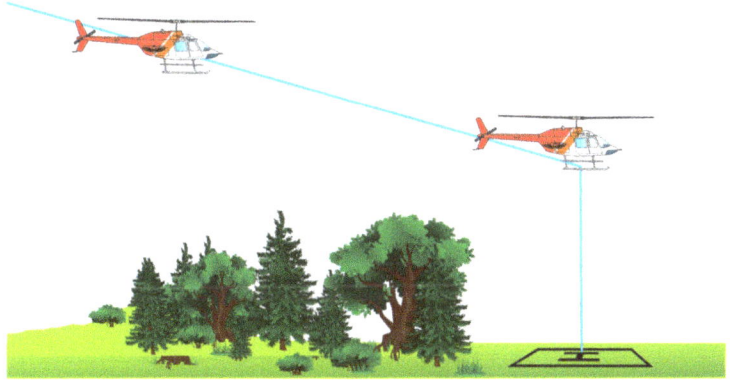

Very Limited Power Take-offs and Landings

When used Very limited power take-offs and landings are used when the power margin available is 1" MAP or 5% Tq or less. These take-offs and landings may be required when operating in high, hot and heavy situations.

They include a:

- Shallow take-off and landing (Cushion Creep), and
- Running Take-off and Landing.

Chapter 10 Limited Power Operations

Shallow Take-off and Approach Profile

Shallow (Cushion Creep) A shallow take-off and approach is used when the power margin between the Power Available and the Power Required is 5% or less (CAT1). It is designed to make the best use of the ground cushion while IGE and to have all the available power in early so that the helicopter can run along the ground if necessary.

If conducting a shallow take-off, it is often referred to as a "cushion creep" take-off as the pilot will creep slowly along the ground using the ground cushion to help pass through translational lift during the transition.

A shallow approach is used for the same reason and is also used during some emergencies where a shallow, slow approach is made to a large open area.

Shallow take-offs and approaches have a glideslope of approximately 3-5 degrees.

When to use A cushion creep take-off may be performed when power is sufficient to just hover in ground effect but not translate in the normal manner into forward flight.

At these power settings, the aircraft may be landed directly on the ground, at zero speed, with no (zero) preliminary hover. Referred to as a zero-zero landing – zero speed, zero hover.

Suitable area The area for a cushion creep take-off must have:

- a smooth surface of sufficient length, and
- shallow approach and departure angles, relatively free of obstacles.

Shallow take-off For a shallow take-off, the helicopter can hover, but the skids are very close to the ground. Gently ease the cyclic forward, staying within ground effect until reaching ETL. Control any sink with collective. On passing ETL, continue to accelerate rather than gaining height until reaching 55 kts, then conduct a shallow climb.

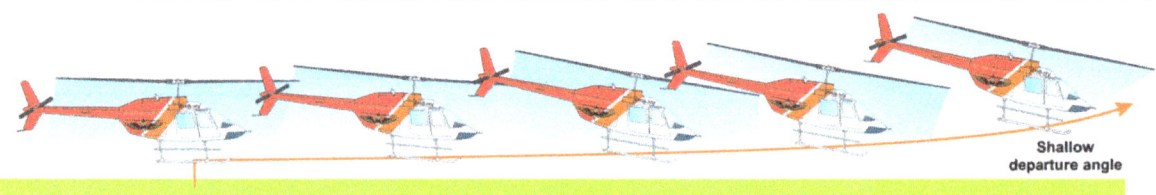

Lift off to Hover. Translate into forward flight making full use of ground cushion

Shallow approach For a shallow approach, set the helicopter up on a longer final and accept a shallower-than-normal approach. The speed will steadily decrease, and the helicopter will enter the ground cushion. Ride the ground cushion until reaching the landing area, then terminate to a low hover.

If the helicopter cannot hover because there is not enough power, the helicopter should gently touch down with a small forward run-on.

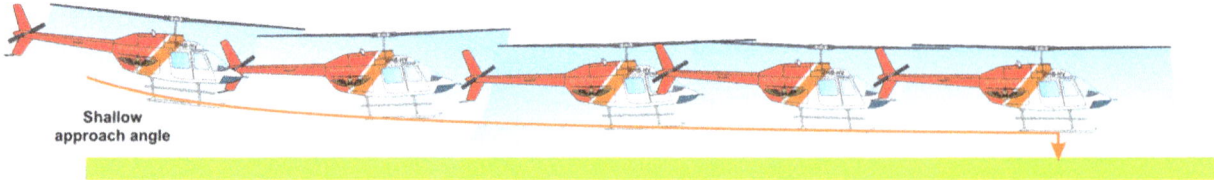

Running Take-offs and Landings

When to use

When there is insufficient power to hover, but the skids feel light on the ground, a running landing and take-off may be used, provided a suitable area is available.

Although very little power is needed for the running landing and take-off compared to other landing and take-off techniques, many areas may be unsuitable.

In the real world, the pilot also has to consider that the helicopter is too heavy and the conditions are unsuitable for the take-off or landing, so another action plan should be considered. Therefore, A running landing is not considered a normal operation but may be required due to other circumstances.

Suitable area

The area for a running take-off and landing must have:

- a smooth surface of sufficient length to allow the helicopter to slide along the ground, and
- shallow approach and departure angles, relatively free of obstacles.

Running take-off

A diagram of a running take-off is shown below.

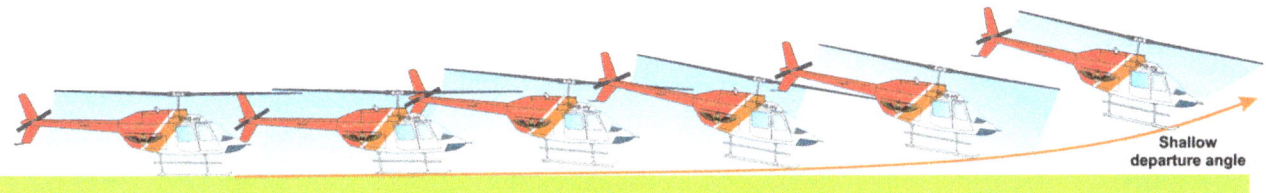

Translate into forward flight making full use of ground cushion

Running landing

A diagram of a running landing is shown below.

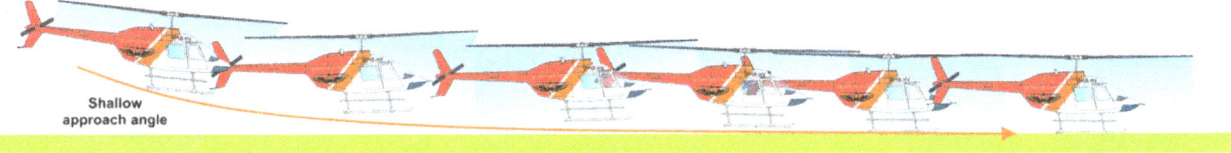

Forward movement on descent

It is always best to have some forward movement on the descent to reduce the risk of inadvertently entering a Vortex Ring State.

- For this reason, Normal, Steep, Constant Angle and Double Angle approaches are preferred when operating in confined areas.
- Vertical approaches are only made in extreme cases to fit into very small confined areas, and the pilot must have a lot of surplus power.
- Shallow approaches are only utilised when power is the primary limiting factor and, by definition, will require a large open flat confined area.

Air Exercises: Limited Power Operations

Introduction

This air exercise involves practising the various take-offs and landings, emphasising cushion creep and running take-offs and landings.

Airmanship

- Memorise the power and RRPM limits
- Memorise the power categories
- Memorise the types of take-offs and landings applicable for each power category
- Be conservative in selecting suitable landing areas
- Maintain situational awareness
- Make best use of the wind

Common faults

The following are common faults of students:
- Overcontrol of the cyclic and collective
- Not maintaining the skids aligned with the direction of travel with pedals
- Losing translational lift too early on the approach
- Exceeding the limits
- Not having patience during the manoeuvre
- Committing to a take-off too early over obstacles

Mike Becker's Helicopter Handbook

Air Exercise 10-1: Determining Power and Wind

Introduction   The instructor will demonstrate the nomination of the Power Category, how to determine where the wind is coming from, and an estimation of the strength to nominate a point on the ground where translational lift will be reached on take-off.

After the demonstration, the instructor will give the student all of the controls, position the helicopter on the ground at the helipad, and allow the student to practice nominating a power margin and determining the wind.

Nominate a Power Category at the Hover

Once in a stable hover into wind, nominate the Power Category for the take-off by conducting the following steps:

Step	Action	Discussion
1	Maintain a stable 5 ft skid height hover into wind.	This will configure the helicopter in a standard hover with the current AUW and atmospheric conditions enabling accurate readings.
2	Note the amount of power that is required to hover.	Power required to hover is _____
3	Calculate the power margin (difference) between the power required and maximum power available.	The difference between the Power Required to hover and the maximum Power Available will give the Power Margin for the flight.
4	Once the power margin is noted, nominate the related Category of Power.	<table><tr><th>Turbine</th><th>Piston</th><th>Power Category</th></tr><tr><td>0-5% Tq</td><td>1" MAP</td><td>CAT1</td></tr><tr><td>6-10% Tq</td><td>2" MAP</td><td>CAT2</td></tr><tr><td>11-15% Tq</td><td>3" MAP</td><td>CAT3</td></tr><tr><td>16-20% Tq</td><td>4" MAP</td><td>CAT4</td></tr></table> *Note: These limits may vary for different helicopters. Check with the flight manual and the Flight School on how they want to reference the power margin.*
5	CRM and the PWPTEM	During the Power, Wind, Plan, Threat Error Management (PWPTEM) brief, state the Category of Power and the type of take-off or approach that will be conducted. **For example:** - *Power* CAT3 - *Wind* from the front right at 10-15 kts - The *Plan* is to conduct a normal take-off with a Standard Airfield Departure, making a right turn and climb to 1500 ft.

Chapter 10 Limited Power Operations

Nominate a Power Category while in Forward Flight

From the cruise configuration, nominate the Power Category for the approach by conducting the following steps:

Step	Action	Discussion
1	While downwind, raise the collective to the maximum power limit.	This may cause the helicopter to climb and accelerate.
2	Note the power limit achieved	In a piston, this will be based on the MAP limit. In a turbine, the limiting factor could be either the Tq, TOT or N1, depending on the Density Altitude. State: Maximum power available is _____
3	Return to normal flight.	Reset the collective for normal cruise.
4	Calculate the power margin (difference) between the Power Required and maximum Power Available.	The difference between the Power Required and maximum Power Available will give the Power Margin for the conditions.
5	Once the power margin is noted, nominate the related Category of Power.	<table><tr><th>Turbine</th><th>Piston</th><th>Power Category</th></tr><tr><td>0-5% Tq</td><td>1" MAP</td><td>CAT1</td></tr><tr><td>6-10% Tq</td><td>2" MAP</td><td>CAT2</td></tr><tr><td>11-15% Tq</td><td>3" MAP</td><td>CAT3</td></tr><tr><td>16-20% Tq</td><td>4" MAP</td><td>CAT4</td></tr></table> *Note: These limits may vary for different helicopters. Check with the flight manual and the Flight School on how they want to reference the power margin.*
6	CRM and the PWPTEM	During the Power, Wind, Plan, Threat Error Management (PWPTEM) brief, state the Power Category, the type of approach and landing that will be conducted, and any other considerations.

Estimating the Wind at the Hover

Estimating the wind at the hover To estimate the wind direction and strength at the hover, conduct the following steps:

Step	Action	Discussion
1	Set a stable 5 ft skid height hover.	Doing this over a surface with some texture, such as grass, is preferable. This exercise is challenging to demonstrate over concrete as you cannot observe the effect of the downwash on the surface.
2	Note where the edge of the downwash is relative to the helicopter.	Looking out the front window, look at the surface, see where the downwash is blowing on the grass, and where it stops having an effect. This is the dividing line between the downwash and the wind.
3	Conduct a slow 360° pedal turn.	As the helicopter turns, note the distance from the helicopter of the dividing line. The following will apply: ■ If the wind is on the **nose**, the dividing line will be very close. ■ If the wind is **from behind**, the dividing line will be far away. ■ If the wind is **from the side**, the dividing line will be between the first two.
4	Stop the 360° turn when the dividing line is closest to the helicopter.	This will place the helicopter directly into the prevailing wind.
5	Note the distance of the dividing line from the nose.	The following will apply: [Diagram showing four helicopters with wind conditions: NIL WIND, WIND 5kts, WIND 10kts, WIND 15kts, with downwash patterns illustrated beneath each]
6	CRM and PWPTEM	Now during the PWPTEM brief, the crew can state the direction the wind is coming from, estimate its strength, and nominate a point on the ground where they estimate reaching translational Lift.

Estimating the Wind while in Forward Flight

Estimating the wind while in forward flight

There are several methods to estimate the wind direction and strength while in forward flight, depending on what is available to the crew at the time. In general, there will be the following options:

1. Obtaining a forecast before departure
2. Listening to the ATIS/AWIS
3. Observing the windsock
4. Observing smoke or dust
5. Observing wind vanes on water
6. Observing the direction trees are being blown
7. Observing drift
8. Observing the difference in ground speed as shown on the GPS versus Airspeed as displayed on the ASI
9. Conducting a constant radius turn about a reference point
10. Assuming the wind has not changed since the take-off from the same HLS

All of these may help the crew determine a wind direction for the approach, but referencing the Airspeed (ASI) versus the Groundspeed (GPS) will always be confirmed when on short finals.

Air Exercise 10-2: Cushion Creep Take-off and Landings

Air Exercise	The student will practise cushion creep take-offs and landings in this air exercise.
Demonstration	The instructor will demonstrate the air exercise first.
Cushion creep take-off	The table below describes how to make a cushion creep take-off.

Step	Description	Discussion
1	Lift off	Complete the Pre-liftoff checks (HEFFR), then lift off the ground to a stabilised, very low hover into wind. There will be minimal opportunity to make clearing turns or to manoeuvre if very limited on power.
2	Power Margin	Conduct the PWPTEM. Note the power margin and decide on a take-off technique (in this case, CAT1 cushion creep take-off).
3	Take-off	From the hover, ease the cyclic gently forward and allow the helicopter to accelerate slowly while in the ground cushion. - Keep straight with pedals - Manage the Power and RRPM Depending on the wind: - The helicopter may sink and any remaining power will be required to maintain height. - If there is not enough power, skids may lightly touch the ground. Keep easing the cyclic forward to accelerate to translational lift.
4	Translational lift	As the helicopter passes translational lift, use some aft cyclic so the nose comes up and the skids break free of the ground, then use some forward cyclic to keep accelerating and allow the helicopter to slowly climb and accelerate away. The climb out will be shallow and slow. Continue until reaching the minimum drag speed for the helicopter (approximately 55 kts).
5	Depart	On reaching the minimum drag speed, the helicopter can now continue to fly away in the normal manner.

Cushion creep landing   The table below describes how to make a cushion creep landing.

Step	Description	Discussion
1	Begin descent	Extend downwind and initiate a normal descent planning to roll out on finals with a shallower approach angle.
2	Finals checks	On final approach, take particular notice of the wind and power available.
		If there is any doubt, go around early and set up again; otherwise, continue the shallower than normal approach, bringing speed back early and bringing the power in early so that this happens slowly and under full control.
3	Maintain slow rate of descent	In the final stages of the landing, maintain a slow rate of descent but maintain speed above translational lift, until making the helipad is assured.
		Only when the helicopter is assured of making the helipad can the pilot reduce speed while entering the ground cushion to come to a very low hover.
		Note:
		With proper handling, cushion creep landings may be terminated to specific touchdown points with no hover.
4	Landing	From the hover, make a normal landing.

Air Exercise 10-3: Running Take-off and Landings

Air Exercise In this air exercise, the student will practise running take-offs and landings.

Demonstration The instructor will demonstrate the air exercise first.

Running take-off The table below describes how to make a running take-off.

Step	Description	Discussion
1	Lift off	Complete the Pre-liftoff checks (HEFFR) and then raise the collective to lift off. With a running take-off the helicopter will not have enough power to lift off the ground, but it should feel very light on the skids. There will be no opportunity to make clearing turns or to manoeuvre.
2	Power Margin	Conduct the PWPTEM. Note the power margin and decide on a take-off technique (in this case, CAT1 running take-off).
3	Take-off	With the skids on the ground, ease the cyclic gently forward and jiggle the pedals left and right (by a very small amount) to help the skids break free of the surface and allow the helicopter to begin a slide along the ground. ■ Allow the helicopter to accelerate slowly while the skids are on the ground ■ Keep straight with pedals ■ Manage the Power and RRPM Depending on the wind: ■ The helicopter may feel like it is getting more friction on the ground, and any remaining power will be required to help it get through this stage into translational lift. ■ If there is not enough power, the skids may not want to break free, and the take-off will have to be aborted.
4	Translational lift	As the helicopter passes translational lift, use some aft cyclic so the nose comes up and the skids break free of the ground, then use some forward cyclic to keep accelerating and allow the helicopter to slowly climb and accelerate away. The climb out will be very shallow and slow. Continue until reaching the minimum drag speed for the helicopter (approximately 55 kts).
5	Depart	On reaching the minimum drag speed, the helicopter can now continue to fly away in the normal manner.

Chapter 10 Limited Power Operations

Running landing   The table below describes how to make a running landing.

Step	Description	Discussion
1	Begin descent	Extend downwind and initiate a normal descent planning to roll out on finals with a shallower approach angle.
2	Finals checks	On final approach, take particular notice of the wind and power available. If there is any doubt, go around early and set up again; otherwise, continue the shallower than normal approach, bringing speed back early and bringing the power in early so that this happens slowly and under full control.
3	Maintain slow rate of descent	In the final stages of the landing, maintain a slow rate of descent but maintain speed above translational lift until making the helipad is assured. - Only when the helicopter is assured of making the helipad can the pilot reduce speed while entering the ground cushion to come to a very low hover taxi and allow the skids to run along the ground. Ensure the skids are aligned with the direction of travel. - Do not lower the collective until the helicopter comes to a stop but small downwards collective can be used to act as a brake by putting more friction contact between the skids and the ground. - Cyclic shall be maintained slightly forward to protect the tail. **Note:** With proper handling, running landings may be terminated to specific touchdown points.
4	Landing	Once the helicopter comes to a stop, lower the collective and follow the landing procedure.

11

Confined Areas

Aim To demonstrate, experience and conduct the techniques to assess, approach, manoeuvre in, land and then depart from an unprepared area with limited space and power available.

Objectives On completion of this lesson, the student will be able to:

- plan a confined area exercise
- state what the difference is between a high and a low recce
- list the symptoms of Vortex Ring State (VRS) and explain a recovery
- recall what causes overpitching and how to recover
- explain recirculation
- recall the PSWATP check
- state two (2) ways to manoeuvre in a confined area while at the hover
- list at least four (4) ways to calculate the wind
- select and assess a landing site, and
- make an approach, land, manoeuvre in and take-off from an unprepared area with limited space available.

Motivation Operations into and out of a confined area are what helicopters are designed for, and it's what we all love to do. Unlike operations around an aerodrome, the confined area operation relies on the pilot's ability to make their own assessments and gather all the information in real-time to make a successful landing and take-off.

There is no ATIS to give weather information, no windsock, no designated runway with a controller advising what to do, and no guarantee that the approach and departure paths are clear of obstacles and hazards. Helicopter performance has to be determined while in flight without access to charts, diagrams and calculators, as the single pilot simply does not have enough arms and legs to fly and administrate.

The ability to process information required for operations in a confined area is a skill that, if understood fully, will make you a much more competent pilot.

Preparation: Confined Areas

Introduction

The pilot needs to recognise potential hazards and threats when operating in a confined area. Some of these are subtle aerodynamic changes that can develop into real problems unless identified and corrected early; others revolve around prior preparation and planning.

Good technique, checks and situational awareness are part and parcel of confined area operations.

Flight Planning

Before departing for a flight, the pilot and crew will always do as much prior preparation as possible. In some cases, when landing in a confined area, the pilot can research the location and make some determinations regarding performance and accessibility into the area. In other cases, the pilot may not be able to collect data on the confined area location, for example, when it is not feasible to identify the confined area before departure.

Gathering any information before the flight will assist the pilot and crew in making better decisions on arrival at the confined area. In today's technological and information age, there are many tools available to assist the pilot in researching a landing area before the flight and may include:

- Google Earth
- Flight School landing site documentation
- GPS or Electronic Flight Bag, and
- other Pilots who have already been to the area.

Performance and Fuel

As part of the flight planning, the pilot can make some assumptions about the helicopter's **performance** and the amount of **fuel** that should be on board.

Enroute

While enroute to the confined area and before arriving, the pilot and crew complete a pre-landing check using the acronym **HEFFRR**.

This check has been covered previously, but there will be three (3) distinct additions.

Heading Home

When operating in a confined area, there may be multiple landing sites, often amongst hills, mountains and remote areas where typical navigation references are not easily recognised. For this reason, it is easy for a pilot to get disorientated and lose all sense of direction. This is particularly relevant in bad weather.

To assist the pilot before entering the area where confined area operations are going to be completed, the pilot should nominate an estimated magnetic compass bearing for a **Heading Home**.

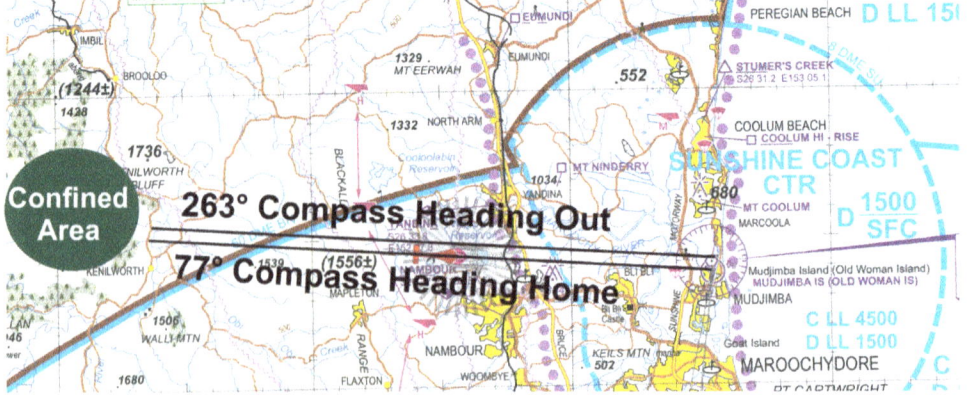

Chapter 11 Confined Areas

Fuel

Fuel needs to be reconsidered and applied to the operation.

Generally, the pilot and crew would have calculated a Fuel on Ground (FOG) time on departure. Based on this, the pilot now needs to reconsider the following:

- time to the confined area from departure
- time to the next refuelling point from the confined area, and
- amount of time available for confined area operations.

To do this, on arrival at the confined area, estimate the flight time to get home or to the next landing or refuelling site, and deduct that from the original FOG. This will give a new time representing when the pilot must stop confined area operations and return home (or to the next refuelling stop) without using any fuel reserves.

The difference between the time the calculation was done and the new return to base time represents the time the helicopter can remain in the confined area.

For example:

On departure at 08:00, the pilot calculated a Fuel on Ground (FOG) of 10:30.

At 08:30, the pilot arrives at the confined area.

Knowing it took 30 minutes to get to the confined area and that the intention is to return to the same departure aerodrome, the pilot deducts 30 minutes from the Fuel on Ground (FOG) time and has a stop time of 10:00 when operations should stop and the helicopter returns to base to arrive with all the fuel reserves intact.

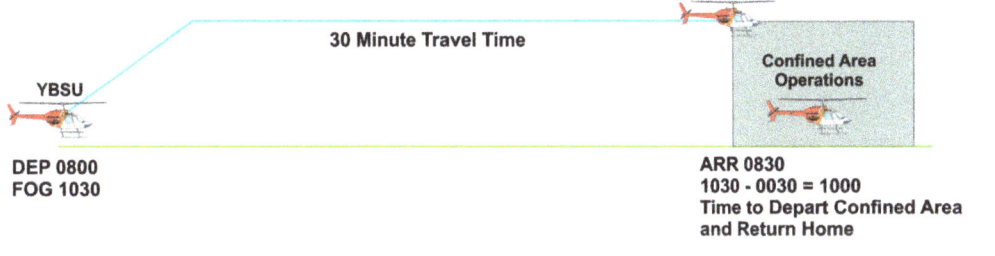

Radio

The final consideration is the **Radio**.

VHF radio works by line of site. Because the helicopter is going to be operated at low level, radio transmissions may not be able to be transmitted or received. If the pilot needs to make radio calls, do so before descending into the confined area.

Reconnaissance

Definition

According to the Dictionary of Aviation Terms, published by McGraw-Hill, reconnaissance as it relates to helicopter operations:

"*is a general examination or survey of the main features, or certain specific features, of a region or area usually as a preliminary to a more detailed survey*"

This means that the pilot and crew will inspect the confined area to determine if it is suitable to land and take-off from.

Reconnaissance can be abbreviated as "Recon" or "Recce." In this book, we have used the "Recce" abbreviation.

In helicopter operations, two types of recce are commonly used:

1. high recce, and
2. low recce.

High Recce

High recce

The high recce is done at 500 ft AGL or above at approximately 45-70 kts IAS, which allows the pilot and crew to look for major obstacles that may be a hazard or threat during the approach or take-off, and to assess or confirm the wind direction and strength.

The shape of the high recce will be determined by the terrain and obstacles that the pilot has to avoid; it may have an irregular shape; however, if at 500 ft AGL or above, in most cases, a standard square circuit should be achievable.

The key to a good recce is to keep the confined area in sight but at a distance where the pilot and crew will have some perspective (angle) to look into the confined area and see what terrain or obstacles may need to be considered. Too close, and the pilot and crew cannot see it and will miss sighting obstacles. Too far away, the pilot and crew cannot see enough of the area to make reliable decisions.

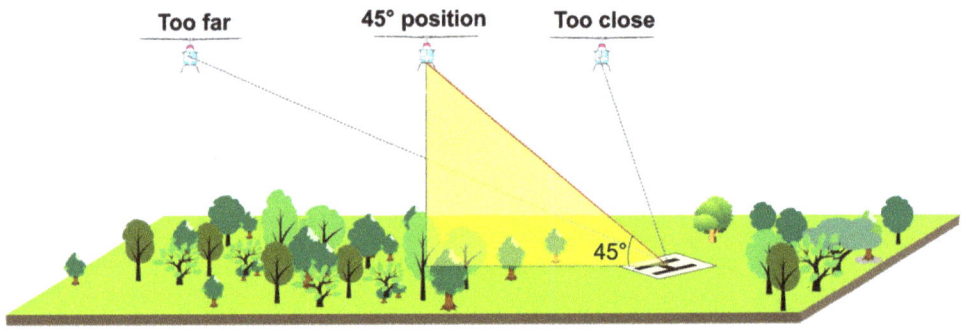

Recce circuit

As a general guide, set the helicopter up so the confined area is on the pilot's side at a 45° angle. It can be observed as the confined area goes past the pilot's door. Once the confined area gets to the back of the pilot's door, turn through 90° so that the confined area is again being observed out the front quarter of the main window, and the confined area again passes down the pilot's right.

Continue this circuit until the pilot and crew are satisfied that all items in the PSWATP check have been considered or confirmed and a suitable plan has been decided upon.

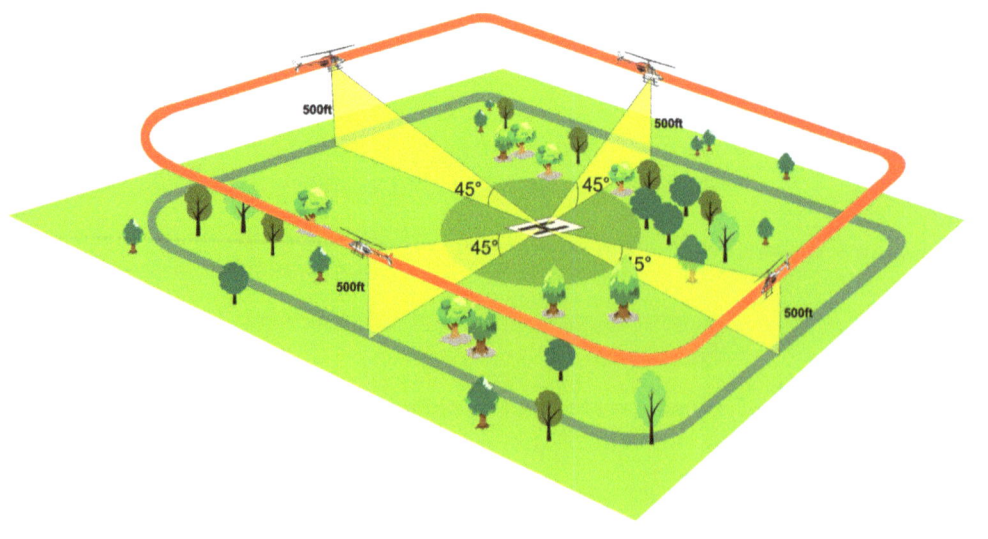

Chapter 11 Confined Areas

Recce circuit overhead

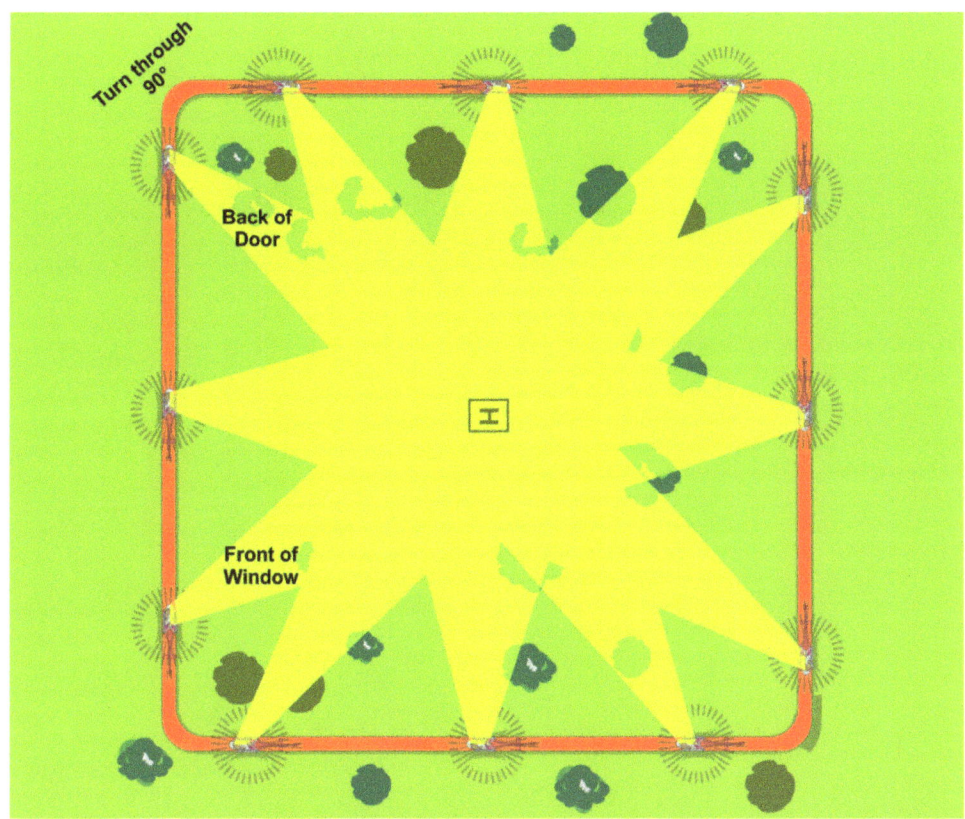

Low Recce

When to fly a Low recce

The low recce is conducted whenever the pilot or crew are not confident that all features have been accurately observed and that a more detailed look at the confined area is required before committing to an actual approach and landing.

Low Recce

A low recce will follow a similar path to the approach; however, it will be off to one side so that the pilot and crew can visually inspect the confined area as they fly past it.

The pilot is conducting a "practice" approach but will not descend below 100 ft AGL as the helicopter flies level past the confined area between 35-50 kts IAS.

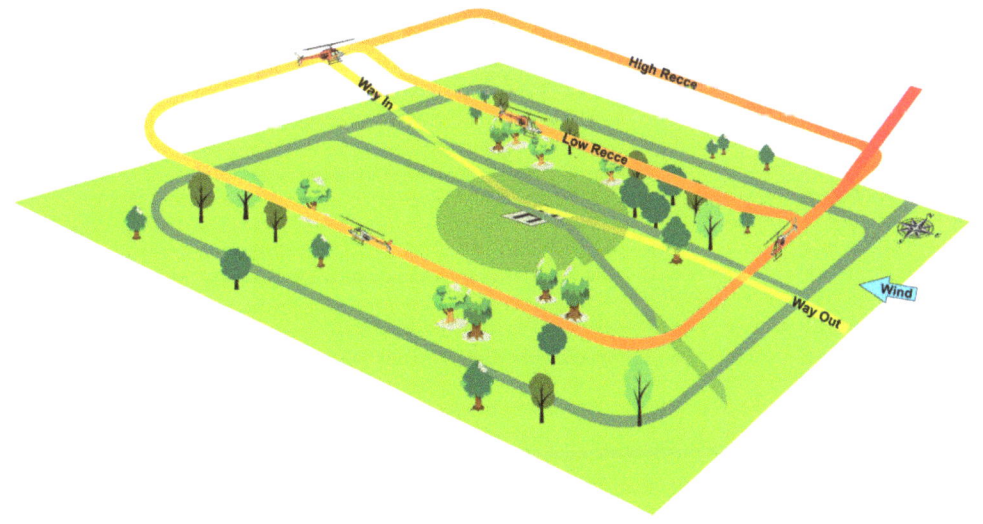

Low recce diagram from the top

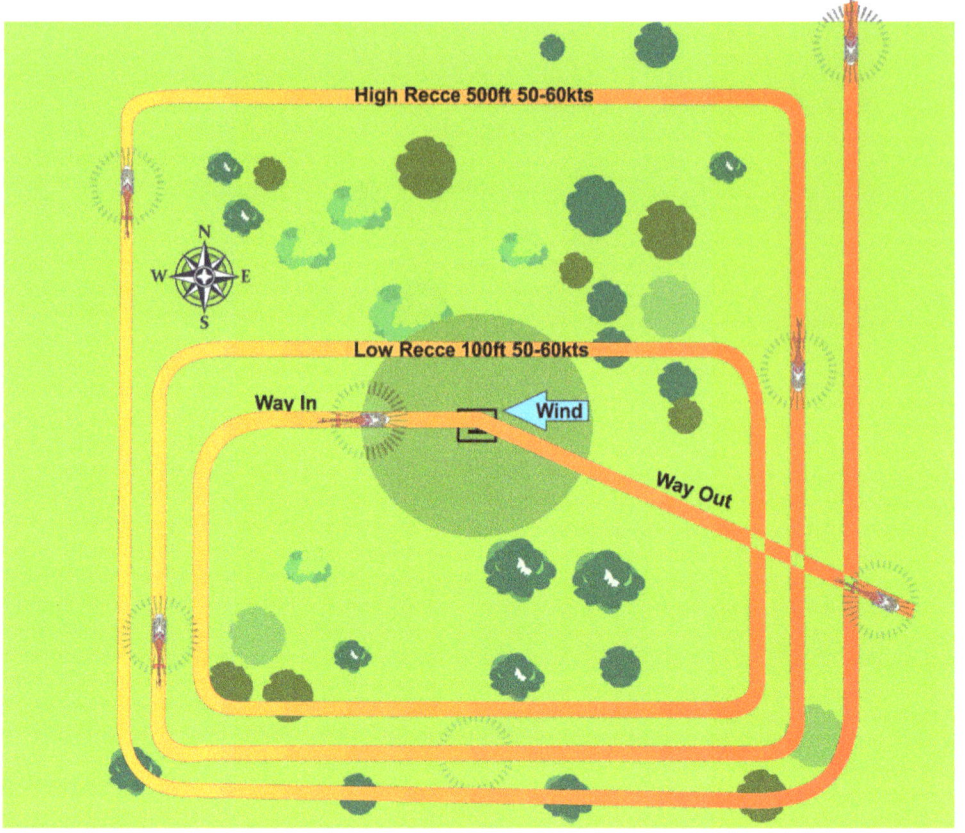

Confirming power

After flying past the confined area during the low recce, the pilot has another opportunity to reconfirm the power by pulling full power available and climbing back to 500 ft AGL before conducting a standard approach and landing in the confined area.

Ground recce

Once inside the confined area, the pilot and crew can conduct a ground recce.

A ground recce can be either conducted:

- while hovering, allowing the crew to observe the area, or
- **when landed and shut down**, allowing the crew to walk around in the confined area and have a good look.

A ground recce is worthwhile if the confined area is going to be used multiple times. The pilot and crew may choose to:

- clear an area
- take down small trees, bushes or branches that pose a threat
- mark out reference points for hovering and landing so that the tail is protected and the best use of the entire area can be considered if also limited in power.

The crew may also be able to make a small bush windsock with a plastic bag or some marking tape for future use.

Chapter 11 Confined Areas

Power Settling and Vortex Ring State

A Little Background

It is generally accepted in helicopter circles that Power Settling (also referred to as Settling with Power) and Vortex Ring State (VRS) describe the same condition. Although this statement is true, it is not 100% accurate.

Learning through invention

To understand Power Settling, we need to understand a little bit of helicopter history. Helicopters (just like fixed wing) were invented by enthusiasts who had a dream, constructed a contraption, and gave it a go.

They were men and women working out of a backyard who were either very brave or very stupid. Thank goodness they did what they did.

Fixed Wing

The advent of fixed wing flying was the same. The Wright brothers (and others like them) had only a rudimentary knowledge of aerodynamics; they simply built it and gave it a go. They failed and tried again, ultimately learning and succeeding from experience.

Before aerodynamics was known

None of these pioneers had proper names for the aerodynamic problems first encountered when flying helicopters. There was no basis to start from, no wind tunnel tests, and no understanding of the effects of induced flow, rates of descent, wind, downwash, and the subsequent changes in the blades' angle of attack regardless of the set blade pitch or Rotor RPM.

Example

For Example

Let's look at an early pioneer who has just lifted to the hover when the helicopter suddenly descends, hits the ground, rolls over and destroys itself. ***What happened Igor?!***

Igor had no idea; the engine was performing normally, the Rotor RPM was stable, and he did not move the collective; everything seemed fine. It just settled to the ground with power on. He didn't think he did anything wrong.

---

[1] Sikorsky Archives, Development Of The Vs-300 Helicopter, viewed 10 November 2013, http://www.sikorskyarchives.com/VS-300_Helicopter.php

[2] Unknown, Foter.com, Public Domain Mark 1.0, Orville Wright&flyer1909, viewed 10 November 2013, > http://foter.com/photo/orville-wright-flyer1909/<.

[3] Sikorsky Archives, Testing Helicopters Becomes A Dangerous Job, viewed 10 November 2013, http://www.sikorskyarchives.com/R-4_Coast_Guard.php

Changing terminology

Over time, with experience and experimentation, we have given many names to a helicopter descending with power, whether it is under control or not under control.

These terms include power settling, over-pitching, vortex ring state, loss of ground effect, loss of translational lift, recirculation, downdraft, updraft, turbulence, normal descent with power on, autorotation, and so on.

Because most aerodynamics encountered on the rotor system are now known and have specific names, the term Power Settling (or Settling with Power) is no longer commonly used as an excuse to cover a myriad of unknown aerodynamic situations. Instead, Power Settling has come to be identified with the early stages of Vortex Ring State (VRS), also referred to as the "*onset of VRS*" or "*Incipient VRS*" (IVRS).

Vortex Ring State (also called fully developed Vortex Ring State) is the ***progression*** of Power Settling to a point where all lift from the rotor system has been removed due to aerodynamic factors.

For this reason, then, when reading texts or completing an examination, you may find that the writer of the text or examination may say Power Settling and Vortex Ring State are different, and they would be technically correct, even though by convention, the people working in the helicopter industry (the pilots) deem them to be the same. There is a difference between the theoretical debate and the practical management of Power Settling and VRS in a real helicopter in any given situation.

Same thing but different!

Think of:

- Power Settling as Baby Bear, and
- Vortex Ring State (VRS) as Papa Bear.

Same thing but different!

How you manage each one can be similar, but depending on the circumstances and which one you encounter can change your response.

Definition: Power Settling

Power Settling refers to the early stages of the Vortex Ring State.

This is when the helicopter starts to descend. The tip vortices and the stalled or nil lift area at the blade root are increasing, but the fully developed vortices in the blade's centre causing VRS are not yet established. The pilot can often power out of Power Settling by raising some collective and increasing speed.

Definition: Vortex Ring State

For this book, the term Vortex Ring State (VRS) refers to a fully developed Vortex Ring State on the main rotor system that will lead to the helicopter sinking towards the ground even though the engine is providing power and driving the main rotors at the correct Rotor RPM.

The helicopter is descending into its downwash with:

- tip vortices at the rotor blade's tip
- inner vortices at the rotor blade's mid-section, and
- a stall or nil lift area at the blade's root.

Regardless of the pilot changing the pitch on the rotor blades with the collective, the helicopter will continue to descend at a very high rate until the rotor disc enters clean air and the vortices are removed.

Chapter 11 Confined Areas

Forces on the Blade

Vortex Ring State can be described as a flight condition like a stall in a fixed wing in that the aircraft sinks suddenly; however, it is less predictable than a stall and cannot be taught in such a precise manner or in such definite terms. It can even be difficult to intentionally enter a Vortex Ring State for demonstration.

To analyse what happens during Vortex Ring State, the forces on the rotor blades need to be considered. The following topics describe the forces on the blade while in three separate flight configurations:

- Hovering OGE in nil wind
- In a shallow vertical descent in nil wind, and
- In a steep vertical descent in nil wind.

Forces on the Blade: Hovering OGE in Nil Wind

While hovering OGE in nil wind

First, consider a helicopter hovering Out of Ground Effect (OGE) in nil wind conditions. The rotational velocity and induced flow are greatest at the tip of the rotor blade.

At the rotor blade's root, although the rotational velocity ratio to induced flow may be similar, the angle of attack is much greater due to the higher blade pitch angle. Remember, washout (the twist built into the blade at manufacture) is built into the helicopter rotor blade giving a higher pitch angle at the blade root compared to the smaller pitch angle at the tip. This is to compensate for the increasing velocity from the blade's root to the tip.

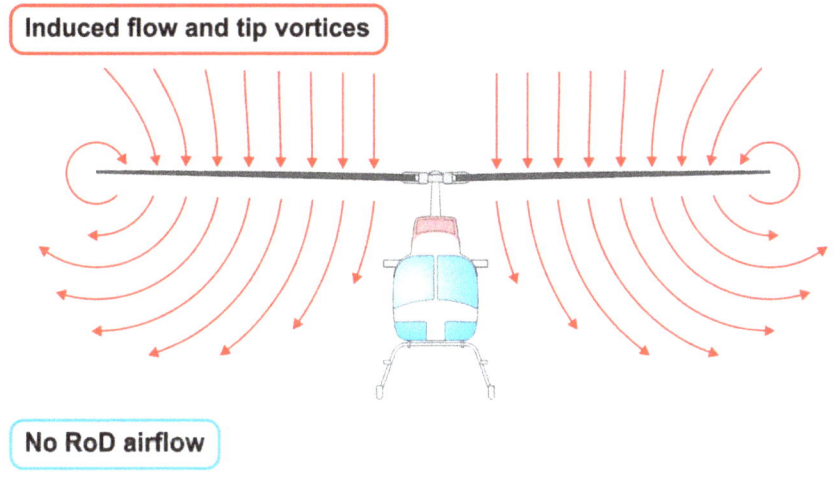

Root end

For this reason, the angle of attack at the root end of the blade is high, to the point there may be a small, stalled area or a section of the blade that is not producing any usable lift close to the rotor head.

Mid-section

The rotor blade's mid-section will produce good positive lift and provide the necessary rotor thrust to oppose weight. The mid-section is typically the workhorse of the rotor blade.

Tip section

The tip section is also producing positive lift, but because of the small tip vortices (air from below the disc moving around to the top of the disc as high-pressure air moves towards the low-pressure air), the induced flow will be slightly greater, and the angle of attack will therefore be slightly smaller.

Vector diagrams

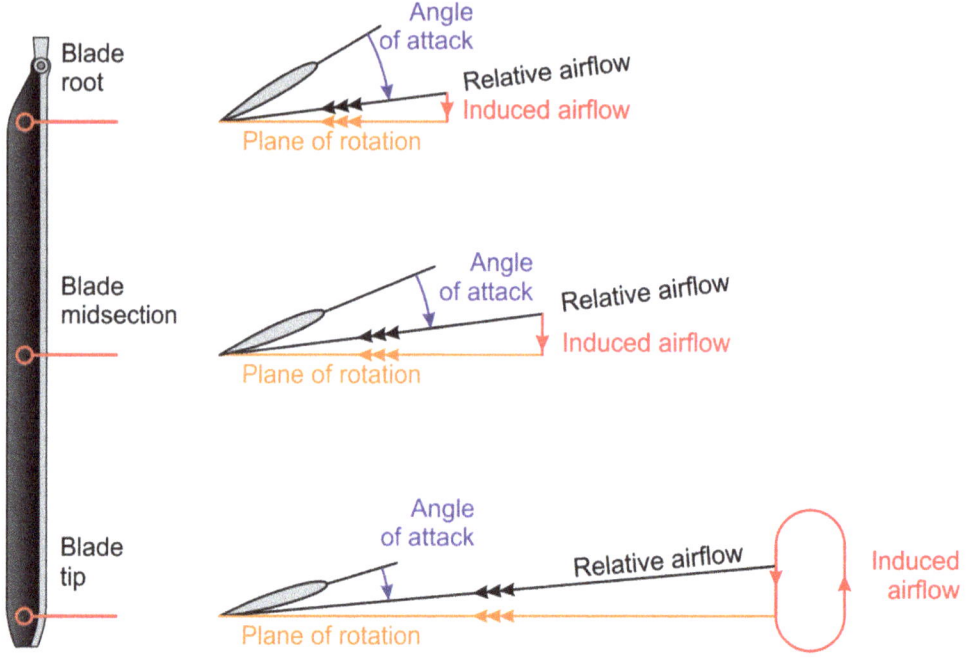

Lift envelope

Lift will be relative to the amount of air processed over the aerofoil, so there will be a direct correlation between the lift produced and the subsequent induced flow created. Both the lift envelope across the blade and the induced flow envelope can be compared, and it can be clearly seen that one correlates to the other.

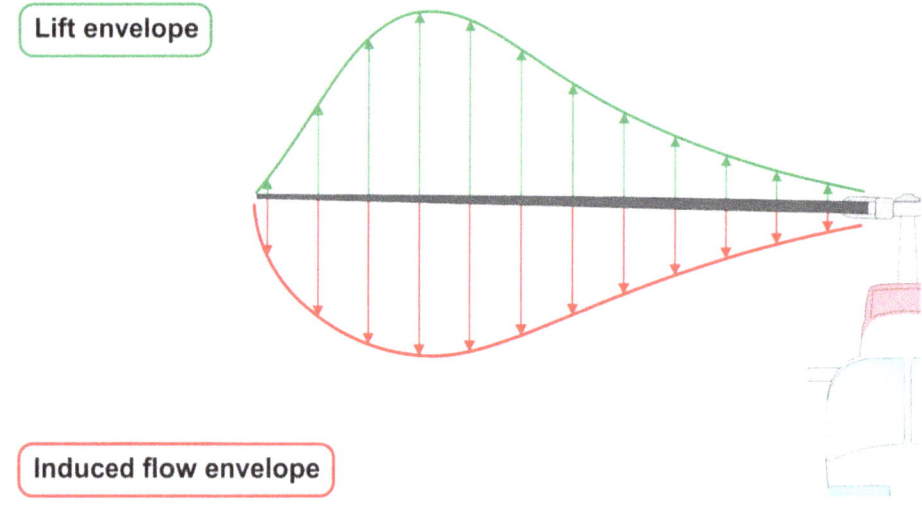

Forces on the Blade: Shallow Forward Descent in Nil Wind

If the helicopter now commences a slight vertical descent, it will experience airflow from beneath the rotor disc due to its downward movement. This airflow is referred to as the Rate of Descent (ROD) airflow and will reduce the induced flow.

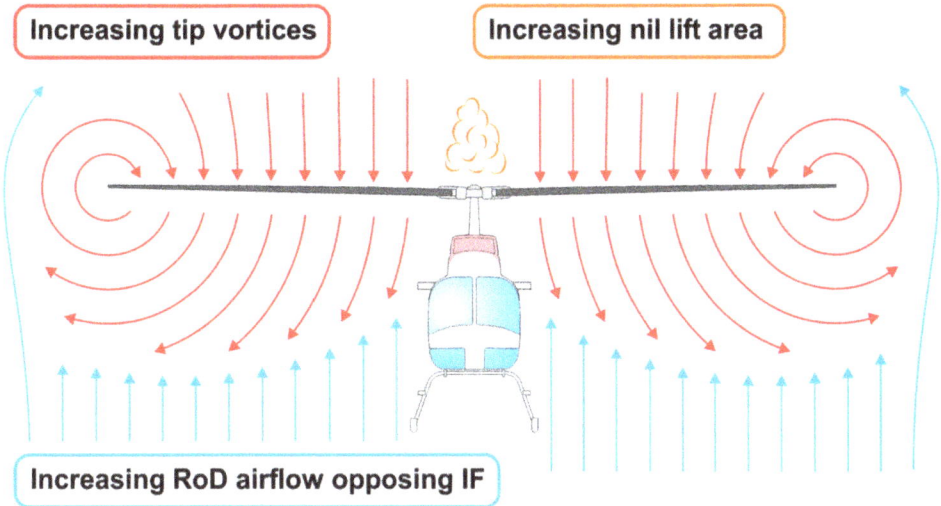

This opposing airflow (whether due to a rate of descent or other cause, such as an updraft when hovering close to a mountain) will affect the rotor blade's angle of attack along the length of the blade because it will oppose and subsequently reduce the induced flow. This variation in the angle of attack will be due to the:

- blades velocity (high at the tip and low at the root end)
- pitch angle of the blade (which is decreasing towards the tip due to washout)
- collective setting by the pilot, and
- reduction of induced flow being produced.

Root end

At the blade's root, the descent airflow rate will oppose the induced flow and further increase the angle of attack. This will effectively increase the small stalled or no lift area, reducing the remaining blade area available for positive lift.

Mid-section

The mid-section of the rotor blade is still trying to produce enough positive lift to oppose weight; however, because the disc area doing this is now reducing the total amount of:

- lift being produced will be reduced, and therefore
- rotor thrust opposing this weight will also be reduced.

Tip section

At the tips, the rate of descent airflow causes a venturi effect for the air that is already trying to move from the high-pressure area under the disc to the low-pressure area above it. This accelerates the tip vortices, making them stronger and moving them inboard on the blade.

Vector diagrams

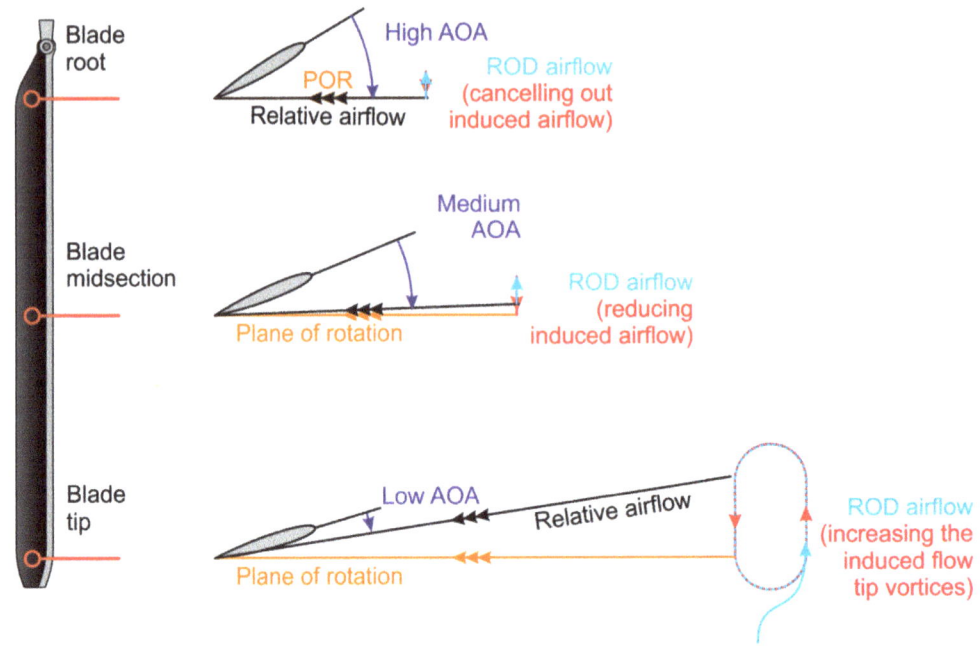

Note: AOA = Angle of Attack

Lift envelope

If you compare the lift envelope to the induced flow envelope, the rate of descent airflow begins to affect the total lift (and, therefore, rotor thrust) produced by the disc.

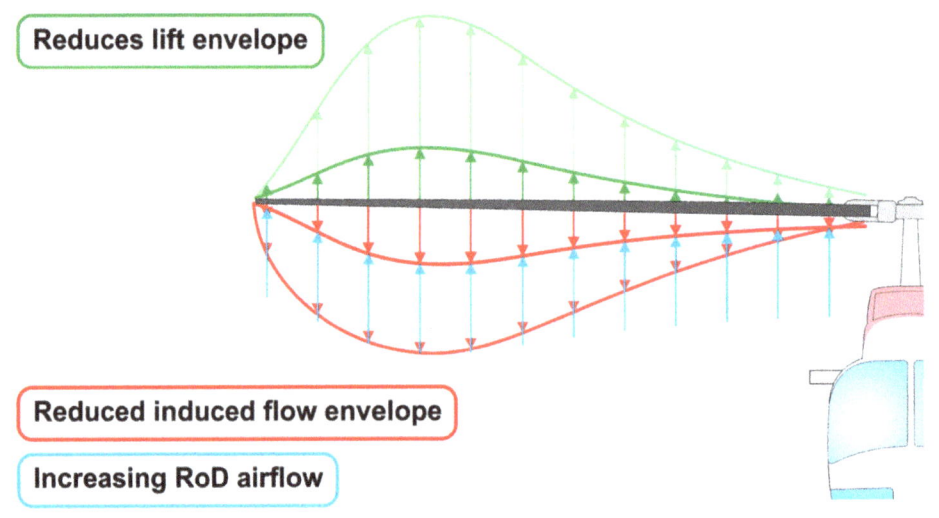

Power Settling

At this stage, the helicopter is not in a Vortex Ring State, but it is Settling with Power and is beginning to descend into its own downwash. This will be noticeable with the pilot feeling a slight vertical vibration, some pitch and roll to no set pattern, and the tail will start to feel sluggish or 'loose' and slow to respond.

It is important to note here that if the pilot decides to pull in more power and increase the induced flow, it may be sufficient to overcome the rate of descent airflow and power out of the descent or power out of the Settling with Power situation.

The aerodynamics described above is typical of a normal descent and are usually quite controllable with power at this point. Any positive airspeed (whether due to forward movement relative to the wind or just a headwind) will reduce the effects of the rate of descent airflow.

Forces on the Blade: Steep Forward Descent in Nil Wind

If the rate of descent is allowed to increase further, the stalled or no lift area will increase and move outboard, and the tip vortices will become larger and move inboard. Also, and most importantly, there will come the point where the mid-section of the blade will have an induced flow going down through the disc and a rate of descent airflow moving up through the disc right beside each other.

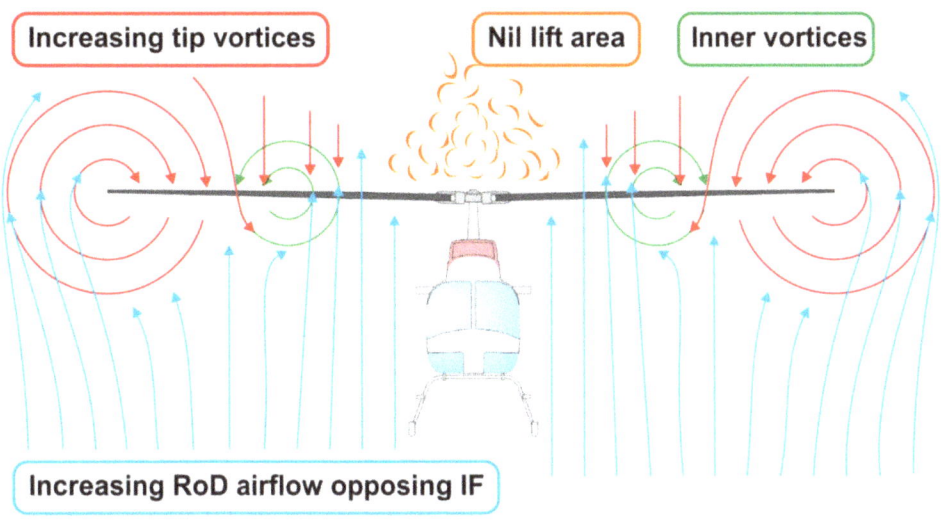

Secondary Vortex or Recirculation This produces a secondary vortex or recirculation in addition to the normal tip vortices. The secondary vortex ring is generated about a point on the blade where airflow changes from up to down. The result is an unsteady turbulent flow over a large area of the disc, which causes a sudden loss of rotor efficiency even though power is still on, and the Rotor RPM is normal. This results in a sudden and large sink and an accelerating descent rate.

Vortex Ring State The helicopter has now entered a **fully developed Vortex Ring State** (VRS) much more severe than Settling with Power.

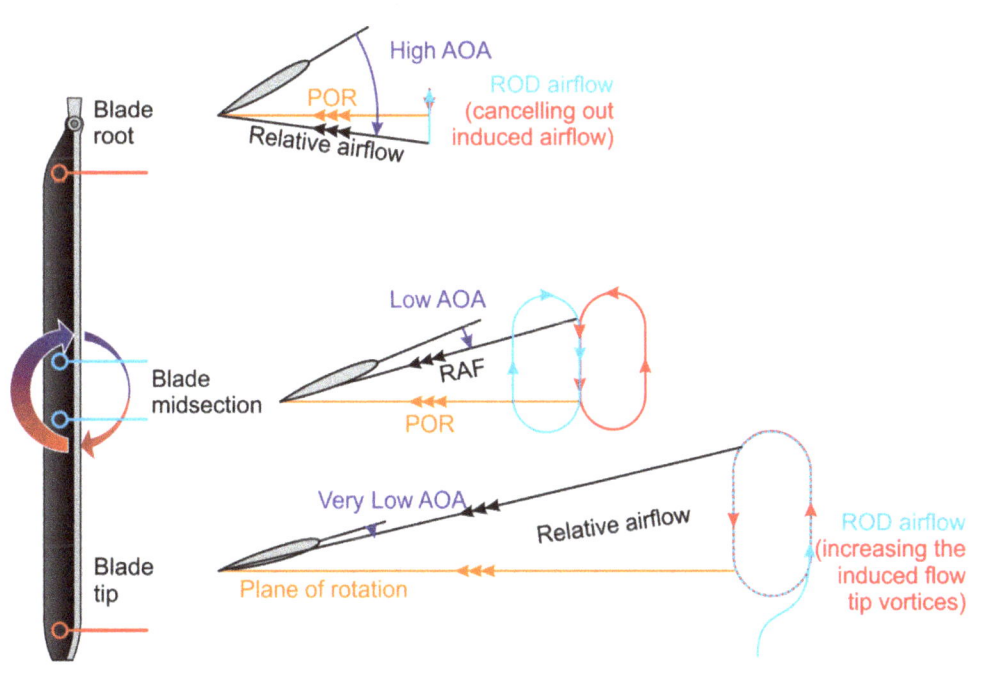

Lift Envelope

If we again look at the lift envelope, it is obvious that the helicopter is now descending with insufficient rotor thrust being produced to oppose weight.

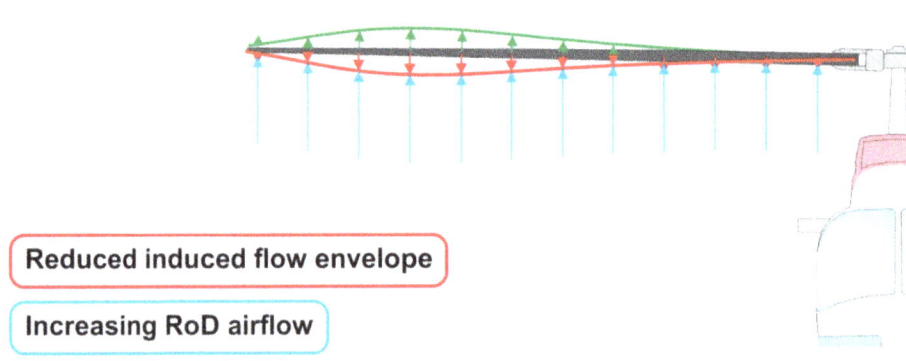

A helicopter in a fully developed Vortex Ring State is descending into a falling column of air. The air outside of this column is unaffected by the helicopter's rotor blades.

Normal recovery techniques will be ineffective such as raising the collective.

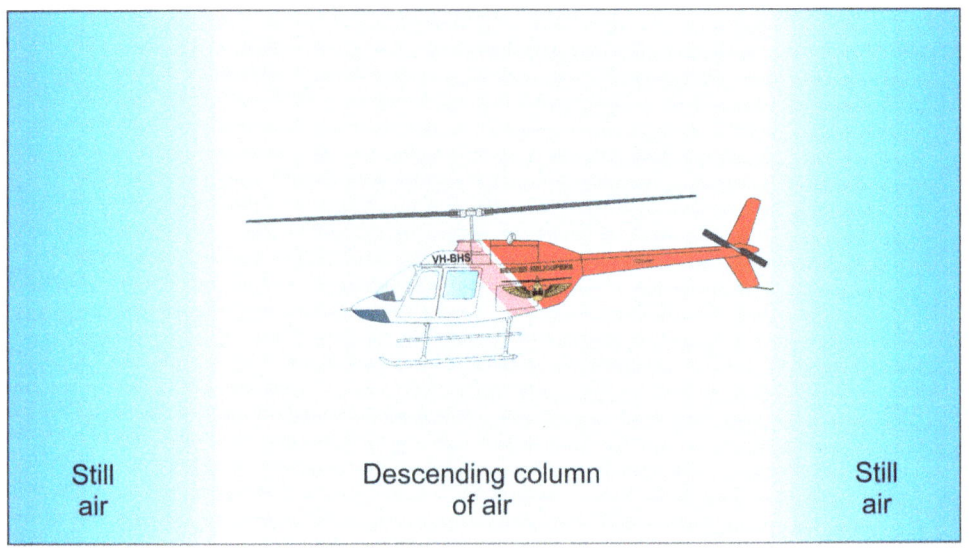

Effect on the Tail Rotor

Another important factor to consider during VRS is the effect on the tail rotor thrust and why the tail becomes "loose" during a low airspeed descent.

Most (not all) manufacturers design the tail rotor so that it rotates with the inboard tail rotor blade advancing up into the descending downwash of the main rotor. This assists the tail rotor in being more efficient by increasing the relative wind passing over it as it turns at the hover.

Chapter 11 Confined Areas

Hovering in nil wind

For example, if the helicopter is hovering in nil wind, the induced flow will be almost vertically downwards, with the inboard tail rotor blade moving up into the downwash.

Moving forward

Once the helicopter starts to move forward and the main rotor disc passes through translational lift, the tail rotor also passes through translational lift, and the whole tail rotor system becomes more efficient like the main rotor system.

Vertical fin

Additionally, the vertical fin can now act as a wing, producing positive lift and unloading the tail rotor system. This means the pilot will be required to put in:

- right pedal in conventional anticlockwise rotating main rotor systems, and
- left pedal in unconventional clockwise rotating main rotor systems.

This is done to remove the need for tail rotor thrust to keep straight as the weather cocking or "streamlining" of the fuselage and the positive lift produced by the vertical fin is now doing all of the work previously done by the tail rotor system.

The advantage of this is extra power from the engine that is now available to the pilot and can be used to increase pitch on the main rotor system and use the extra rotor thrust to climb. All this is normal and happens every time we take-off.

On approach and transition to the hover

Now, consider a helicopter in nil wind transitioning back from forward flight back to the hover. Because the helicopter is at a very low airspeed and transitioning back to the hover, the tail rotor will lose the benefits of translational lift, the vertical fin will no longer be producing positive lift, and the fuselage will lose its streamlining effect.

As the tail rotor is in a transition between losing the effects of translational lift and not yet gaining any benefits from the downwash, it will not be as responsive or efficient as required.

Therefore, the helicopter will naturally want to yaw in the direction torque is pulling it, so to prevent this, the pilot will need to input some more pedal to increase the tail rotor pitch to keep the helicopter straight. This is normal and, in normal circumstances, can be done and controlled smoothly, even though the amount of pedal required can sometimes be relatively large.

Inadvertent entry into VRS

A sudden, large increase in the pitch on the tail rotor will require more power from the engine, and this extra demand can result in a decay (droop) in the main Rotor RPM. This droop in Rotor RPM can also lead to an additional increase in the ROD. To prevent this, the pilot will typically pull some more collective.

Depending on how close the pilot is to entering VRS, this increase in collective may either arrest the ROD or suddenly cause the helicopter to sink by inadvertently entering a fully developed VRS. The inadvertent entry into VRS can be solely dependent on where the wind is coming from.

The Effect of a Tail Wind

If the helicopter encounters a small tailwind while the helicopter is at a very low airspeed while in descent, this will influence two things:

1. the helicopter will now start to descend into its own downwash, and
2. the tail will act more like a weather vane trying to push the tail away from the wind.

This will make the tail "waggle" from side to side. For the pilot, the helicopter will feel "loose" as the wind affects the tail, and the pilot tries to correct it by using and sometimes overusing the pedals. This tells the pilot that the helicopter has a tailwind and should immediately abort the landing and effect a go-around.

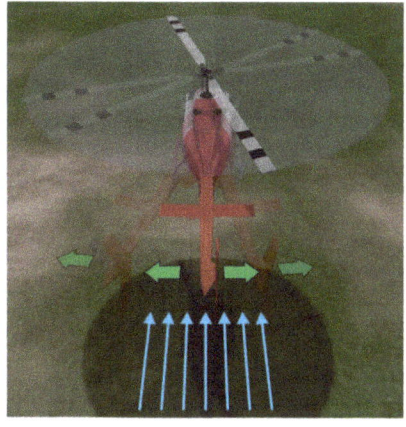

Tailwind

Having a tailwind makes the helicopter extremely susceptible to entering a Power Settling or a fully developed VRS condition instantly. While an experienced pilot may be able to read all of these signs and still be able to fly the helicopter to the ground, the inexperienced pilot is setting themselves up for a hard landing.

Preventative measure

This is why it is vital the pilot always:

- ensures any approach is into a headwind, and
- maintains directional control of the helicopter and does not allow the helicopter to yaw unintentionally with changes in power or wind effect. The large pedal inputs and high-power requirements to fix it could help cause the helicopter to suddenly enter a VRS condition.

Effect of Collective during VRS

What the pilot does with the collective is also very important in minimising any height loss during a VRS condition. Regardless of where the collective was when entering the VRS, the pilot will probably automatically start raising it to reduce the ROD. This is a natural reaction for the helicopter pilot in controlling an unwanted ROD. If in the early stages of settling with power, the pilot may be able to power out of it and maintain the altitude or climb.

If the settling with power has advanced, the increase in collective could worsen the situation, and the helicopter could enter a fully developed VRS and the ROD suddenly increases. Often the inexperienced pilot is unaware of how close the helicopter is to entering a VRS, so the right decision may not be clear.

Over-pitching If the helicopter suddenly sinks, it is not uncommon for the pilot to continue pulling the collective in the vain hope of powering out of the VRS and avoiding hitting the ground. However, the reality is the pilot could induce an "*over-pitching*" situation causing the Rotor RPM to decay, again making the whole situation worse by bringing in another problem (over-pitching) where there is no longer enough engine power to overcome the increasing drag on the main rotor system.

Chapter 11 Confined Areas

Conditions Which Produce Power Settling and VRS

The conditions initially required for power settling to occur, leading to a fully developed Vortex Ring State, include:

- **low forward indicated airspeed** (in the vicinity of 5-10 kts or less, in fact in most cases, it requires a tailwind with nil IAS),
- **an airflow directly opposing the induced flow** (usually a rate of descent of more than 300 ft per minute), and
- **the use of some power**, giving an induced flow passing through the disc.

To enter power settling, the above conditions must all be present to some degree; the amount of each can vary depending on the helicopter type. Some helicopters enter power settling more easily than others; just as fixed wing stalling characteristics vary from type to type.

Generally, two-bladed, high-inertia helicopters are reasonably stable in a VRS condition compared to a multi-bladed, low-inertia system which can be much more volatile.

From experience, this author has found that VRS tends to occur in stages.

Initial Sink

The initial sink rate commences with a tailwind at medium power ranges (18-21" MAP for piston engines or 50-75% torque for turbine engines) and a rate of descent of 500 ft per minute. At this point, the helicopter feels loose, the controls feel sloppy, the tail becomes less efficient, and the helicopter wants to pitch and roll to no set pattern.

Accelerated Sink

If allowed to continue and the rate of descent gets to 1000 ft per minute, the helicopter experiences an **accelerated sink** (which is very noticeable to the pilot), and the rate of descent may then increase rapidly to over 3000 ft per minute (that is as high as the VSI we have in our helicopter goes).

Recovery from Vortex Ring State

In the early or developed stages of power settling, recovery may be made by removing any of the three factors, which make up the condition, that is:

- power
- low airspeed, or
- rate of descent.

There are two possible recovery techniques, depending on the height above the ground at the time of entry into VRS.

Option 1: Enter Autorotation

Although not usually an option, the pilot could enter autorotation, stopping any further conflict between descending and ascending airflows.

If the helicopter has altitude, then this recovery action will result in exiting the vortex ring state, but the helicopter will still be experiencing a very high ROD, and the pilot will then need to affect a power recovery back to normal flight from the autorotation. This is not a great recovery plan, but it is an option.

Option 2: Use Cyclic	The most common recovery option is to get back into clean air, and this is done by moving the cyclic and controlling the yaw of the helicopter to move the helicopter out of the descending column of air and use the relative wind to change the airflows over the disc. This effectively wipes out the vortices, restoring the normal flow of air through the rotor disc and onto the tail.
Use of collective	The collective is also an essential part of the recovery, but whether it is raised or lowered will depend on where it is at the time. The secret to the collective is to **set climb power** (not full power, but the power setting used for a steady climb). So, if the collective is: ■ low, raise it to climb power and leave it there. ■ high, and the possibility of an overpitch exists, lower the collective to restore the Rotor RPM and then set climb power as soon as possible.
Use of cyclic	**Moving into clean air is done by** moving the cyclic **forward** or **sideways**, thereby changing the main rotor disc orientation and allowing the helicopter to fly out of the affected area.
Disadvantage of forward cyclic	The disadvantage of using forward cyclic is that the helicopter is already moving forward with the moving column of descending air, so it may take longer, and more height may be lost in effecting a recovery.
Lateral cyclic	Some experienced pilots, especially those operating with a sling load, prefer to use lateral cyclic and move sideways. This also puts the helicopter into clean air and allows at least a partial recovery very quickly so that when the sling load arrives at the ground, and there is a sudden reduction in aircraft weight, the collective, along with entering ground cushion, can produce enough thrust to reduce the rate of descent and cushion the landing.

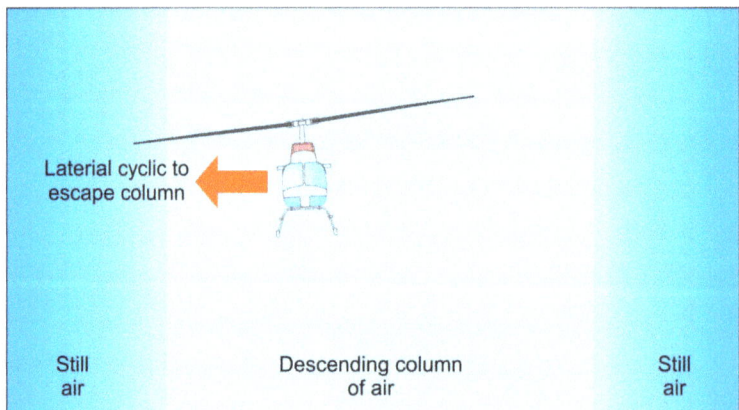

The use of lateral cyclic to affect a quicker recovery is now referred to as the **"Vuichard method"**.

Chapter 11 Confined Areas

Advantage of lateral cyclic

The advantage of moving **laterally** is that the helicopter will move into clean air sooner, and less height will be lost in the recovery.

Which one you choose to use (forward or lateral recovery) will depend on your experience, the conditions (such as current height and speed, etc.) and the environment (close to trees, buildings or the ground, etc.).

In anticlockwise rotating main rotor systems, it is always best to use right cyclic while at the same time maintaining directional control with the tail, usually with left pedal. In this manner, tail rotor thrust will also help the helicopter move to the right and again assist in a quicker recovery by moving smoothly into clean air.

In clockwise rotating main rotor systems, the opposite will apply.

How Does Wind Affect VRS

Into wind

Let's return to the helicopter descending in a descending column of air. If the helicopter is moving forward INTO the wind, then this column of air is always moving away from the helicopter. In other words, the helicopter is always moving into clean air, and a power-settling situation followed by a VRS is less likely to occur.

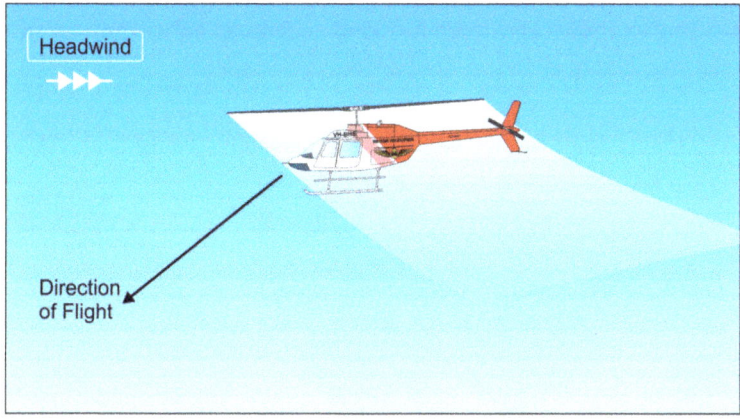

Tailwind

If the helicopter is moving forward with a tailwind, then the descending column of air will remain with the helicopter, and the helicopter's downwash may make the situation worse by pushing and accelerating the air in front of it. This is when power settling and VRS are most likely to occur.

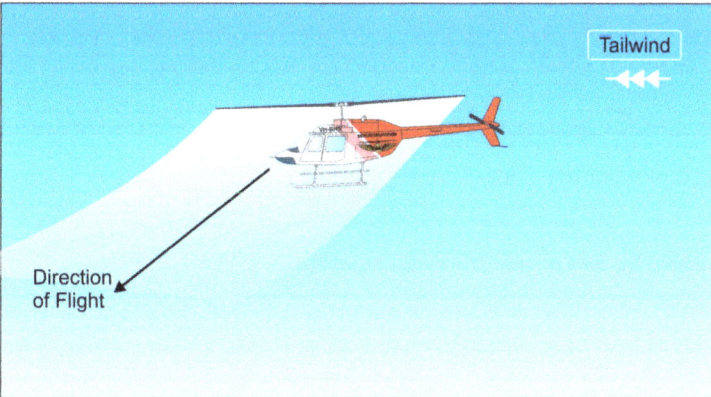

The two most common flight scenarios for a helicopter pilot to inadvertently enter a power settling and VRS situation are:

- in light and variable winds, and
- at a high hover.

Approach in light and variable wind

When the wind is light and variable, at a high all-up weight, the wind may change direction at the wrong time, or the pilot may choose the wrong way in (because it can be hard to determine the wind when it is light and variable) and therefore, inadvertently place the helicopter into a power settling or VRS situation. This is why early detection and recovery is important.

Wind from a steady direction

If the wind is from a steady direction, whether weak or strong, then the pilot is less likely to be making a downwind approach, and the chances of getting into power settling and VRS are much reduced. (Unless, of course, the pilot makes a mistake and unwittingly chooses to approach downwind!)

Short Finals checks

Completing the Short Finals Check is designed to help recognise the symptoms leading to VRS early. (The Short Finals Check is explained in the Circuits chapter.)

The finals check includes:

1. Airspeed over ground speed
2. ROD is under control, and
3. Power in hand.

If any of the above is not where the pilot expects it to be, the pilot should abort the landing and conduct a go-around early.

Summary

Every approach or high hover has the potential to lead to a Power Settling and to a full VRS condition. Pilot awareness of the situation, surroundings, demands for power, wind and familiarity with the aircraft can all contribute to good techniques, which reduce the risk or aid in early recognition and recovery from VRS.

Maneuvering in a Confined Area

Introduction

Once in a confined area, the helicopter may have to be manoeuvred. This may be to:

- find the best landing site
- avoid obstacles, or
- position the helicopter for the take-off giving the shallowest take-off path to avoid obstacles.

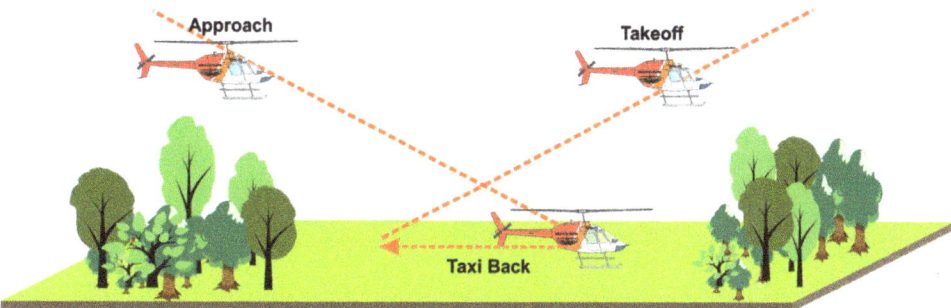

Manoeuvring in a confined area requires that the pilot:

- looks after the tail rotor, as the pilot cannot directly see it
- maintains the rotor blades a safe distance from obstacles, and
- minimises the effects of the downwash on the surface to prevent recirculation of debris through the rotor system

This manoeuvring requires a combination of turning, hover taxiing, sliding sideways and adjusting the hover height.

Turns

Each confined area is different, but the technique is the same. Hover turns are made into areas the pilot or crew can see, and the tail is kept in the middle of the pad, away from the major obstacles.

The pilot cannot directly see the tail, so care must be taken to protect it. The pilot can see the tips of the main rotor blades, so it is easier to judge the distance from any obstacles to the main rotor blades.

Conducting turns about the tail and taxiing forward is the safest technique in a confined area.

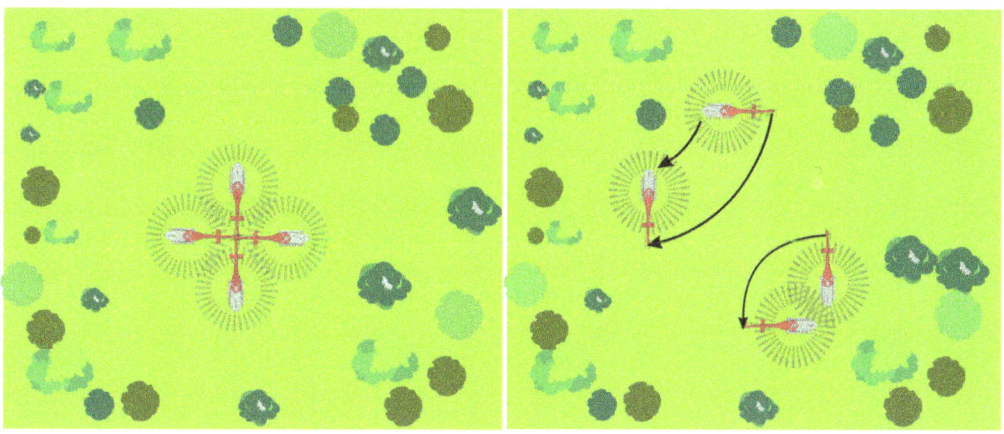

Hover height	If operating in a confined area and sufficient power is available, hover at 10 ft AGL. The extra hover height will allow:

This is a helicopter in a high hover, staying clear above the debris blown around by the downwash.

- extra distance from any hidden or unnoticed obstacles on the surface, particularly under the helicopter.
- reduce the recirculation effects of the main rotor downwash so less debris will be circulated through the rotor blades, and
- give the pilot and crew a better observation point to manoeuvre and select the landing site.

Trees When manoeuvring in a confined area surrounded by trees, be aware of overhanging branches and the possibility that the helicopter's downwash may cause branches to break or dead trees to collapse.

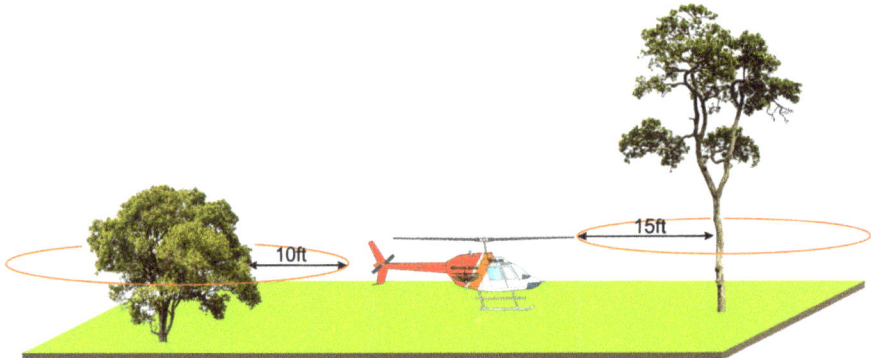

When hovering IGE, the helicopter's main rotor blades are closer to the base of the tree, but big trees branch out with height.

It is easy for a pilot not to realise that the helicopter rotor blades are under overhanging branches which can fall onto the helicopter, or the pilot climbs vertically, and the main rotor blades inadvertently strike the branches.

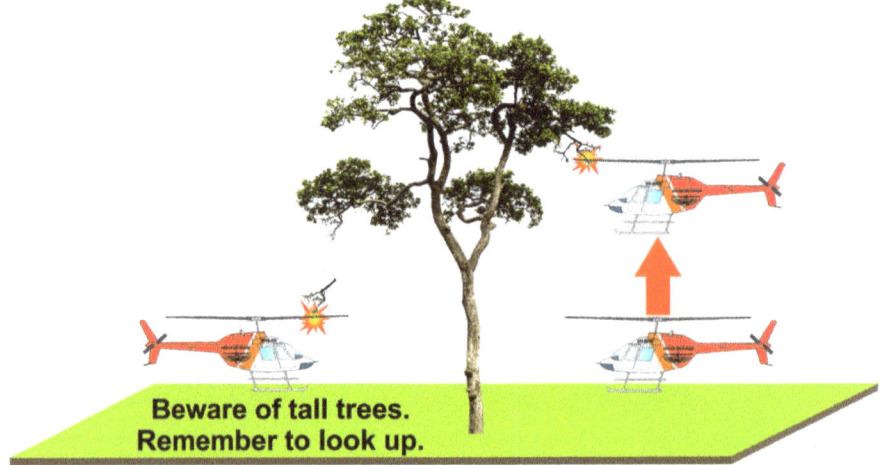

To avoid this, the pilot and crew need to have a good lookout, especially above, and communicate with each other.

---

[4] Marines from Arlington, VA, United States [Public domain]

Recirculation

Recirculation When hovering near the ground, some air passing through the disc is re-circulated, increasing in velocity as it passes through the disc for the second time. The local increase in induced flow near the tips causes a loss of rotor thrust. Some recirculation is always taking place, but over a flat, even surface, the loss of rotor thrust is more than compensated for by ground cushion.

Recirculation will increase when any obstruction on the surface or near where the helicopter is hovering prevents the air from flowing away—for example, hangers, tall bushes, trees, tall buildings, etc.

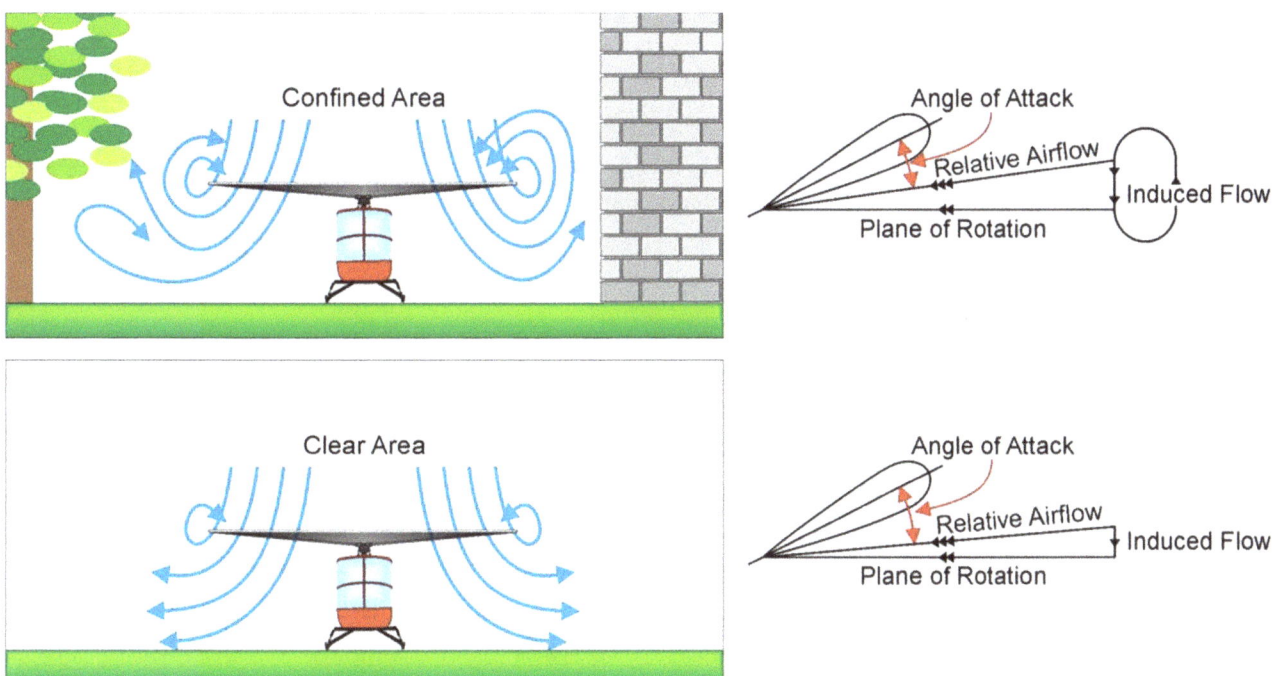

The effect of recirculation is a loss of lift at the tips, therefore if you are hovering next to, say, a hanger, or a cliff face or similar, where only one side of the disc experiences recirculation, the helicopter will tend to drift or appear to be pulled towards the object causing the recirculation. This is undesirable, and the pilot must maintain a good awareness to recognise recirculation early and take measures to avoid it.

The PSWATP

Introduction

During a recce, the pilot and crew need to consider many different things. To help remember what to look for, the acronym **PSWATP** is used.

There is a lot of information in the **PSWATP,** and it is unreasonable to think that the pilot and crew can articulate every single item every single time.

Instead, the **PSWATP** is used as a memory jogger so that the pilot and crew consider the relevant items, which may vary on any given day. As a crew, it is not necessary to tick off each item individually but be continuously aware of all the factors as the recce unfolds. Much of the **PSWATP** can be conducted during the planning stage or while enroute to the confined area.

Pilots under training

For a pilot under training who has never done a confined area before, the instructor will go into a lot of detail explaining every item in the **PSWATP** check.

This may require four (4) or more high-recce circuits to allow time for the information to be discussed. As the student pilot gains more experience with the items in the **PSWATP**, detail can be determined more quickly to the point that a quick glance at the confined area can cover multiple items of the **PSWATP** without the crew having to articulate each one.

By the end of the second confined area lesson, the goal is for the student to conduct one high recce and then make an approach.

Only those items relevant for the day need to be discussed by the crew, emphasising the overall **PLAN,** not necessarily the entire detail in the **PSWATP**.

PSWATP

The PSWATP check is broken down into the following elements:

P	POWER, PILOT, PAYLOAD
S	SIZE, SHAPE, SLOPE, SURFACE, SUN/MOON, SHADOWS, SURROUNDS, STOCK
W	WIND, WIRES-WIRES-WIRES, WAY IN, WAY OUT
A	APPROVAL, APPROACH, ABORT
T	TURNING, TERMINATION POINT AND THREATS
P	**PLAN**

P POWER, PILOT, PAYLOAD

Power

Power margin available

The Power Available check may be conducted enroute with a quick confirmation on arrival.

Power available check: During the orbit of the confined area and while assessing all other aspects, pull in full power available and note what was achieved.

Power required. On lift-off, the pilot will have carried out a hover check, including a check of "power available at the hover". This information should have been noted, so now you can estimate the power you are likely to require when coming into the hover in the confined area (assuming similar conditions). If the conditions are different, the pilot will need to take this into account. From this, the pilot can make a judgement.

Power margin in excess of that required to hover IGE				
				CAT4 4" MAP 16-20%Tq **Options**
			CAT3 3" MAP 11-15%Tq **Options**	Vertical
		CAT2 2" MAP 6-10%Tq **Options**	Constant Angle or Towering Steep	Constant Angle or Towering Steep
	CAT1 1" MAP 0-5%Tq **Options**	Normal	Normal	Normal
	Cushion Creep Shallow Running	Cushion Creep Shallow Running	Cushion Creep Shallow Running	Cushion Creep Shallow Running
Limited Power Operation				

Pilot

Pilot's capability

The pilot and crew are to discuss and determine if the confined area is within their current capabilities and experience to use. It is important that a pilot and crew know their own personal limits based on their experience.

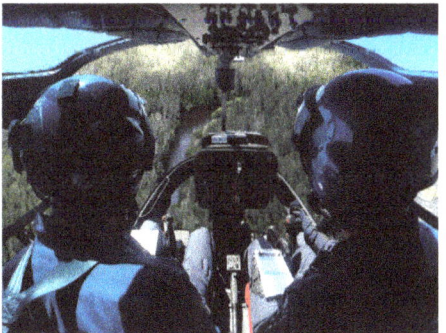

This may be discussed enroute with a quick confirmation on arrival.

Payload

Is the weight going to change?

Consider changes to the payload (passengers and/or cargo) and how this may affect the approach and departure performance. The crew will consider if the helicopter is coming into the confined area is:

- **full** (performance decreases), but are departing empty (performance increases);

OR

- **empty** (performance increases), but departing full (performance decreases);

OR

- **with no change** to the payload (no change in the performance).

This may be discussed enroute with a quick confirmation on arrival.

S SIZE, SHAPE, SLOPE, SURFACE, SUN/MOON, SHADOWS, SURROUNDS, STOCK

Size

Is the area big enough to get into and big enough to get out of

The pilot needs to consider if the confined area is big enough to get into and also big enough to get out of. This is based on the **power** category, the **physical size** of the area to accommodate the helicopter and allow it to manoeuvre, and the **experience** of the pilot. The size of the confined area may be discussed enroute, but it will be essential to make the final decision on arrival.

Shape

Shape of the area

Does the shape of the confined area lend itself to a particular approach and departure path? Is it long and skinny? Does it have a bend in it? Is it an open area that is circular or square? Can you imagine what the area may look like on final approach?

Whatever the shape of the confined area, the pilot and crew need to consider how this will affect the approach and departure path. The shape of the confined area may be discussed enroute, but it will be essential to make the final decision on arrival.

Chapter 11 Confined Areas

Slope

Is there any evidence of a slope?

This can be very difficult to observe while conducting the recce at 500ft, so it is more likely to be considered on very short finals or after terminating at the hover in the confined area.

Generally, consider **every** landing and lift-off in a confined area as a slope landing and lift-off.

If this confined area is:

- **new**, where the slope and surface is **unknown**, only go in if there is a good power margin as this will allow the pilot to manoeuvre.
- **known**, where the slope and surface have been recceed, then limited power operations may be conducted at the discretion of the crew.

Sometimes the slope can be hidden by what is on the surface, such as long grass or snow. This will require the pilot to conduct the landing without knowing what the slope may actually be. Although slope can be discussed enroute or during the recce, it can only be confirmed once in the confined area.

Surface

Surface type

The pilot needs to consider what type of surface is within the confined area, how it will affect helicopter performance, and the ability to hover and land without causing any damage to the helicopter or the surrounding areas.

For example:

- Hovering over long grass can hide solid obstacles that may damage the helicopter.
- Trying to land on a rocky surface can be risky as the skids slide on the rocks.
- Hovering over sand or snow can reduce visibility.

Grass	Rocks	Sand

This may be discussed enroute with a quick confirmation on arrival.

---

[5] http://mauinow.com/files/2014/06/waihee-helicopter-hard-landing-photo-by-wendy-osher.jpg
[6] http://weaselzippers.us/wp-content/uploads/248905_10151341538548092_1695922555_n-550x412.jpg
[7] Sgt Tammy Hineline [Public domain]

Sun (Moon)

Visibility

During the day, the sun's position relative to the approach and take-off path is to be considered. During night-time operations, then this consideration relates to the moon. If the sun or moon is very low on the horizon and the approach and take-off path is directly into it, it can be difficult to see and avoid any obstacles.

If the sun is very low on the horizon, select approach and departure paths with the sun off to one side. This may be discussed during the planning stage or enroute with a quick confirmation on arrival.

Bad Option

Better Option

Shadows

Visibility

Both the sun and the moon can create shadows. This is most obvious when operating in mountainous terrain. If flying on the sunny side of a valley and trying to look into the side in shadow, it can be impossible to see into it at times.

Unless the area has already been surveyed, an area of shadow is considered an obstacle. To avoid the shadow, either come in lower or enter the shadow area from an angle that will guarantee ground clearance.

Once inside the area of shadow, it again becomes easier to see.

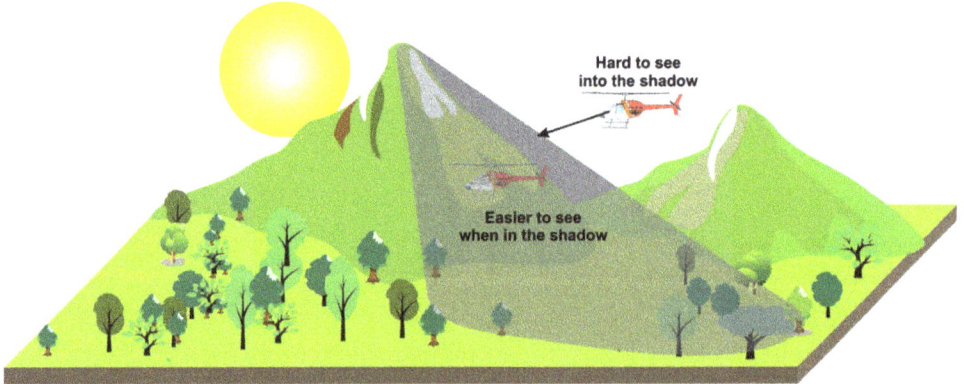

This may be discussed during the planning stage or enroute with a quick confirmation on arrival.

Chapter 11 Confined Areas

Surrounds

Obstacles and hazards

Consider what is surrounding the confined area that may be affected by the helicopter's downwash, which can create a threat or a hazard to the helicopter or those on the ground.

For example:

Trees, buildings, vehicles, other helicopters, tents, debris, fences, open hanger doors, refuelling equipment and people are all examples of considering the surroundings of a confined area.

This may be discussed during the planning stage or enroute with a quick confirmation on arrival.

Stock

Livestock and wild animals including birds

Animals generally do not like helicopter noise or what appears to be a huge bird coming to attack them from the sky. For this reason, stock (animals) must be considered when assessing a confined area.

The following applies if there is stock in the vicinity of the confined area:

- Do not fly directly over the top of stock, as they will scatter in all directions.
- If possible, land downwind of the stock so that the wind blows any noise or dust away from the stock.
- If possible, in the planning stages - when getting permission from the landowner - ask if they have any stock in the area that may be affected and ask for them to be secured or moved.

This may be discussed during the planning stage or enroute with a quick confirmation on arrival.

---

[8] http://www.bestgore.com/wp-content/uploads/2013/05/british-soldier-beheading-woolwich-uk-03.jpg
[9] http://www.centralstation.net.au/wp-content/uploads/2014/06/1.3.png

W WIND, WIRES-WIRES-WIRES, WAY IN, WAY OUT

Wind

Wind direction and strength

Determine or confirm the wind direction and strength. It is always preferred to approach and depart a confined area with the wind on the nose or, at the very least, with a headwind component within 10° either side of the nose.

Because confined areas come in all shapes and sizes, the best approach and departure path may not allow for the wind directly on the nose; instead, the wind may have to be coming in at a slight angle. This is acceptable within certain limits as follows:

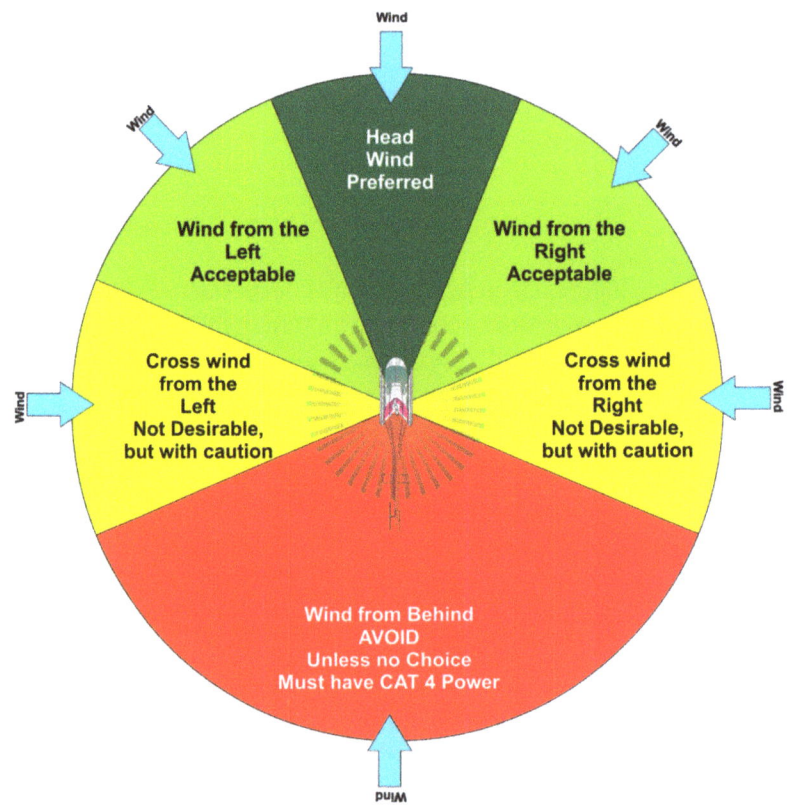

Ways to determine wind direction and strength

The following information can be used to help determine the wind direction and strength.

Item	Description
Weather forecast	Obtain a weather forecast in the planning stages and relate it to the confined area operations. This will indicate the prevailing wind for the day.
ATIS / AWIS	If available and applicable to the confined area, listen to the ATIS or a local AWIS.
ATC	Advice from the tower: Air Traffic Control.
Another pilot	If there are other helicopters in the area, ask another pilot on the radio for any relevant wind information.
Ground report	If operating from a remote base, there is often ground support who can indicate the wind on the radio.

Chapter 11 Confined Areas

Windsock

There are manmade and natural windsocks that indicate to the pilot and crew where the wind is coming from and indicate its strength. These wind indicators are the best way to determine the wind when operating away from an aerodrome into a confined area, and they include:

Aerodrome windsock

Bush windsock

Tying a piece of tape or a plastic bag to a tree or bush is a quick way for a pilot to create a "bush" windsock. This is useful if going into the same confined area on multiple occasions.

Smoke

Smoke from fires creates a natural windsock.

Dust

Dust created by cars or trucks on dirt roads shows up very well and acts as natural windsocks. In the image below, the wind is blowing from right to left.

Water

When there is no wind, the water will look glassy and smooth. The stronger the wind, the rougher the surface of the water gets.

To determine the direction, look for an area of smooth water by the edge of the water source. It is smooth because the edge of the bank protects it. The wind will, therefore, be moving from smooth water towards rough water.

---

[10] © Copyright Mick Garratt and licensed for reuse (CC BY-SA 2.0)

[11] http://www.curiousanimal.com/wp-content/uploads/2014/10/15Day-3-Road-Train-travelling-along-the-Fairfield-Leopold-Road-Western-Australia-DSC_0033.jpg

Trees and vegetation

Wind will blow leaves and branches back, usually exposing the underside of the leaves, which look much lighter in colour compared to the top of the leaves.

Other

The pilot and crew can use anything that indicates wind direction and strength. Clothes on a clothesline are a good example.

Drift

While enroute, it may be noticeable due to the difference in track and heading just where the wind is coming from.

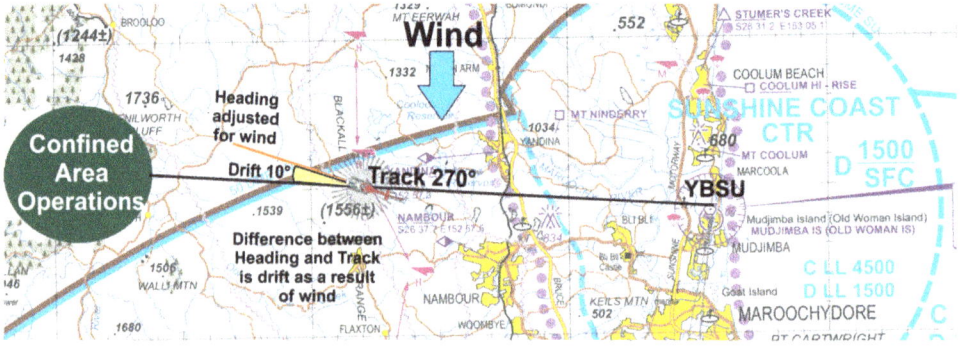

Drift is a good indication of the prevailing wind.

Groundspeed vs Airspeed

Calculating or observing the difference in airspeed to ground speed indicates where the wind is coming from.

Airspeed	Groundspeed	Wind
High	Low	Headwind
Low	High	Tailwind
Same	Same	No wind or crosswind

---

[12] HoremWeb [CC BY-SA 4.0]

[13] Image by Michael Coghlan / Clothes blowing in the Wind (CC BY-SA 2.0)

The airspeed is indicated on the ASI or the glass cockpit display; the ground speed can be calculated by the pilot or displayed on the GPS.

ASI

ASPEN Tape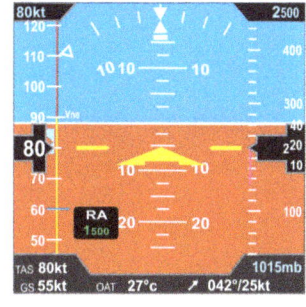

Garmin430 GPS Display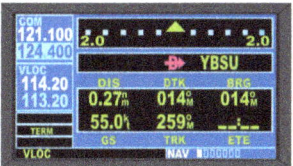

For example:

If the ASI and ASPEN Tape are showing 80 kts and the Ground speed on the GPS is showing 55 kts. The difference is a 25 kts headwind.

Turns

Constant angle and constant radius turns indicate where the wind is coming from, if there are no other obvious signs. These are two different techniques that rely on pilot accuracy to work properly. Although not used often, they are just another tool a pilot can have up their sleeve if needed.

Constant radius turns

Pick a feature and conduct a 360° turn around it at a constant height, speed and distance.

This will require the angle of bank to be increased or decreased as required to maintain a constant distance away from the chosen obstacle. Where the angle of bank is greatest indicates where the wind is coming from.

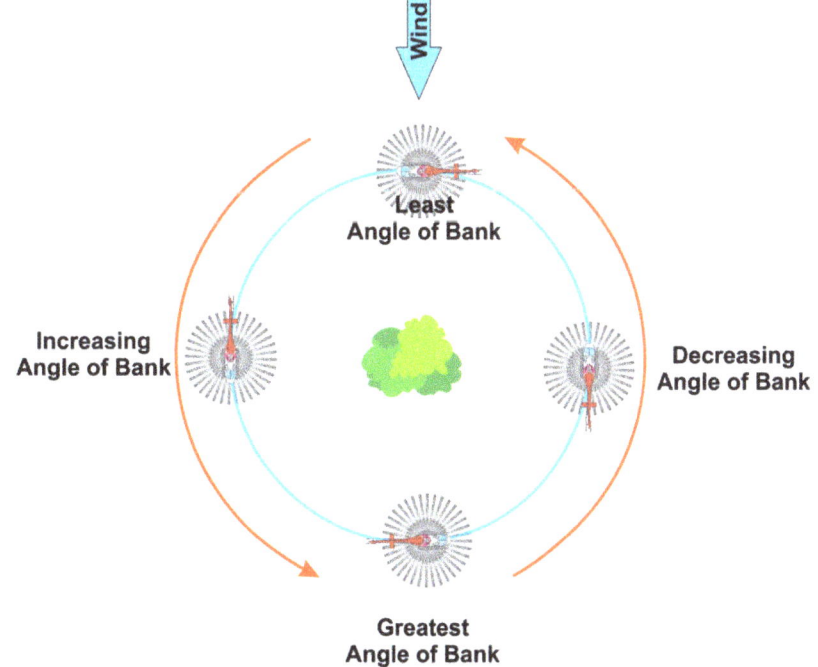

Constant angle of bank turns

Pick a feature and conduct a constant angle of bank turn through 360°. Allow the helicopter to be affected by the wind. On completion of the 360° turn, note where the helicopter is in relation to the chosen feature. Whichever side the helicopter is on indicates where the wind has caused it to drift.

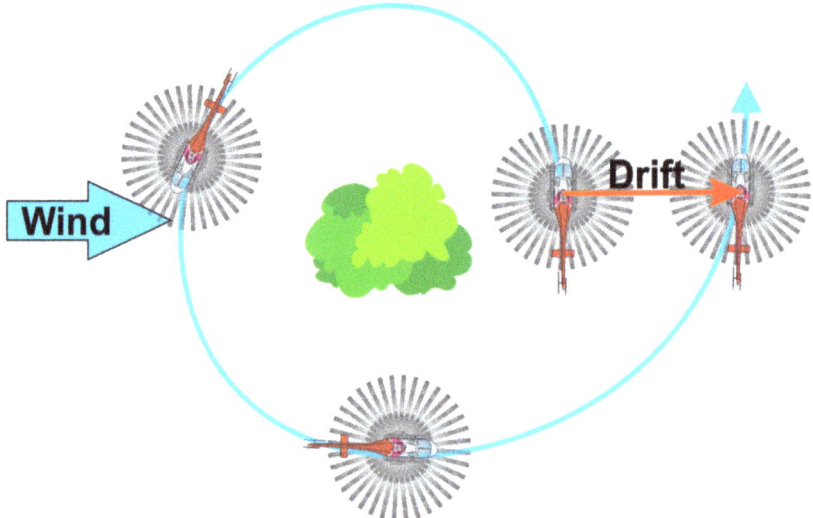

In the example above, the helicopter is conducting a left-hand constant radius turn, and the wind is blowing the helicopter away from the reference point.

The different ways to determine wind direction and strength may be discussed during the planning stage or enroute with a quick confirmation on arrival.

Wind shear

Wind shear is defined as a change in wind speed or direction at a specific location.

Wind shear or wind gradient can be dangerous to helicopter pilots during an approach or take-off. If approaching a confined area with a known wind direction and strength, the power required will be consistent as long as the wind is consistent.

If, for some reason, the wind disappears or changes its direction significantly, this can affect the amount of power the pilot requires and can lead to an over-pitching or VRS situation.

Pilots are to expect the wind to decrease the moment the helicopter drops below the tree line or descends below the height of the obstacles.

Wind turbulence

Wind can become turbulent as you approach a confined area, and the helicopter may experience some downdrafts.

The pilot needs to be vigilant on the final approach and watch for any significant changes or abrupt requirements for more power. Maintaining forward movement to the termination point will help reduce the effects of wind shear.

Wires, wires, wires

Wires, wires wires

Helicopters flying into wires contribute to many helicopter accidents when operating low level. This is because wires can be difficult to see. There are four rules when considering wires:

- All roads, railway lines and water courses have wires.
- All buildings have wires going to them.
- All valleys and rivers have wires strung across them.
- If in doubt, there is a wire there.

The only sure way to know there is no wire in front of you is if you have been through the same area before. Even that can sometimes be misleading, as a wire may have been strung up overnight or since your last visit. Never take wires for granted, and never just assume.

Recognising wires

To identify a wire, do not look for the wire itself but for the evidence that there is a wire there; this includes looking for:

- Roads, railway lines, rivers and water courses
- Buildings
- Valleys
- Cut lines, and
- Poles.

Roads, railway lines and rivers

Assume that all roads and railway lines have wires running parallel to them and that all water courses (rivers) have wires cutting across them until you confirm otherwise.

Road   Railway line   River

Buildings

Assume all buildings have wires running to them. It is often a good bet that if there is a road nearby, there will be a wire running from the road somewhere to the building. Assume wires are running between adjacent buildings.

All towns, villages, factories, mills etc., will have wires running to them and away from them in various directions.

Valleys

If flying inside a valley (below the ridge tops) that you have no local knowledge of, you can only assume there is a wire strung across it somewhere, and you should not be there until you can confirm otherwise.

Cutlines

Look for evidence of vegetation being cut down, showing a clear path across the ground where a wire might be.

---

[14] https://commons.wikimedia.org/wiki/File:Telephone_line_by_road.jpg See page for author [Public domain]

[15] https://pxhere.com/en/photo/998878 (CC0)

[16] © Copyright Hugh Venables / Pylons crossing the lark (CC BY_SA 2.0)

[17] https://pxhere.com/en/photo/1492555 (CC0)

[18] https://c1.staticflickr.com/3/2746/5708010455_804301a76d_z.jpg

Poles

Assume every pole has a wire. Once you have seen a pole, fly over it or at least ensure the helicopter is higher than the pole, as it is unlikely that a wire will be higher than a pole. When you find a pole, look at its cross members and conductors. This is called the "equipment" or the "hardware". If you carefully read the equipment or hardware, you can see where the wire may go.

This may be discussed during the planning stage or enroute with a quick confirmation on arrival.

Way in and Way out

Way in and way out

As the recce is conducted, look for the best way in and out with a headwind component. This may be a compromise between avoiding obstacles, following the natural contours of the terrain, achieving the headwind component within a 45° arc of the nose and the power available to achieve a normal to shallow approach if possible.

As the recce circuit is conducted, look for an approach path where the confined area opens up and exposes an easy path in. It will become apparent when it is seen.

It is not a good approach if the area is hidden behind trees, buildings, wires or other obstacles.

A APPROVAL, APPROACH, ABORT

Approval

Permission to land

Ensure that the land owner has given prior permission to use the confined area. This should be completed during the planning stage before departure.

Approach

Type of approach

Taking into account the power category, the wind, the way in and way out, the pilot's ability, and all the other considerations already discussed above, the pilot will need to nominate an approach and departure type as either:

- Vertical
- Steep
- Constant Angle
- Normal
- Shallow, or
- Double angle.

---

[19] http://blackcombaviation.com/images/subpage_images/hydro.jpg

Abort

The Decision Point

The pilot will nominate a point on the approach and the take-off legs where a *decision* is made to either continue with the approach/take-off or to abort (not do it).

Although we call it the abort point, it is the *"DECISION"* point.

Abort on approach

On approach, the abort point is usually on very short finals just before the helicopter passes the tallest obstacle on approach and before the loss of translational lift.

If the pilot decides to abort the approach for whatever reason, doing so early will give the pilot the necessary time to bring in power and increase speed so the helicopter can fly away without hitting any obstacles.

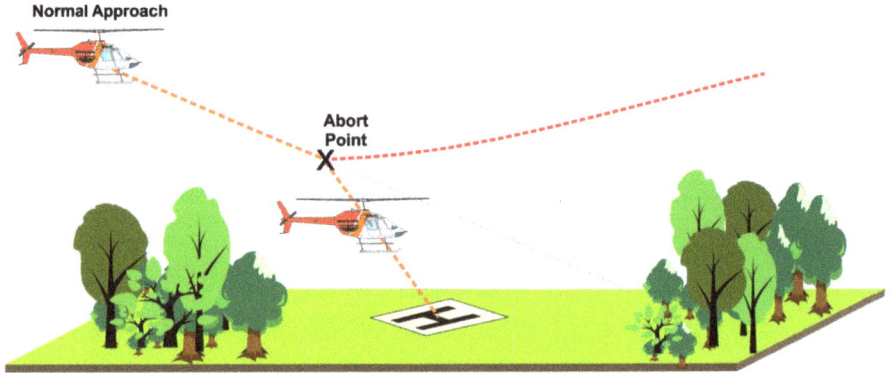

The abort on approach can be summarised as follows:

Item	Description	Action
ASI vs G/S	ASI high, G/S low	Continue
	ASI low, G/S high	Abort – Go around
ROD	Under control	Continue
	Not under control	Abort – Go around
Power in hand	Sufficient power in hand	Continue
	Insufficient power in hand	Abort – Go around

Chapter 11 Confined Areas

Abort on take-off

The abort on take-off is determined by how the helicopter performs and its ability to clear the obstacles in front of it. The pilot and crew will have articulated a PWPTEM take-off plan and nominated a type of take-off. That is vertical, steep, constant angle, normal or shallow.

Based on the wind and obstacles, the pilot will have nominated an abort point based on the power, wind, size of the confined area and identified where the Last Safe Point of Hover (LSPH) is.

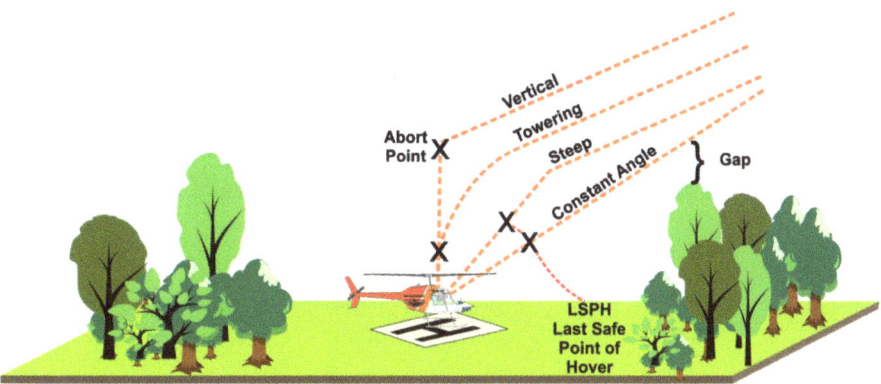

If conducting a vertical steep or constant angle take-off, the pilot is looking for a steady ROC and the ability to clear the obstacles. This is evidenced by the pilot identifying a gap between the tips of the rotor blades and the tops of the obstacles.

The take-off can continue if this gap is achieved and the ROC is maintained.

If at any point the gap cannot be achieved, or the helicopter stops climbing, then the take-off shall be aborted.

Gap with a ROC - GO	No gap and no ROC – NO GO

On take-off, the pilot will nominate a point on the ground: this usually relates to the point where translational lift is expected to be reached or a point that, if conducting a constant angle or steep approach, if it is not progressing as expected, the pilot will have enough space to bring the helicopter back to the hover before hitting any obstacles.

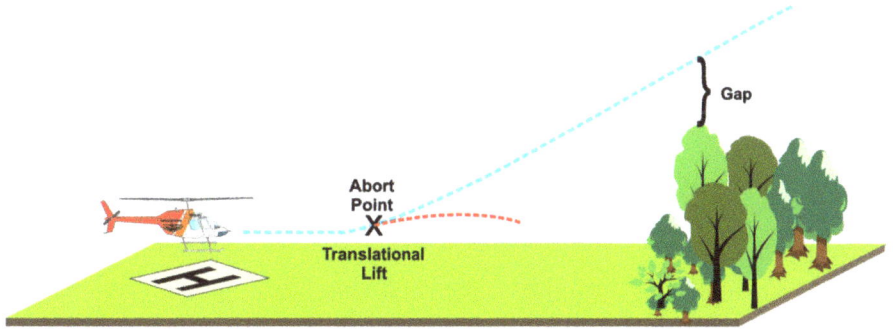

Page 11-41

If the decision to abort is made on take-off, the helicopter will keep moving forward for a short distance before returning to the hover. It is important that when selecting the abort point, the pilot and crew look beyond it and determine where the helicopter can safely come to a hover after passing the abort point. This is referred to as the Last Safe Point of Hover (LSPH).

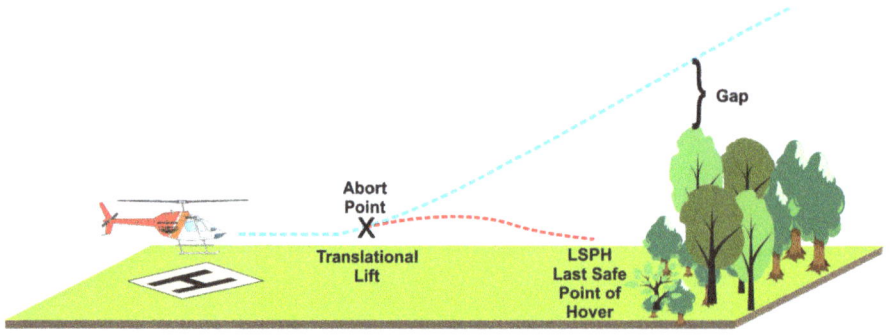

The abort on take-off can be summarised as follows

Item	Description	Action
ROC	Helicopter has a ROC sufficient to clear the obstacle	Continue
	Helicopter does not have a ROC sufficient to clear the obstacle	Abort
Gap	There is a gap between the tips of the blades and the obstacles and a steady ROC as viewed by the pilot	Continue
	There is no gap between the tips of the blades and the obstacles and no steady ROC as viewed by the pilot	Abort
Translational lift (Normal and shallow only)	The helicopter passes through translational lift at or before the abort point	Continue
	The helicopter does not pass through translational lift at or before the abort point	Abort

T TURNING and TERMINATION POINTS. THREATS

Turning

When conducting a confined area recce, it is very easy to get disorientated as there will be no runway or known features. Instead, there will be trees, rivers, obstacles, mountains, roads, houses, wires etc. Therefore, it is important to pick some key features that the pilot and crew can identify to help them find the confined area and keep the crew orientated to where they are in relation to the confined landing area.

Pick a particular tree, a bend in a river, a road or some other prominent feature as a base point. If other features act as **lead-in features** showing where the confined area is or even a **too far feature** that can indicate that you have gone past the confined area, this will help.

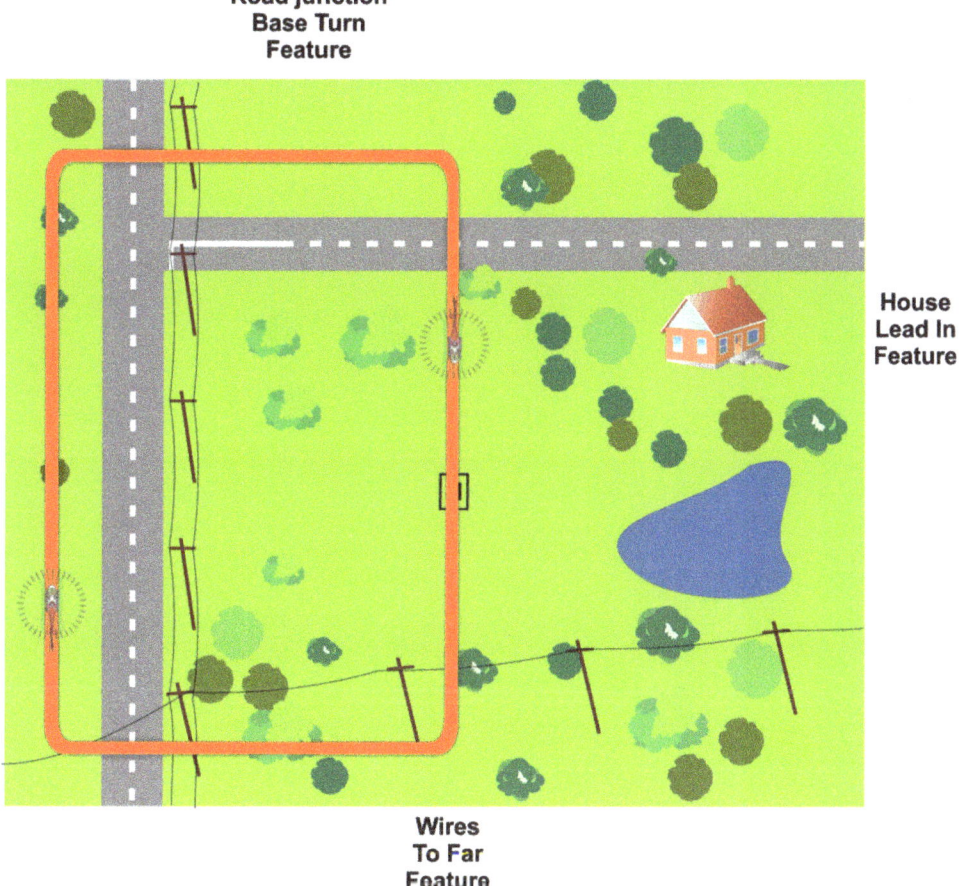

Turning points, **lead-in features**, **too-far features** and **alignment features** do not have to be selected for every confined area. Because no two confined areas are the same, the pilot and crew will decide what turning points they want to nominate based on what will help them get into and out of the area on the day.

Termination

As a general rule, plan to terminate (come to the hover) at least two-thirds of the way into the confined area. This will give the best approach angle while looking after the tail and allowing for some overshooting.

The termination point may not be where the helicopter will land, but it is the point to aim for on the approach.

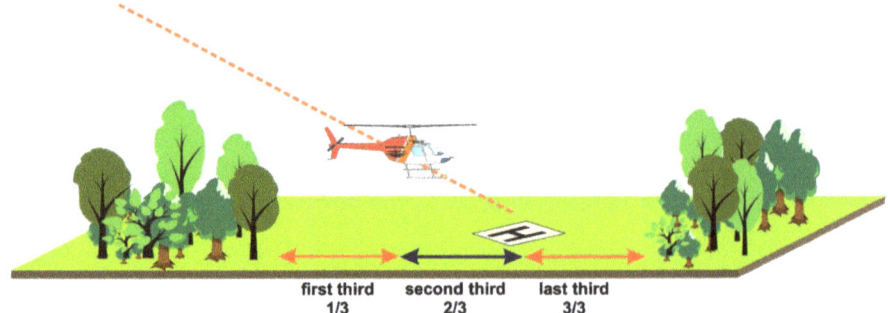

Landing shorter can put the tail rotor much closer to the trees or obstacles when on approach, and the pilot could inadvertently put the tail into the trees on approach.

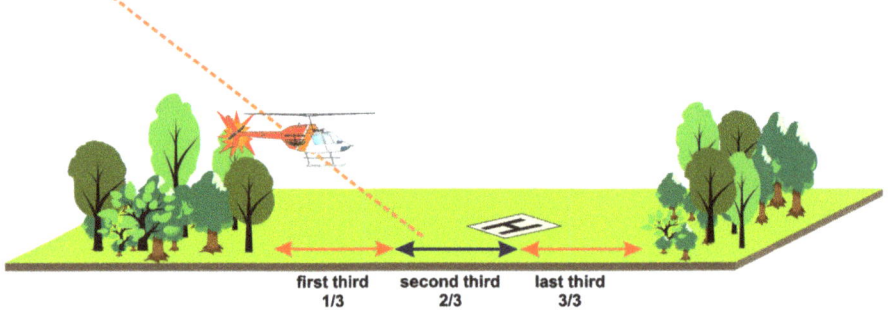

On very short finals, while the termination point is in front and can be viewed by the pilot out the front window, project the helicopter onto the pad and imagine where it will come to the hover.

This allows the pilot to nominate a reference point within the termination area, allowing the pilot to check that the tail is going to be in the clear and the rotor disc will fit before arriving there. This will allow the pilot to focus on making the final approach to the ground rather than coming to a high hover.

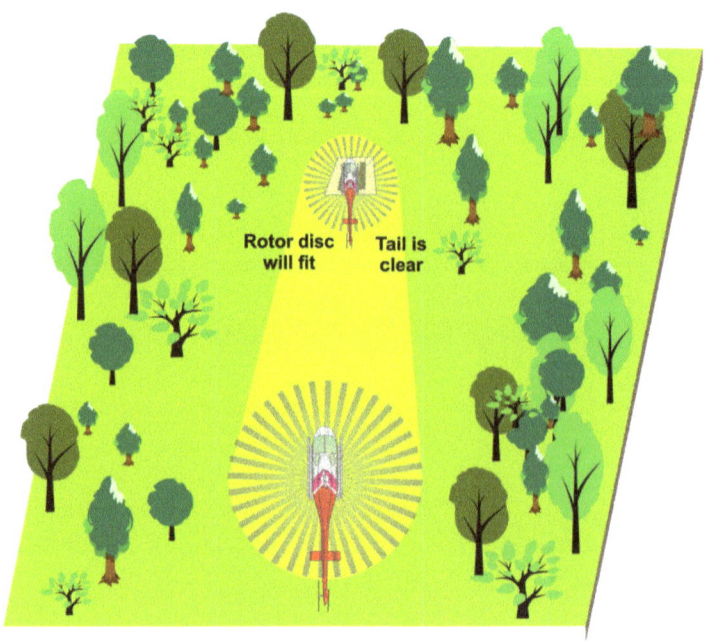

Threats

This should be a general discussion between the pilot and crew for any threats and hazards that could pose a risk to the helicopter or to those on the ground. Most items would already have been covered within the **PSWATP**; therefore, **Threats** is to cover anything relevant to the operation that has not yet been covered.

For example:

- Other helicopters operating in the same area
- Remote area, wildfire, volcano or other natural threats
- Approaching bad weather
- Lose items on the ground
- An approaching vehicle

P PLAN

Plan

The plan is the main part of the **PSWATP**. This is where all the items considered in the **PSWATP** can be summarised and articulated to the crew in the approach and departure Plan.

The plan is articulated during the Power, Wind, Plan (PWP) briefing on downwind and before take-off.

Piston example:

A plan for an approach (in a piston helicopter) may be articulated as follows:

- 3" MAP power margin
- Wind is from the southeast at 10-15 kts
- The confined area is suitable
- The plan is to conduct a 500 ft circuit and a normal approach with a double angle on finals to the open grass area
- The abort point is at the tree line
- Any questions?

Turbine example:

The plan for a take-off (in a turbine helicopter) may be articulated as follows:

- 15% Tq power margin
- Wind is from the southeast at 10-15 kts
- Plan is to conduct a constant angle take-off and, once clear of the obstacles, climb straight ahead to 500 ft before departing to base
- Abort point is the small green bush 20 ft in front; the LSPH is the open grass area before the trees
- Major obstacle is the set of wires running along the tree line
- Any questions?

Air Exercises: Confined Areas

Introduction

Confined Area operations put into practice all the previous lessons, including circuits, limited power, low flying and slope lift-offs and landings.

Using the PWSATP and flying at least one (1) circuit pattern around an area you have never landed in before is paramount to a successful outcome.

An inexperienced pilot may need a lot more time for the evaluation, planning and execution. This is normal, so do not be tempted to rush it, even if your instructor seems to be able to do it all in the blink of an eye. It is important that a pilot work within THEIR capabilities for a particular area, not try to work within the capabilities of others.

Common faults

Common faults to guard against during the air exercises

- Lack of preparation
- Overestimation of individual capabilities and that of the helicopter
- Losing sight of the confined area landing area during the recce
- Disorientation
- To much focus on the landing site and ignoring the big picture, in particular not sighting wires and threats
- Not maintaining a safe height AGL when conducting the recce
- Flying too fast or too slow
- Not using the PSWATP
- Terminating at a very high hover
- Not considering the tail rotor and its distance outside the rotor disc
- Getting too close to obstacles
- Practising with an instructor who does not know how to do the exercise

Airmanship

- Maintain a good lookout and constantly be moving your head to look around, not just your eyes
- Articulate (speak) what you are doing
- Be conservative

Chapter 11 Confined Areas

Air Exercise 11-1: Power Settling an VRS

Air Exercise

In this air exercise, the student will experience the symptoms and practise:

- entering a power settling situation that may develop into a full VRS, and
- recovering from power settling or VRS.

The techniques of initiating and recovering from power settling and VRS will be practised at an entry height of 2000 ft AGL with a recovery height of 500 ft AGL.

The instructor will first demonstrate, while the student follows through on the controls, to understand what power settling feels, sounds and looks like when encountered. Obviously, there is no danger in this happening at altitude, but it is most commonly encountered on short finals into a landing area close to the ground, so knowing the signs and recovery becomes very relevant. Having an experienced instructor, confident in the exercise and able to help set up the scenario and understand the environment is important.

Power Settling

Step	Action	Discussion
1	PREPARATION	Prepare as for a confined area. ■ Set the helicopter at 2000 ft AGL into wind straight and level into wind. ■ Conduct the HASEL check.
2	Initiate Power Settling	From Straight and Level flight: ■ Commence a deceleration (small flare) by lowering some collective and cyclic to return to the hover while at the same time turning downwind. ■ Try to maintain height. ■ Maintain balance or as speed reduces keep straight.
3	Downwind	Once downwind, allow the helicopter to descend at approximately 500 ft per minute ROD: ■ This may require the collective to be set below hover power. ■ Allow the helicopter to drift with the wind. ■ IAS should be reading zero (0). ■ The tail will waggle and feel loose. ■ Maintain direction. ■ The helicopter may want to pitch and roll to no set pattern. It may vibrate.
4	Increased descent rate	If entering Power Settling, the helicopter will be descending. This may be constant but 'turbulent'. This is usually 500 ft per minute ROD or less.
5	Power Settling Recovery	If in the very early stages of Power Settling, a recovery may be initiated by: ■ Set Collective to cruise power. ■ Some forward cyclic to gain speed. ■ Keep straight by looking outside. Because the helicopter will feel unstable, the balance ball may not be a good indicator. ■ Fly away.
6	Height loss	Height loss may vary, but with a straight-ahead recovery expect to lose 200-500 ft in altitude.

NOTE: In an attempt to recover from Power Settling and setting the collective to climb power, the pilot may actually make the situation worse. Raising of collective can force the helicopter from Power Settling into VRS and experience a sudden increase in the ROD. In any Power Settling and VRS recovery, the use of the collective should be conservative, with the primary purpose to maintain Rotor RPM.

Vortex Ring State

Step	Action	Discussion
1	PREPARATION	Prepare as for a confined area. ■ Set the helicopter at 2000 ft AGL into wind straight and level into wind. ■ Conduct the HASEL check.
2	Initiate Power Settling	From Straight and Level flight: ■ Commence a deceleration (small flare) by lowering some collective and cyclic to return to the hover while at the same time turning downwind. ■ Try to maintain height. ■ Maintain balance.
3	Downwind	Once downwind, allow the helicopter to descend at approximately 500 ft per minute ROD: ■ This may require the collective to be set below hover power. ■ Allow the helicopter to drift with the wind. ■ IAS should be reading zero (0). ■ The tail will waggle and feel loose. ■ Maintain direction. ■ The helicopter may want to pitch and roll to no set pattern. It may vibrate.
4	Increased descent rate	■ If entering Power Settling, the helicopter will be descending. This may be constant but 'turbulent'. This is usually 500 ft per minute ROD or less. ■ If entering a fully developed VRS, the helicopter will suddenly fall out from underneath you, and the descent rate will significantly increase. ■ Descent rates seem to go in stages and can vary with helicopter type and design. However, each 500 ft increment, the helicopter(and pilot) will experience an acceleration point in the ROD, so you will feel it falling. ■ Descent rates can typically be between 1000 ft and 4000 ft per minute ROD.
5	VRS Recovery	In a fully developed VRS, a recovery may be initiated by: ■ Set Collective to cruise power. ■ Some lateral (sideways) cyclic to move the helicopter sideways into clean air (typically to the right in anticlockwise rotating systems and to the left in clockwise rotating systems). ■ Keep straight by looking outside. Since the helicopter will feel unstable, the balance ball may not be a good indicator. ■ As the helicopter enters clean air, the ROD will stop almost instantly. ■ At that point, use some forward cyclic to gain speed and fly away
6	Height loss	Height loss may vary, but with a recovery to the left or right, the pilot can recover the helicopter in as little as 50 ft. This is important to know if experiencing VRS on short finals into a confined area.

Air Exercise 11-2: Confined Area Operations

Air Exercise In this air exercise, the student will learn the judgement required to:

- select a suitable confined landing and take-off area
- make an assessment, approach, landing and departure, and
- make best use of power, approach and departure paths.

Demonstration The instructor will demonstrate a confined area recce, approach, landing and take-off. After the demonstration, the instructor will give the student all of the controls and position the helicopter where there are confined areas.

The instructor will limit the power as required depending on the students' level of ability.

Step	Action	Discussion
1	PREPARATION	PRIOR PREPARATION PREVENTS POOR PERFORMANCE - Flight Planning - Fuel - Performance - Personal preparation
2	Enroute	Discuss: - Power - Pilot - Payload - Sun - Wind - Wires - Approval Conduct a HEFFR check before arriving adding: - Heading home - Fuel calculation and nominate a time to depart the area - Radio calls before descending
3	Recce	Select a suitable area, then conduct a high recce and, if necessary, a low recce. Confirm: - Power - Size, shape, slope, surface, sun (moon) surrounds, stock - Wind, wires, way in and way out - Approach type - Abort points - Turning and termination points

Step	Action	Discussion
4	Plan	Based on all the relevant information, give a PWP brief before landing (downwind leg of the circuit). For example: - 3" MAP (or 20%Tq) power margin - Wind will be on the front right at 10-15 kts - Plan is to conduct a normal approach until the treetops, then a double angle to the termination point - Abort point is at the tree line - Threats are the wires running across finals - Any questions?
5	Approach	Make the approach. On short finals before the abort point confirm: - Aairspeed vs Groundspeed - ROD under control, and - Power in hand. On short finals project forward and imagine where the helicopter will land ensuring: - the tail rotor is in the clear - the main rotor disk will fit, and - confirm the surface and slope.
6	Landing and lift off	If possible, land and lift-off in the normal manner.
7	Prior to lift off	Consider the PSWATP items.
8	Plan	Give a PWP prior to lift-off. For example: - 3" MAP (or 20%Tq) power margin - Wind will be on the front right at 10-15 kts - Plan is to conduct a towering take-off and once clear of the obstacles a normal departure - Abort point is passing the small tree 30 ft on the nose, and the last safe point of hover is over the flat grassed area beyond it - No threats to be concerned about - Any questions?
9	Take-off	Conduct the take-off as briefed.

Summary

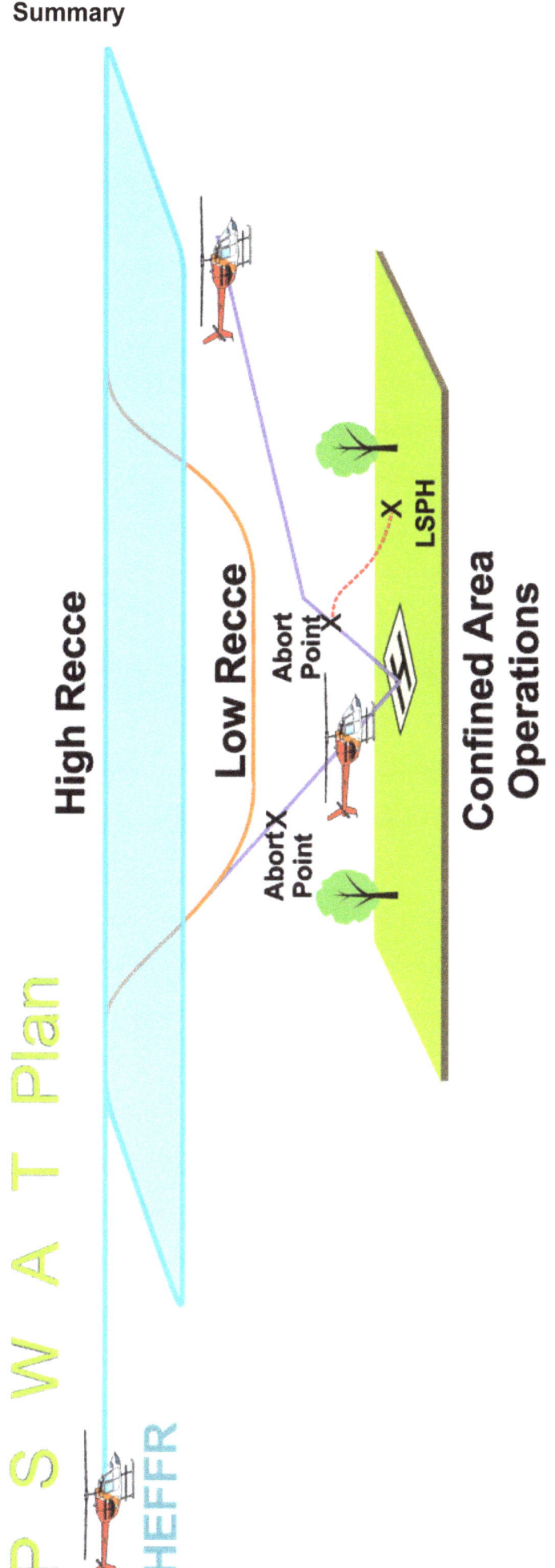

12

Low Level Flying

Aim To manoeuvre the helicopter close to the ground.

Objectives On completion of this lesson, the student will be able to:

- conduct a risk analysis for a low level operation
- explain the height velocity graph and how it applies to low level flying
- describe the effect of wind on low level flying
- explain how control effectiveness can change with speed
- list at least three low level manoeuvres
- perform low level manoeuvres in the helicopter
- conduct low level circuits with maximum performance takeoffs and zero-zero landings in the helicopter, and
- manage an engine failure while flying at a low level.

Motivation Low-level flying is an essential aspect of helicopter operations and goes beyond being just an additional rating. It is crucial for any basic helicopter training to encompass a thorough understanding of how to fly at low altitudes.

Helicopters generally offer excellent visibility and manoeuvrability, allowing them to operate in close proximity to the ground and navigate around obstacles. Unlike fixed-wing aircraft, helicopters are not as constrained by airspeed requirements, enabling them to perform unique tasks.

In narrow valleys, helicopters can execute tight turns, while climbs and descents can be carried out near trees, obstacles, and hills. The versatility of helicopters allows them to be positioned in close proximity to the terrain, surpassing the capabilities of other aircraft types.

Various operations such as sling loads, aerial agriculture, fire fighting, surveys, hunting, supply drops, photography, and search and rescue heavily rely on the pilot's fundamental skills and judgment in low-level flying. These skills are vital for executing these types of missions effectively.

Preparation: Low Flying

Risk Analysis

Introduction

Prior to conducting any low level operation, it is important that the pilot and crew conduct a Risk Analysis.

This means that information is gathered, and a plan is put in place so that the operation has a level of safety and predictability.

The Risk Analysis does not necessarily have to be onerous; it is simply an element during the planning stages that shows there has been some thought put into the task.

Believe it or not, you have already been doing this in some previous lessons. The use of the PSWATP acronym in confined areas is actually a Risk Analysis tool that can also be applied to low flying.

The big difference in modern low level operations is a Risk Analysis is a legal requirement with a formal process and sign-off so that the regulator (FAA, EASA, CASA, CAA etc.) has evidence of what you have planned for.

Helicopter Companies will have a formal process to follow. A private pilot may not, so below is a general outline of what a Risk Analysis looks like and then gives an example based on conducting a Low Flying Training exercise in an approved low flying training area.

Risk Analysis

A Risk Analysis involves identifying potential hazards, assessing their associated risks, and implementing appropriate mitigation measures to enhance safety.

Step 1: Identify

The process begins by identifying all potential hazards that may pose a risk during the low level operation. These hazards can include natural or man-made obstacles, adverse weather conditions, airspace considerations, human factors, equipment failures, and other relevant factors.

If able and relevant, consult historical incident and accident data, industry guidelines, and local regulatory requirements to comprehensively identify hazards.

Step 2: Assess

Assess the probability and severity of the identified hazards.

Assess the probability of the occurrence and severity of the consequences associated with each hazard. Probability refers to the likelihood of a hazard materialising (happening), while severity reflects the potential impact or harm it may cause.

Use a Risk Matrix or a similar method to categorise hazards based on their probability and severity levels. This categorisation will aid in prioritising risk mitigation strategies.

Example Risk Matrix

Below is an example of a common Risk Matrix.

Likelihood			Potential Consequences				
			Minor injuries or discomfort. No medical treatment or measurable physical effects.	Injuries or illness requiring medical treatment. Temporary impairment.	Injuries or illness requiring hospital admission.	Injury or illness resulting in permanent impairment.	Fatality
			Not Significant	Minor	Moderate	Major	Severe
	Expected to occur regularly under normal circumstances	Almost Certain	Medium	High	Very High	Very High	Very High
	Expected to occur at some time	Likely	Medium	High	High	Very High	Very High
	May occur at some time	Possible	Low	Medium	High	High	Very High
	Not likely to occur in normal circumstances	Unlikely	Low	Low	Medium	Medium	High
	Could happen, but probably never will	Rare	Low	Low	Low	Low	Medium

Step 3: Analyse

Analyse existing controls (already in place to reduce the risk) and evaluate their effectiveness in mitigating the identified hazards. This can be done by reviewing company Operations Manuals, Standard Operating Procedures (SOPs), Maintenance Protocols, Check and Training Programs, and any other relevant documentation you can access. Determine if the current measures adequately address the identified risks or if additional controls are necessary.

Step 4: Rate

Combine the probability and severity assessments to rate the overall risk level associated with each hazard. This can be done using the Risk Matrix above. Where the intersecting cells represent different levels of risk, such as low, medium, or high then assign risk levels to each hazard based on this rating.

Step 5: Mitigation

Decide on the Mitigation Strategies you will use based on the identified risks and their associated levels. These mitigation strategies are designed to reduce or eliminate the identified risks. Consider a hierarchy of controls, including elimination (don't do it), substitution (go somewhere else or do something else), engineering controls, administrative controls, and Personal Protective Equipment (PPE). Prioritise strategies that effectively reduce risks to an acceptable level while considering operational feasibility (cost vs reward).

Step 6: Implement

Implement and monitor the mitigation measures. To implement the identified mitigation measures and ensure proper integration into the operation, they must be communicated to relevant personnel, participants and stakeholders to accommodate the necessary training and any changes to operational manuals and procedures.

Report back once the flight has been completed to monitor the effectiveness of the implemented measures and gather feedback on their effectiveness to continuously improve safety performance.

Step 7: Review

Periodically review and update the risk analysis process and its outcomes to ensure its ongoing relevance and effectiveness. This review should consider changes in operational procedures, regulations, technology, and lessons learned from incidents or accidents. Update the Risk Analysis as needed to account for new hazards or changes in risk levels.

Example Risk Analysis

1. **Identify hazards**
    (a) Natural obstacles (trees, hills, water, animals, etc.)
    (b) Man-made obstacles (buildings, power lines, roads, towers, etc.)
    (c) Adverse weather conditions (high winds, reduced visibility, storms, cloud etc.)
    (d) Airspace (controlled airspace, other aircraft, local procedures etc)
    (e) Human factors (pilot error, fatigue, distractions, experience of instructor etc)
    (f) Equipment failures (engine, avionics, navigation etc.)

2. **Probability and Severity Assessment**

    Using a risk matrix, assess the probability and severity of each hazard.

    (a) Natural obstacles: Moderate probability, moderate severity
    (b) Man-made obstacles: Low probability, high severity
    (c) Adverse weather conditions: Low probability, moderate severity
    (d) Airspace congestion: Low probability, low severity
    (e) Human factors: Moderate probability, high severity
    (f) Equipment failures: Low probability, high severity

3. **Analyse existing controls**
    (a) Low-level flight procedures and limitations
    (b) Pre-flight planning and briefing on potential hazards
    (c) Training on obstacle avoidance and emergency procedures
    (d) Weather monitoring and decision-making protocols
    (e) Communication and coordination with air traffic control
    (f) Regular maintenance and inspections of helicopter systems

4. **Rate the level of risk**

    By combining the probability and severity assessments, the level of risk can be determined:

    (a) Natural obstacles: Moderate risk
    (b) Man-made obstacles: Low risk
    (c) Adverse weather conditions: Low risk
    (d) Airspace congestion: Low risk
    (e) Human factors: Moderate risk
    (f) Equipment failures: Low risk

5. **Mitigation Strategies**
    (a) Natural obstacles: Conduct thorough pre-flight planning, including detailed route analysis and briefing on potential hazards. Emphasise proper scanning techniques and the importance of maintaining situational awareness during low-level flight. Use the instructor's prior experience; use google earth to look at the area before leaving.
    (b) Man-made obstacles: Identify and mark obstacles within the training area. Emphasise the use of obstacle clearance techniques and maintain a safe distance from structures during training exercises.
    (c) Adverse weather conditions: Establish specific weather minimums for low-level flight training. Ensure pilots get a weather report and receive weather briefings before each training session and implement strict adherence to weather-related limitations.
    (d) Airspace congestion: Coordinate with air traffic control to schedule training exercises during low-traffic periods. Emphasise the importance of scanning for other aircraft and maintaining proper spacing during low-level maneuvers. Use the radio.
    (e) Human factors: Implement a fatigue management program for flight instructors and students. Emphasise the importance of adherence to standard operating procedures, effective communication, and managing distractions during low-level training exercises.
    (f) Equipment failures: Ensure regular maintenance and inspections are conducted on helicopters. Implement a robust emergency procedures training program to address various equipment failure scenarios, such as engine failures and radio failures.

6. **Implementation and Monitoring:**

    Communicate the identified mitigation strategies to instructors and students. Provide training sessions and materials to ensure understanding and compliance. Establish a reporting system to monitor the effectiveness of the implemented measures and encourage feedback from instructors and students to continuously improve safety performance.

7. **Review and Update:**

    Periodically review the risk analysis process to account for changes in procedures, regulations, or training requirements. Consider feedback from instructors, students, and safety reports to identify areas for improvement. Update the risk analysis as necessary to ensure ongoing relevance and effectiveness.

Height Velocity Diagram

Introduction

The Height Velocity (HV) Diagram (also called the Deadman's Curve) is produced by the helicopter manufacturer during type certification flight testing and shows those combinations of height (on the left) and speed (on the bottom) that if the helicopter experienced an engine failure, the likely hood of conducting a safe autorotation without damaging the helicopter is reduced.

For this reason, flight within the HV Diagram shaded areas should be done consciously with an understood elevated degree of risk and, if possible, to be avoided.

Sample HV Diagram

The HV Diagram below is an example of the Bell206 Jet Ranger.

It is a straightforward diagram; however, over the years, regulators (FAA, EASA etc.) have changed the rules on how these diagrams are created, and more modern helicopters can have HV Diagrams that consider more information and look more complicated.

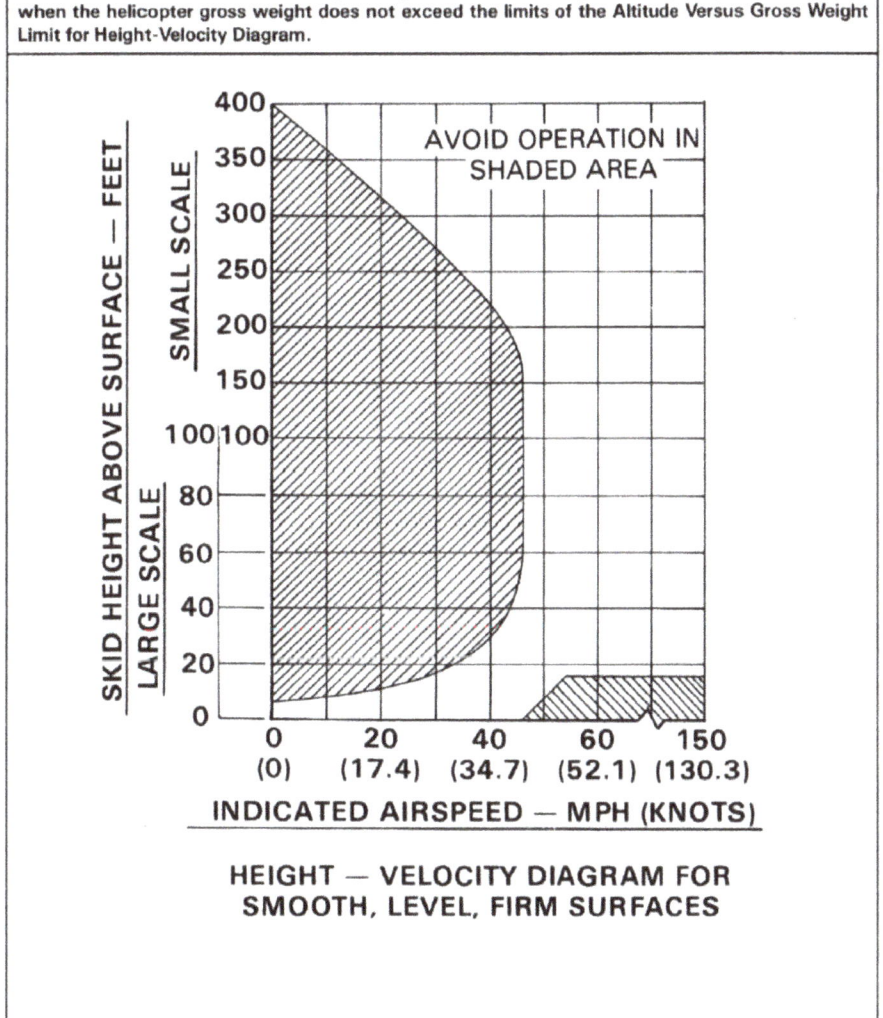

Chapter 12 Low Level Flying

RFM

It is important to note that the HV Diagram is located in the **PERFORMANCE** section of the RFM, which means the pilot can use it for planning purposes.

It is **NOT** in the **LIMITATIONS** section; therefore, the **HV Diagram is not a limitation** and can be flown into deliberately by the pilot when operating low level, but it must be considered in the operation's performance planning.

Basis for the chart

As previously stated, the manufacturer plots the HV Diagram during flight testing. The rules for creating it state that it assumes nil wind with an average pilot at the controls who will delay in responding to the engine failure by approximately 1.5 seconds over a smooth flat hard surface. This simulates a pilot's delay in real life in realising there is an issue.

This means an above-average pilot with some wind and less of a delay in responding can do much better than what is described within the HV Diagram, and we all know a real engine failure while low flying does not happen over a nice flat runway!

Example

The HV Diagram for the Bell206BIII has two (2) shaded areas that indicate those combinations of height and speed that can be dangerous in an engine failure below 400 ft AGL.

A	A large, shaded area that indicates high hovers between 10 ft and 400 ft AGL at low speeds from 0 kts to 40 kts are areas to be avoided for prolonged periods of time.	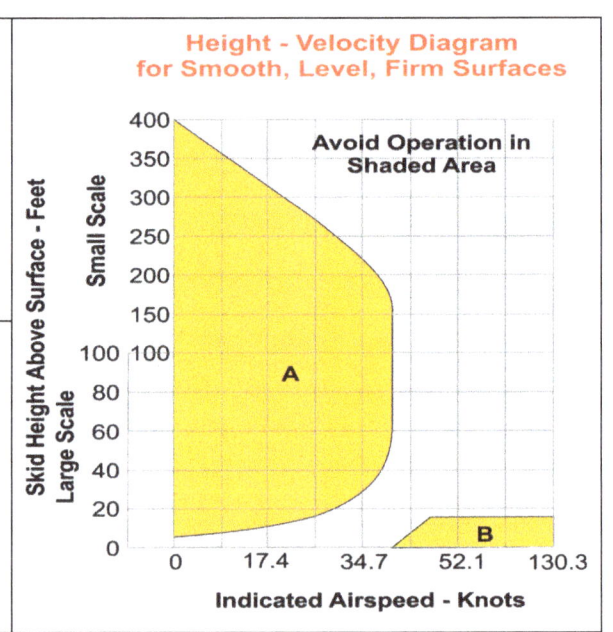
B	A smaller shaded area indicates high speeds above 40 kts when below 20 ft AGL are areas also to be avoided for prolonged periods of time.	

Planning

If planning a low-flying operation, part of the Risk Analysis is to try to fly at heights and speeds that reduce the helicopter's exposure to operations within the HV Diagram shaded areas.

If the helicopter must be operated within the HV Diagram shaded areas, then plan to minimise the time in the area to only what is needed to complete the task.

Remember, the HV Diagram is only relevant if the engine fails; if the engine is functioning, the helicopter will not struggle to operate in these areas.

Summary

In summary, when conducting a low level operation, the HV Diagram is to be **considered**, but it is **not a limitation**.

Wind

Introduction

When flying low level, the wind is a significant consideration. It can be both an advantage and a disadvantage depending on where it is coming from in relation to the helicopter's movement and how strong and consistent it is blowing.

Drift

Due to their lower forward speeds, helicopters are affected more by wind and the resulting influence on performance and drift.

Drift is very apparent at low levels and must be constantly allowed for when:

- maintaining a set track
- turning downwind
- maintaining a constant radius turn about an object on the ground, or
- performing turns about a line feature, such as a straight section of road.

Wind on the nose

If the wind is on the nose, it will:

- reduce the power required
- reduce the time in the Deadman's Curve, and
- give a lower ground speed for a higher airspeed.

In general, the stronger the wind, the greater the advantage when into wind.

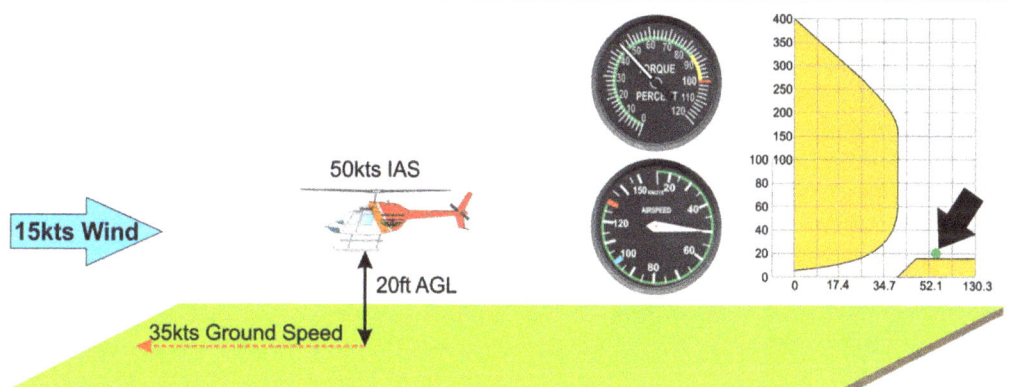

Wind on the tail

If the wind is on the tail, it will:

- increase the power required
- increase the time in the Deadman's curve, and
- give a higher ground speed for a lower airspeed.

In general, the stronger the wind, the greater the disadvantage when downwind.

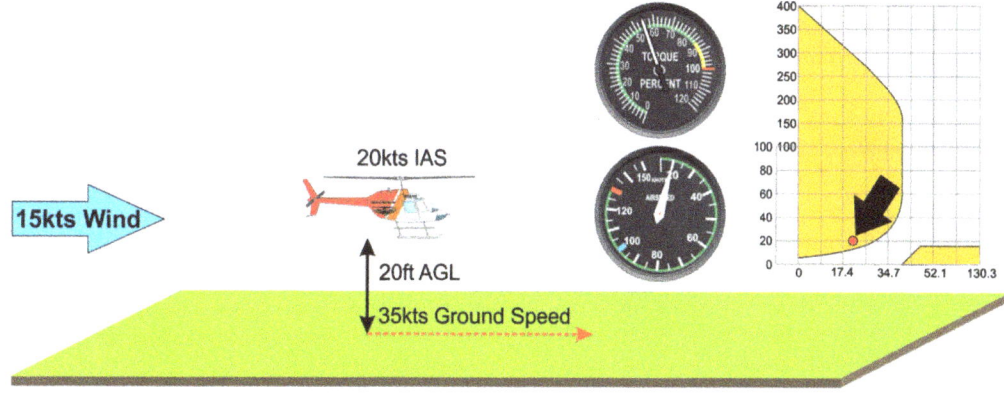

Chapter 12 Low Level Flying

Speed illusion when turning downwind

When flying low level, the pilot should be looking outside and estimating speed by the movement of the ground underneath the helicopter with small glances inside to ensure the helicopter is maintaining a constant nominated airspeed. Instead, the indicated airspeed (IAS) is often ignored, and the pilot incorrectly focuses only on the ground speed to the detriment of managing the airspeed.

When flying into the wind, the ground speed (G/S) will appear slow, but when turning downwind, the ground speed will start to increase. To prevent this, the pilot tends to automatically slow the helicopter down so that the same ground speed sight picture is achieved. Unfortunately, this can cause a further reduction in **airspeed.**

If the pilot is not careful, the airspeed can reduce to the point that the helicopter starts to sink towards the ground, which the pilot will want to stop by raising collective. This can spiral into a dangerous situation where the helicopter speed reduces to below translational lift, and there is not enough power to hold the height downwind.

Unless recognised early, this could lead to an over-pitching situation and loss of control.

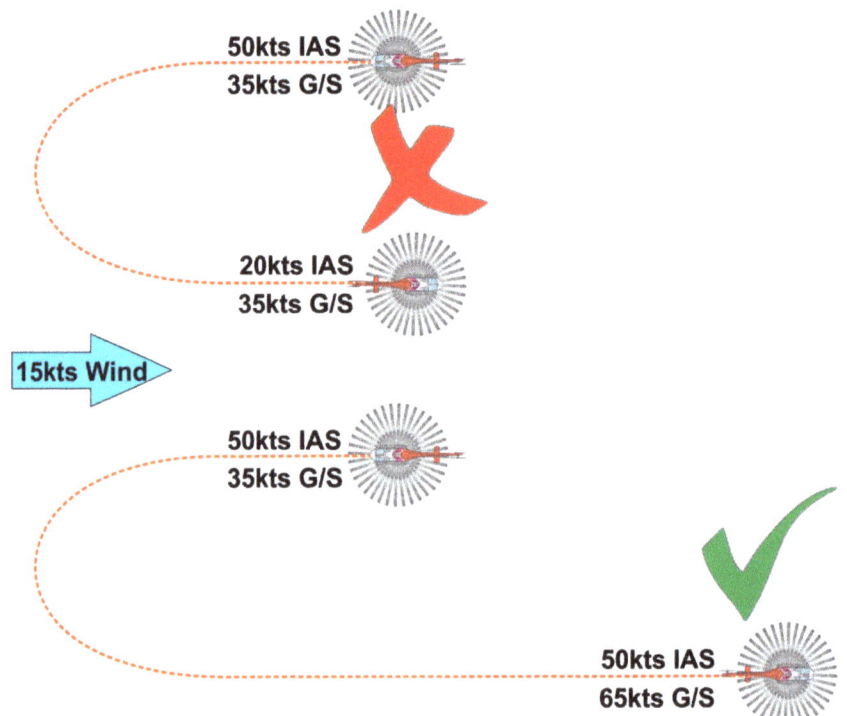

Instead, when turning downwind, the pilot should attempt to *maintain airspeed,* which will mean accepting an increase in ground speed.

This large variation in ground speed with operations into wind, then downwind, is not the preferred way to operate efficiently and safely at low level. Instead, the smart low level pilot will always try to operate and conduct turns with a crosswind.

Crosswind

When operating low level, it is always best to operate crosswind with all turns made into wind. This practice will give the pilot and crew the most consistent visual cues for airspeed and ground speed but will always have the helicopter positioned only 90 degrees away from the prevailing wind in the event of any emergency or engine failure.

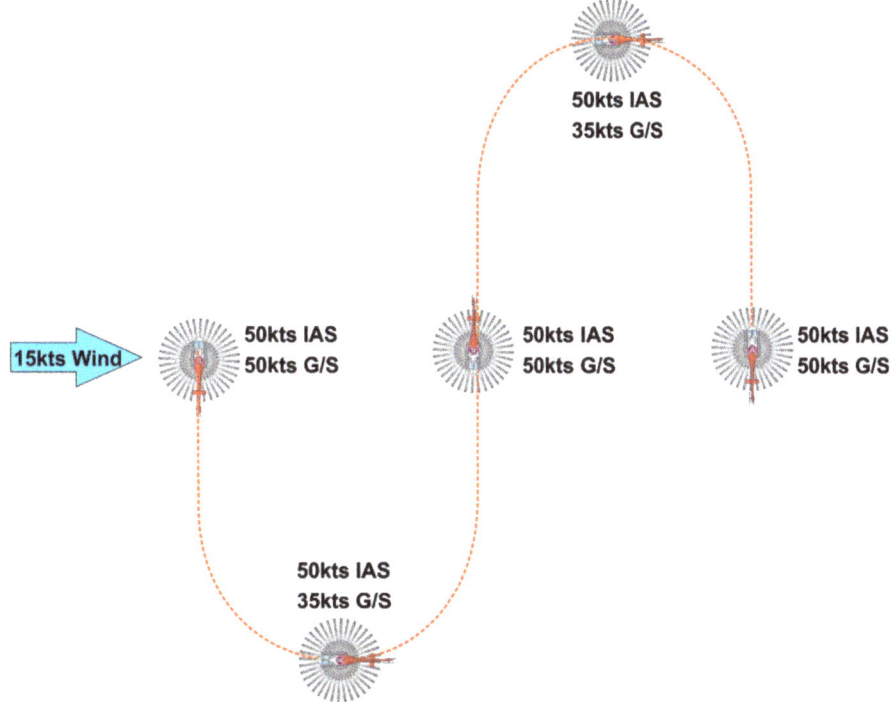

This does not mean that the helicopter cannot operate into wind or downwind; all this does is reduce the exposure of the helicopter to downwind flight and maximises the opportunity for turns into wind. It gives the pilot more control over the low level operation.

Control Effectiveness

Cyclic

In normal flight, the:

- **primary effect** of cyclic is to control attitude, and the
- **secondary effect** of cyclic is a change in altitude (more correctly referred to as height above the ground when low level).

When operating low level, the **secondary effect** of cyclic to change altitude is used more often as the helicopter will respond faster, and a change in altitude or height above the ground will happen faster if using the cyclic instead of the collective.

Collective

In normal flight, the:

- **primary effect** of collective is to control altitude by increasing or decreasing power, and the
- **secondary effect** of collective is a change in attitude.

When operating low level, because changes in power are slower to make a noticeable change, the collective is used early in anticipation of the manoeuvre and to assist the use of the cyclic as the primary control.

Coordination

When operating low level, the pilot will have to coordinate the cyclic, collective and pedals differently to get a timely change when operating close to the ground. The way these controls are coordinated will change with the helicopter's speed.

Chapter 12 Low Level Flying

Control effectiveness

The use of cyclic and/or collective to manipulate **height** above the ground will change with airspeed.

As airspeed **increases**, the **cyclic** becomes **more effective** in changing **height** faster.

As airspeed **decreases**, the **collective** becomes **more effective** in changing **height** faster.

The diagram below displays this relationship between the two controls.

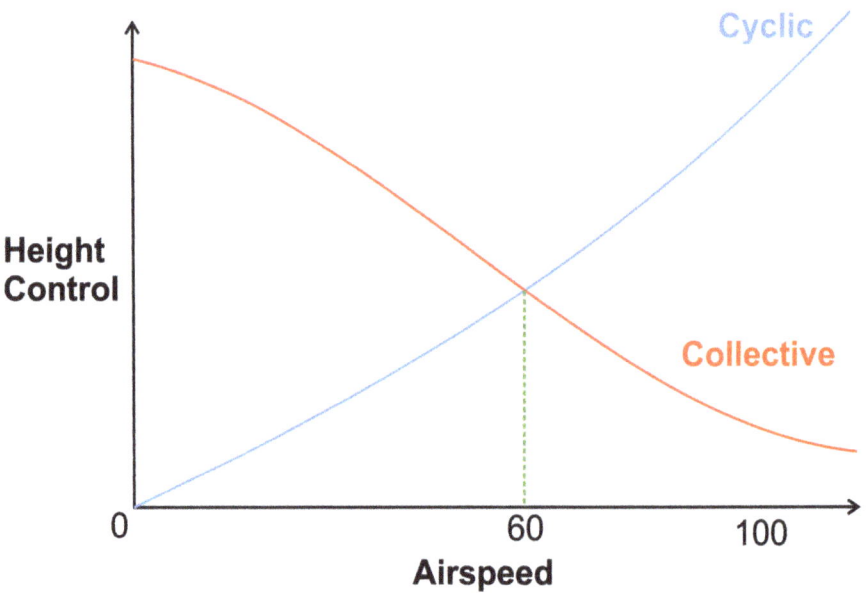

For example:

Consider a helicopter at the hover. If the pilot wants to increase **height**, the collective is raised. Cyclic will have little to no effect on **height** at the hover; instead, it will affect speed.

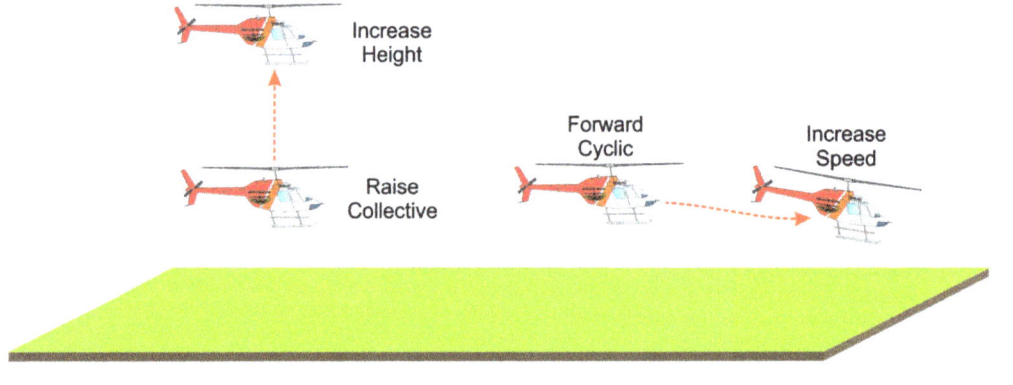

Options to gain height

Now consider a helicopter cruising at 80 kts straight and level. If the pilot wants to increase **height,** there are two options:

1. Raise collective and maintain speed; the helicopter will respond slowly and start a climb in the conventional manner; or

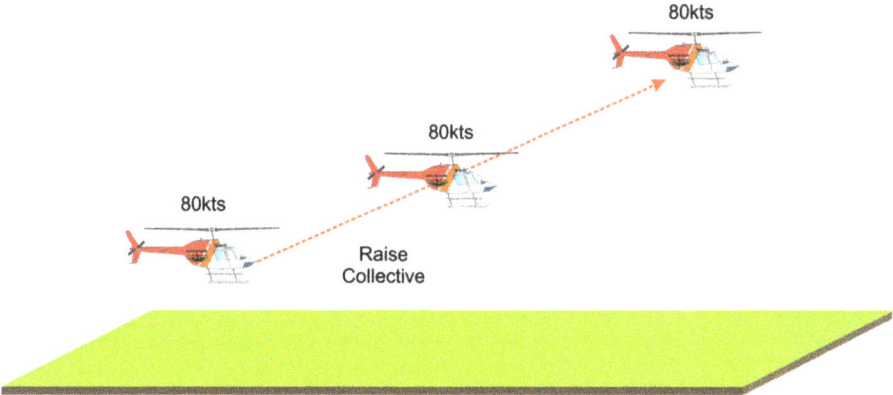

2. Use some aft cyclic to change the attitude, and the helicopter will immediately start a climb even without using the collective, even though the helicopter will lose some airspeed. This is using the effects of a flare to sacrifice speed for height when needed quickly.

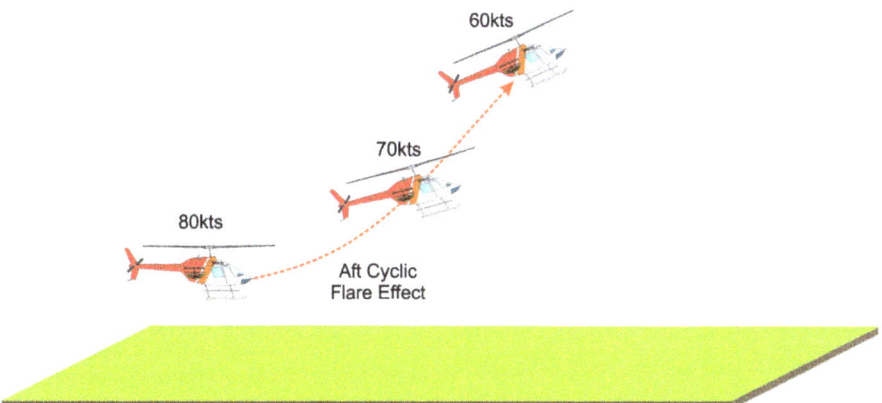

When operating low level, the pilot can choose which technique is required to gain height, depending on how urgently it is required or use a coordinated combination of the two.

Chapter 12 Low Level Flying

Low Level Manoeuvres

Introduction During training, an instructor will teach the student a series of standard low level manoeuvres. This does not cover all the possibilities that may be available to fly a helicopter low level, but they do help a student gain the skills, knowledge, and attitudes to safely manage a helicopter close to the ground. Competency in these manoeuvres is the building block for flying in the mountains, in tight terrain and conducting a commercial operation.

Low Level Turns

S-Turns Although S-turns are not commonly used during actual operations, they are a good exercise to initially allow the student pilot to develop the coordination and anticipation required to manage the helicopter when low level. The exercise will teach how to manage wind, allow for the inertia of the helicopter and how to lead with collective and stay in balance while maintaining airspeed.

To conduct a series of S-turns, first pick a straight-line feature such as a tree line, a road, a fence, or something similar that is orientated 90 degrees to the prevailing wind. Cross the straight-line feature at approximately 45 degrees at 45 kts IAS at no more than 50 ft AGL. As the helicopter crosses the feature, commence a turn back the other way and continue this until reaching the end of the feature.

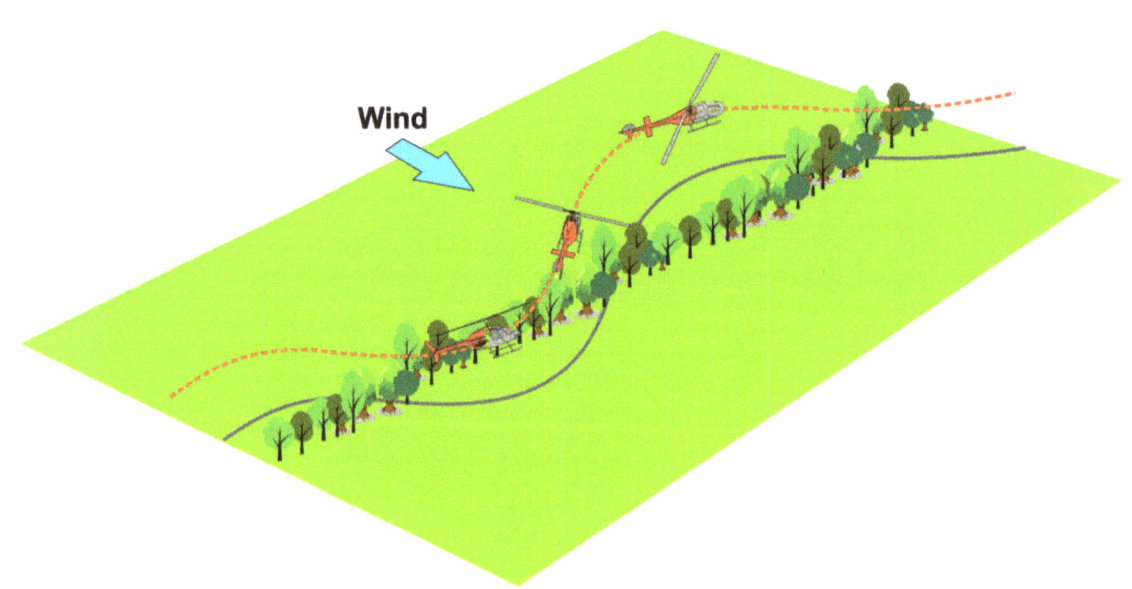

Page 12-13

Changing wind position	When turning, the rotor disc will change its orientation to the crosswind, which will affect rotor thrust.

When the helicopter banks and the wind is positioned on top of the rotor disc, the helicopter will want to sink towards the ground due to the reduction in rotor thrust. This will require the pilot to raise some collective, maintain balance with pedals and maintain the attitude to maintain height, speed, and ground track.

When the helicopter banks the other way, the wind is positioned under the rotor disc, and the helicopter will want to climb due to the increase in rotor thrust. This will require the pilot to lower some collective, maintain balance with pedals and maintain the attitude to maintain height, speed, and ground track.

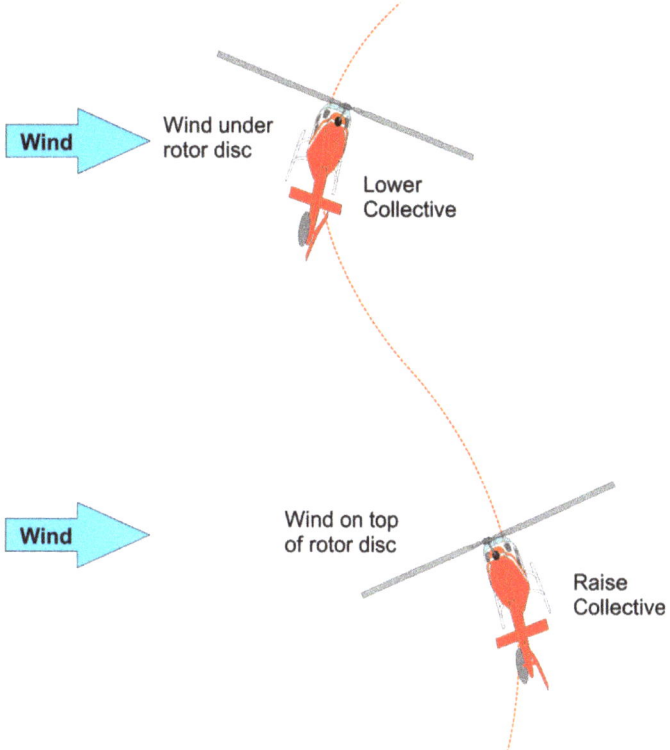

Chapter 12 Low Level Flying

180-degree turn

Once the helicopter comes to the end of the straight-line feature, the pilot needs to initiate a 180-degree turn to head back the way the helicopter has come from (later on in this chapter referred to as a Reversal Turn).

To do this, initially commence a turn towards the downwind with cyclic while raising some collective to commence a small ROC. Once the helicopter starts climbing, turn into wind through 180 degrees. Halfway through the turn, the pilot may lower a small amount of collective so that the helicopter is positioned at 45 degrees, 50 ft AGL and at the selected speed, ready to commence the S-turns again in the opposite direction.

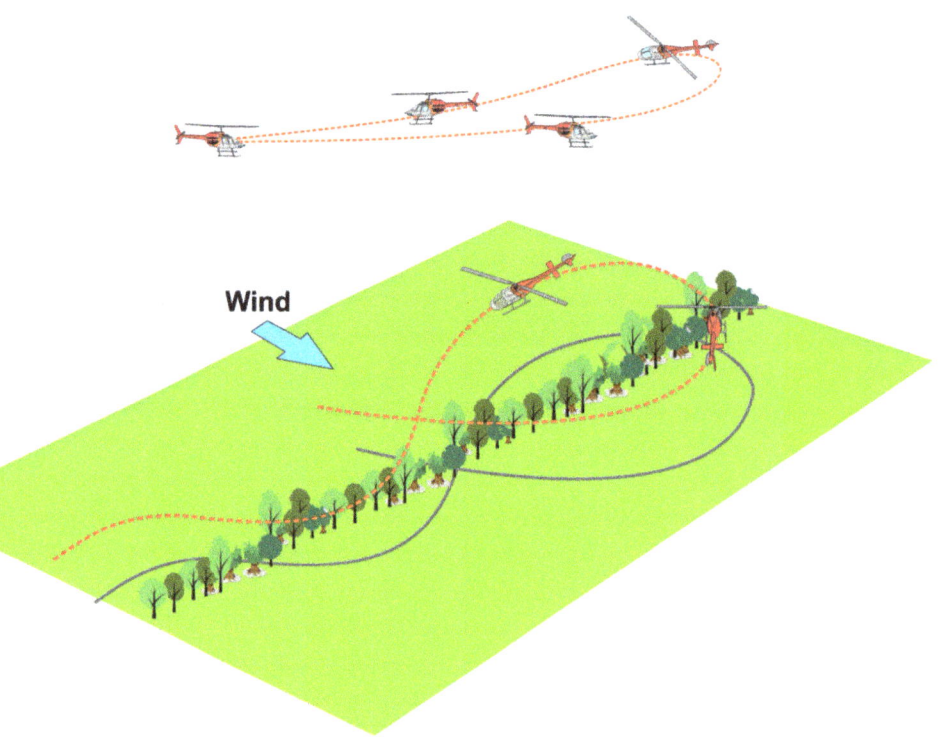

Putting it all together

Putting all this information together, the pilot should be able to conduct S-Turns along a straight-line feature at a constant height and speed and then do a reversal turn to conduct the same S-turns in the opposite direction.

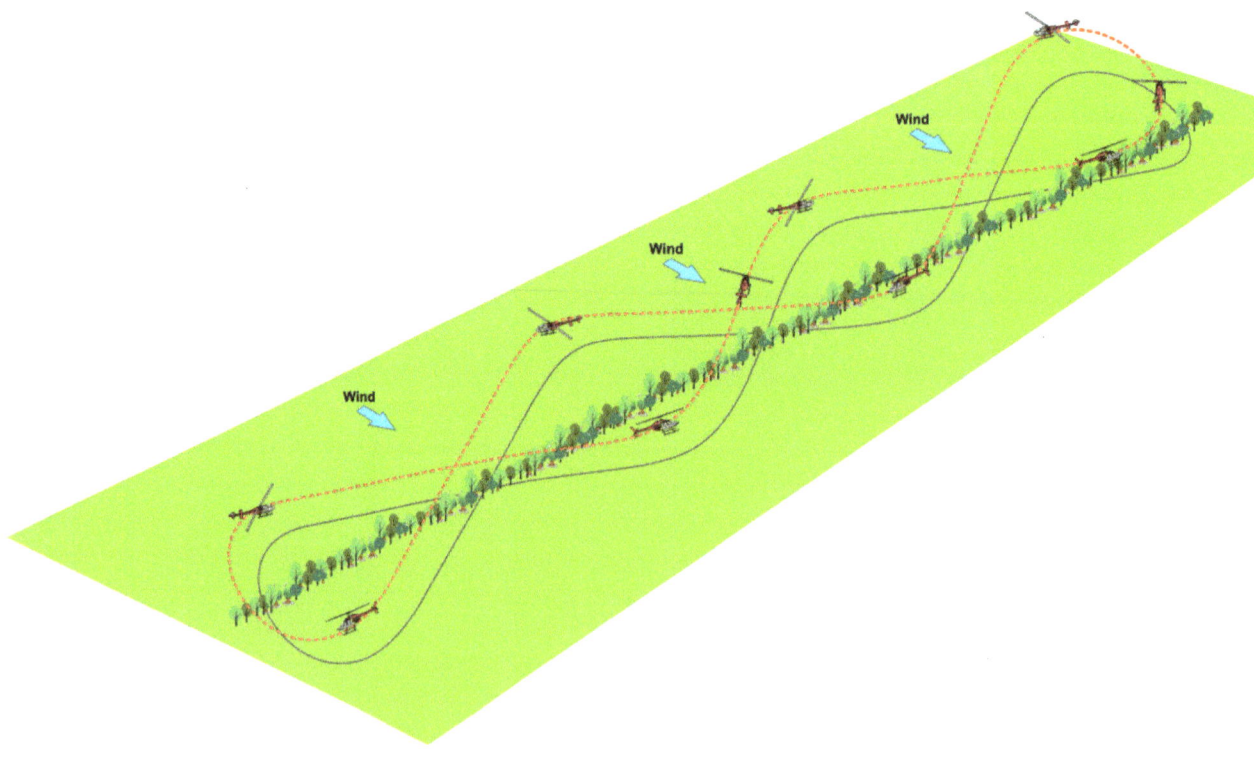

Figure of 8

The Figure of 8 manoeuvre is simply a method of conducting a 180-degree turn (Reversal Turn) so that the pilot can overfly the same feature multiple times. Again, it is a coordination exercise making the pilot allow and adjust for wind, speed, and obstacles while low flying.

To conduct a series of Figure of 8-Turns, pick a central feature to fly over, such as a helipad, a tree, a small structure, or something obvious. Plan to orientate the Figure of 8 pattern 90 degrees from the prevailing wind. Cross over the feature at approximately 45 degrees at 45-60 kts IAS and 50 ft AGL. As the helicopter crosses the feature, commence a 180-degree turn in the same manner as conducted at the end of the S-Turns.

Keeping it simple, a figure of 8 simply puts the two turns done at the end of each S-Turn together.

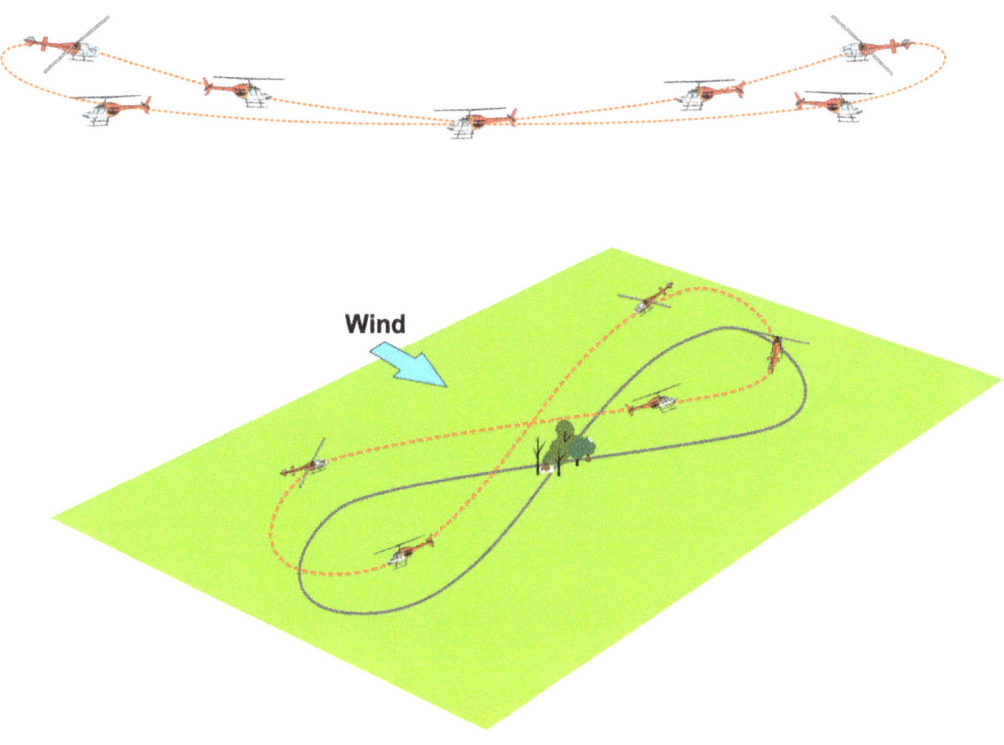

Quick stop

The purpose of a quick stop is to **decelerate the helicopter rapidly at a constant height**. It does not mean that the flight controls are moved quickly or that the manoeuvre is done quickly, but it does mean that the helicopter will respond quickly.

The manoeuvre reduces forward speed and terminates at the hover to avoid obstacles or change the helicopter's flight configuration.

A pilot can conduct several types of Quick Stops; each will depend on where the wind is coming from, as the quick stop should always finish with the helicopter hovering into the wind, usually OGE.

Chapter 12 Low Level Flying

Straight ahead quick stop

When flying directly into the wind, initiate a quick stop straight ahead by lowering some collective, then flaring with cyclic. Terminate at the hover, remaining into wind.

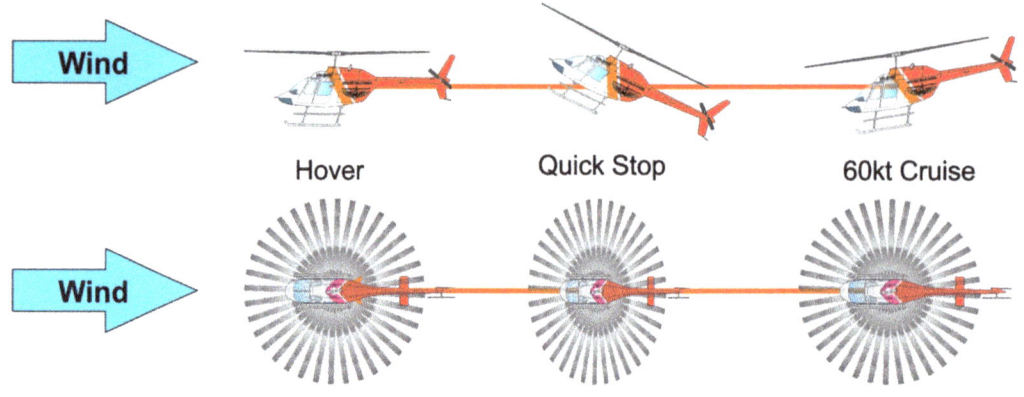

90-degree quick stop

When running 90 degrees to the wind, initiate a quick stop by lowering some collective, then flaring and turning. Terminate at the hover into wind.

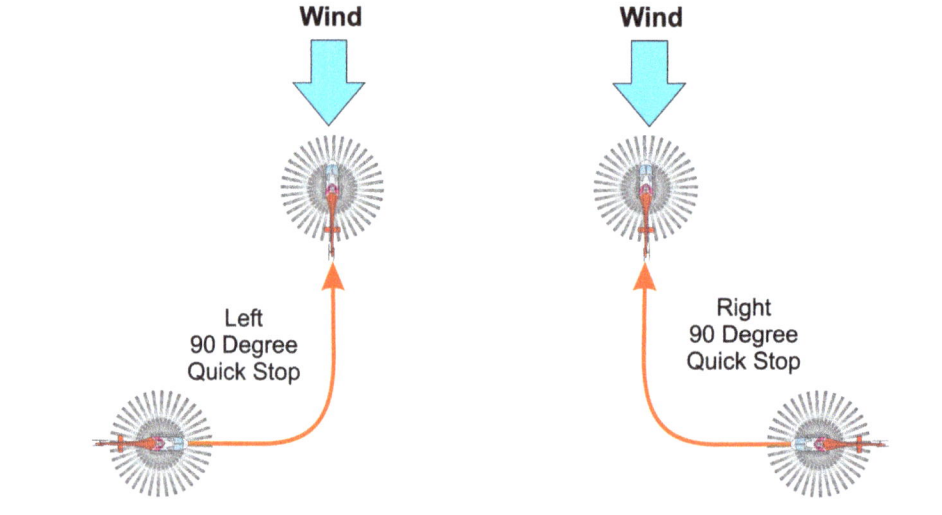

180-degree quick stop	When running downwind (the wind is on the tail), initiate a quick stop by first turning into the wind. As the helicopter comes 90 degrees to the wind, start lowering the collective and flare the helicopter. Terminate at the hover into wind.

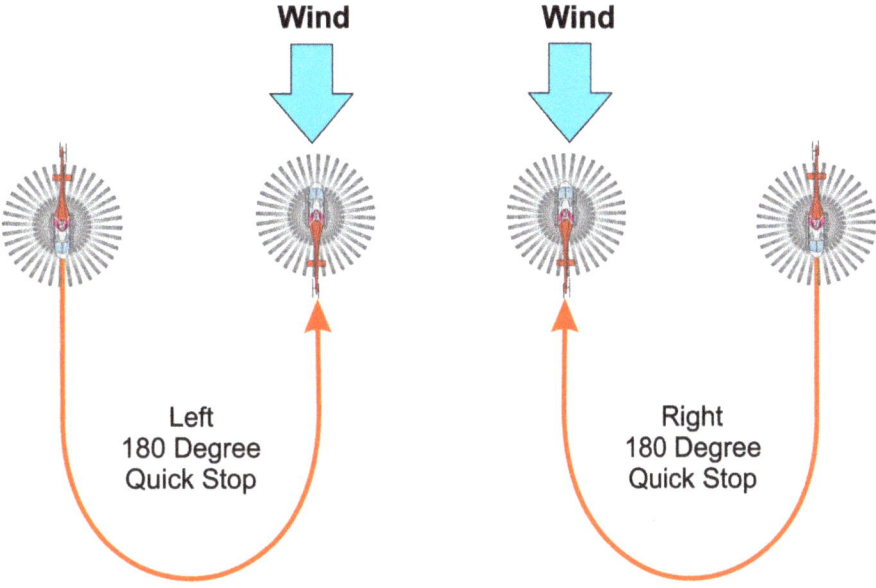

A 180-degree quick stop can be more problematic as operating downwind and low level can be dangerous if not done correctly.

Techniques	When initiating a quick stop from downwind, there are two techniques: 1. Turn and flare method, or 2. Flare and turn method. Which technique is used will depend on the: - Current IAS - strength of the wind - AUW of the helicopter, and - Density Altitude.
Turn and flare	The Turn and Flare method is the preferred method to use for 180-degree quick stops. From downwind, initiate a turn before flaring when passing 90 degrees to the wind. This method is best when the wind is strong (above 15 kts), the helicopter has a high AUW, and the Density Altitude is high regardless of the speed. This method was described above previously.

Chapter 12 Low Level Flying

Flare and turn

The flare and turn method is used when the helicopter is running downwind at a high speed (above 60 kts), and the pilot needs to commence the **deceleration** first before turning.

From downwind, commence a **small flare** to start decelerating and lower collective to control height and maintain balance with pedals. As the helicopter passes through no less than **60 kts IAS**, commence a turn with cyclic. Passing through 90 degrees to the wind, the collective will need to be raised to maintain height. Terminate at the hover into wind.

It is important that the initial flare is small and treated as a deceleration to set the helicopter up for the turn and additional flare to come. Doing a harsh flare with the wind from behind, thinking that the helicopter will come to a hover downwind, will only lead to VRS and overpitching.

Wind

The flare and turn method is not preferred with students or when the wind is strong (above 15 kts), the helicopter has a high AUW and the Density Altitude is high.

It is a good method if the wind is light (less than 15 kts), the helicopter is light, the Density Altitude is low, and the airspeed prior to entry is high, and an experienced pilot is using it for an operational purpose.

Other methods

There are other Q-stop techniques that experienced pilots may utilise as part of an operation, but they do not need to be taught at this stage of training.

Reversal turns

A reversal turn is simply a turn through 180 degrees allowing the helicopter to change direction and go back in the direction it has just come. There are many different ways to complete a reversal turn, and the technique used will depend on the:

- amount of area available
- type of terrain or local obstacles
- wind
- amount of power available
- AUW of the helicopter
- Density Altitude,
- Type of operation and
- Decision and experience of the pilot.

Having completed S-turns, Figure of 8s and 180 Quick stops, the student pilot has already learnt three (3) different types of reversal turn.

Other common types of reversal turn are the:

- flat 180-degree turn
- cyclic 180-degree turn, and
- Pedal and Torque turns.

Flat turn: The advantage of a flat turn is that the helicopter stays close to the ground throughout the turn. This may not be necessary for normal operations but is an advantage during military operations when the helicopter is trying to stay hidden below the tree line or behind a hill or obstacle. During the turn, the pilot must avoid allowing the helicopter to sink towards the ground.

From straight and level flight low level, between 45 and 80 kts IAS, commence a left or right-hand turn with cyclic. Do not gain or lose height but coordinate the cyclic and collective to achieve a flat turn. Maintain balance with pedals.

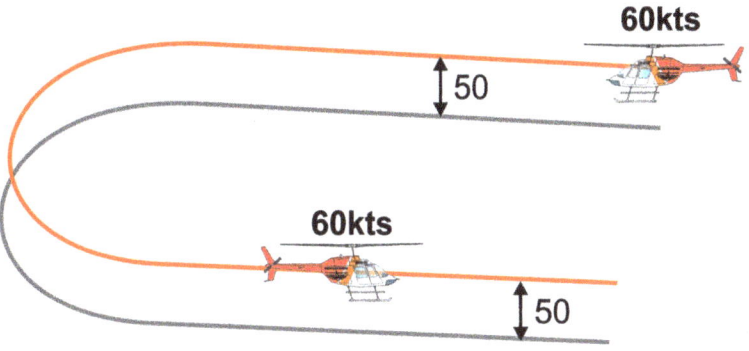

Cyclic turn: A cyclic turn uses a combination of sacrificing speed for height while conducting a turn. The collective does not need to be used except to fine-tune height, if required, on the exit of the turn.

The advantage of a cyclic turn is that the helicopter gains height, so it can avoid obstacles while turning in a tighter circle at a slower speed. It also can be done without changing the power.

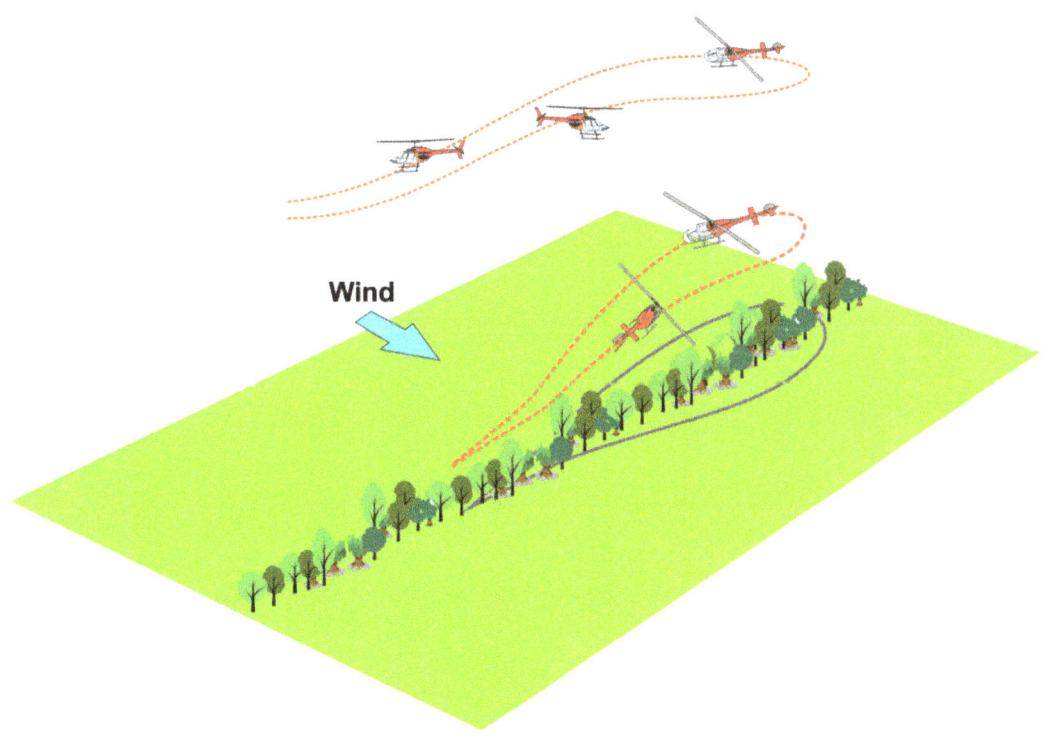

Chapter 12 Low Level Flying

Pedal and Torque turns

Pedal turns are a useful manoeuvre in tight areas or when a quick turnaround is required, such as when agricultural spraying. It enables the helicopter to turn virtually about the axis of the rotor mast. The turn is made mainly in the yawing plane and always into wind.

Torque Turn (turn to the right)

In a conventional helicopter, a turn to the right is assisted by torque, known as a torque turn.

Pedal turn (turn to the left)

A turn to the left requires left pedal and is against torque, referred to as a pedal turn. (The opposite would apply for a helicopter with the blades rotating clockwise).

Caution

Unless the pilot has carried out advanced training, such as specific low flying training, a pedal or torque turn should be avoided as it is too easy to do incorrectly, putting the helicopter (and you) in a potentially dangerous situation.

Maximum Performance Takeoff and Zero-Zero Landings

Maximum performance takeoffs

A maximum performance takeoff describes the use all the power available to get away from the ground as quickly as possible.

Used when lifting off from a surface such as sand, dust, snow or anytime when the pilot does not wish the downwash to have an adverse effect on the surface or visibility may be reduced because of the effects of the downwash on the surface.

A maximum performance takeoff commences while the helicopter is on the ground with the collective full down. To lift off, raise the collective positively and do not hesitate at the hover. Instead, continue a vertical climb to approximately $1/2 - 2/3$ of the rotor diameter or until clear of the rotor wash or obstacles, then transition into forward flight.

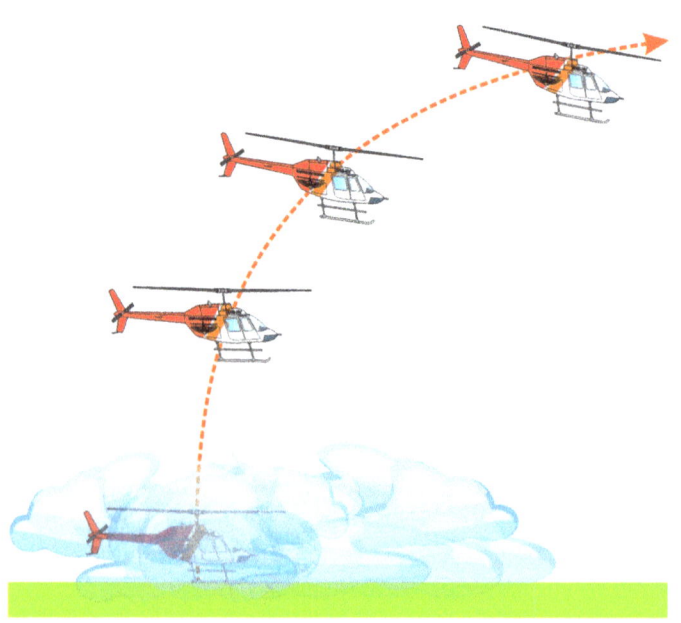

Zero Zero landing   A zero zero landing is used in a similar manner to the maximum performance takeoff. It is used to get the helicopter onto the ground as quickly as possible without first coming to a hover.

Used when landing on a surface such as sand, dust, snow or anytime when the pilot does not wish the downwash to have an adverse effect on the surface or visibility may be reduced because of the effects of the downwash on the surface.

At the end of a normal or steep approach, continue the approach all the way to the ground, levelling the skids at the last moment so that the helicopter lands level with zero speed and zero hover.

Once the skids are on the ground, lower the collective to reduce the downwash.

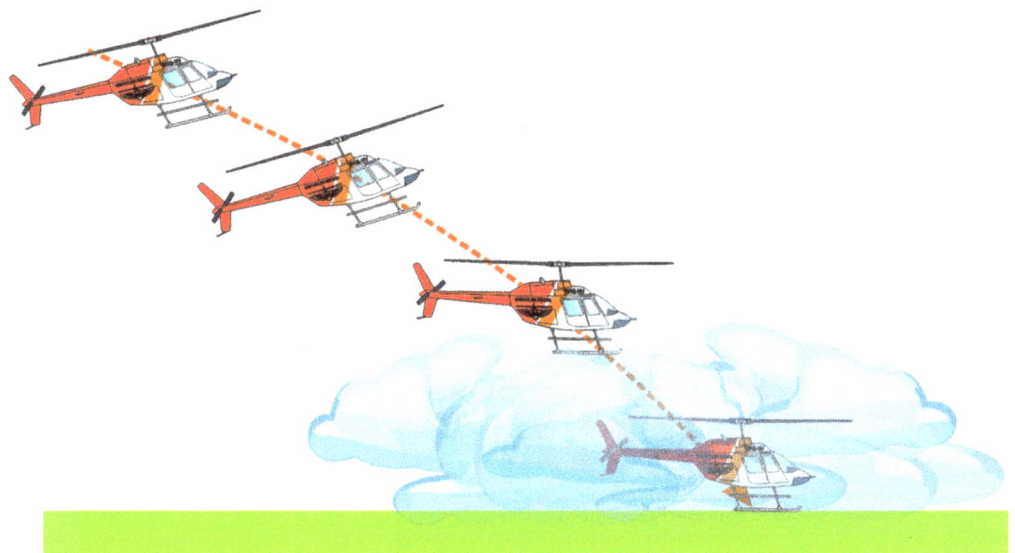

Low Level Circuit

Low Level circuit A low level circuit is a normal racetrack pattern circuit flown not above 300 ft AGL.

The lower the circuit, the tighter and faster it will be. Any nominated height from 50 ft AGL to 300 ft AGL is classified as a low level circuit.

Often the circuit height will be adjusted to consider the surrounding areas, so it will be at the pilot's discretion on what height the circuit is flown.

Lifting off from the ground, transition forward until reaching 55 kts IAS, then turn downwind. Abeam the pad, commence a descending turn to terminate over the helipad.

It is very common to combine the Maximum Performance takeoff and Zero Zero Landing with a Low Level Circuit.

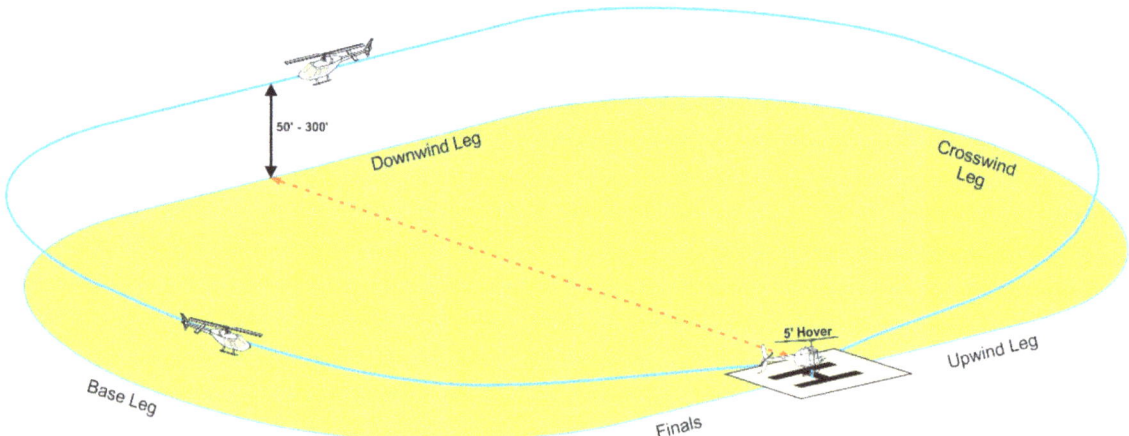

Decelerating Climb and High Hover

Decelerating climb and high hover

When operating very low level, especially in a remote area, it is easy to get disorientated and lose the sense of direction or the sense of where the objective is.

For this reason, the crew will often need to gain some height quickly to look at where they are to maintain situational awareness. Because the HV Diagram should be considered, the height gained should be greater than 400 ft AGL.

From the lift-off, conduct a normal low level circuit, but once abeam the helipad, use flare effect to climb the helicopter. As speed reduces, start a left or right hand turn terminating at the hover into wind. This is almost like a climbing quick stop, except the collective is used as required to fine-tune the height gain.

When at the high hover OGE, the pilot should be able to see the helipad in front of the helicopter.

The high hover will give the crew the opportunity to:

- observe and reconfirm what they will do next, and
- provide a platform as a radio or video link for other operations.

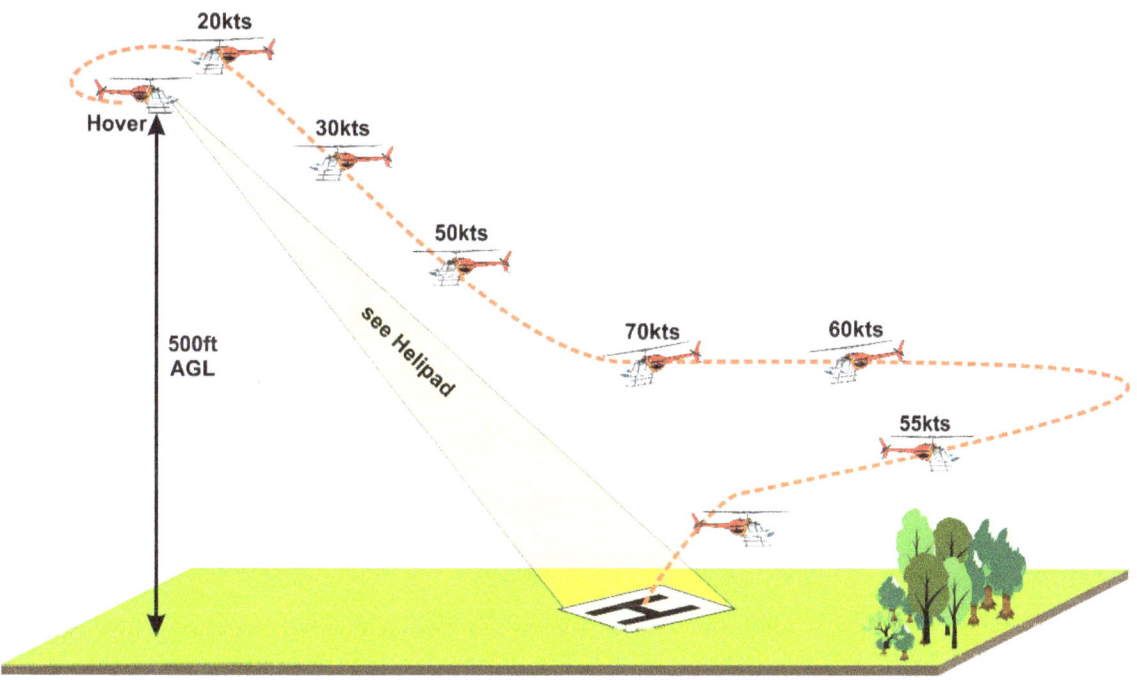

Chapter 12 Low Level Flying

Reference points

To maintain a high hover, the pilot will need to select reference points in front and to the side so that any movement of the helicopter can be detected and adjusted for. It is also important to monitor the height as any sink could result in a VRS developing. The VSI, altimeter or radar altimeter will need to be scanned often to confirm that the altitude is constant. This is most relevant if conducting a left or right pedal turn while at the high hover as the helicopter will experience a tailwind at some point during the turn, and this is when it is most likely to sink and enter VRS.

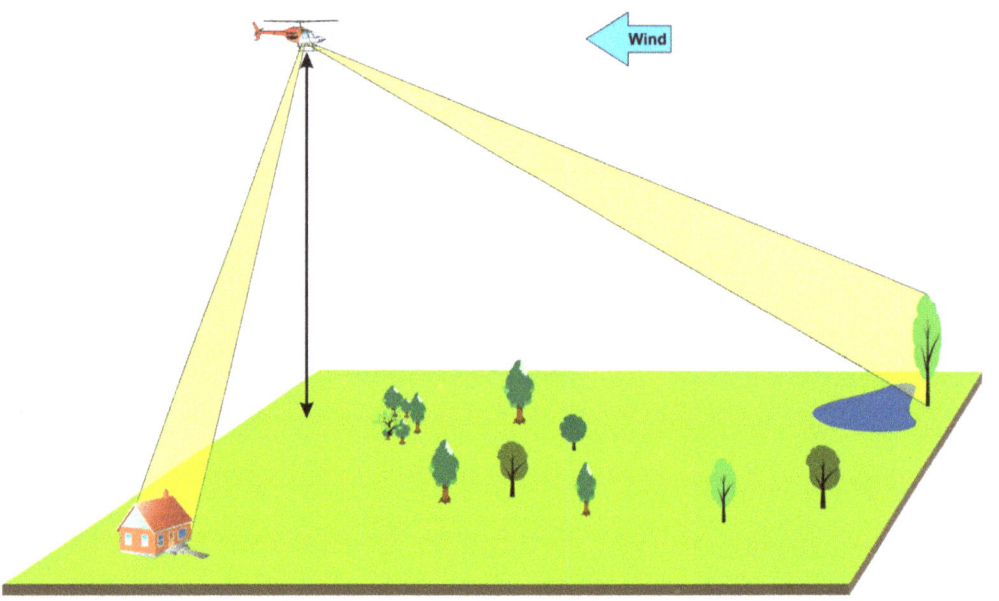

Movement detected

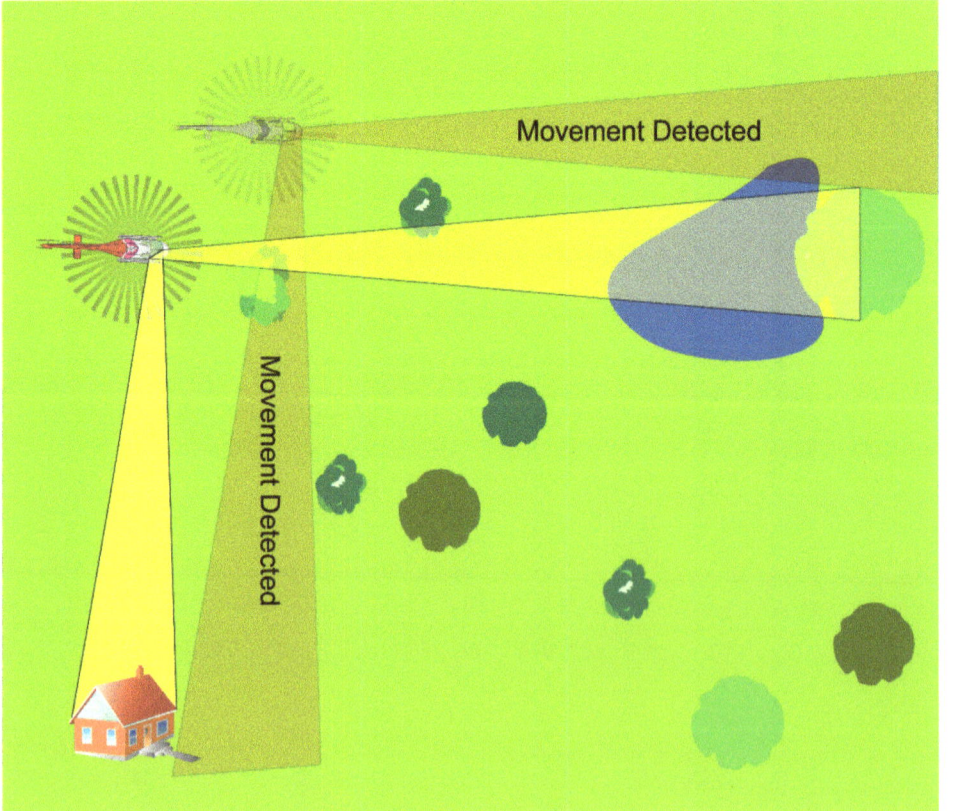

If the pilot detects movement or sink it shall be corrected immediately.

Rapid Descents

Rapid descent

From the high hover, the pilot has several different options on how to descend again. It can be done in a slow conventional manner or when operating low level it may be more appropriate to conduct a rapid descent.

This can be done by conducting S-turns or a 360-degree turns while descending. At the end of the descent, the goal is to either continue operating low level between 45-80 kts or terminate at the hover over the nominated helipad.

For this exercise, we will plan to terminate at the hover over the nominated helipad.

To conduct a rapid descent from a high hover, first lower some collective, allow the nose to drop and the speed to build up to approximately 30 kts IAS. At that point, turn left or right while allowing the helicopter to descend and build speed up to 45-60 kts IAS. As the helicopter rolls through 360 degrees, the helipad should come into view and conduct a steep approach terminating at the hover over the helipad.

360

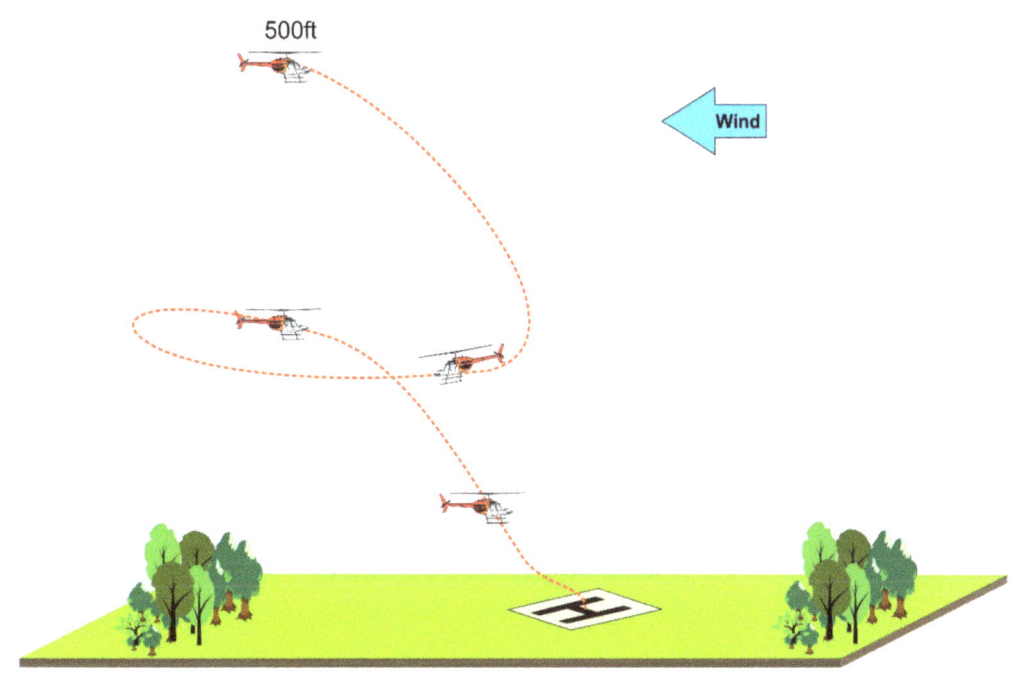

S Turns

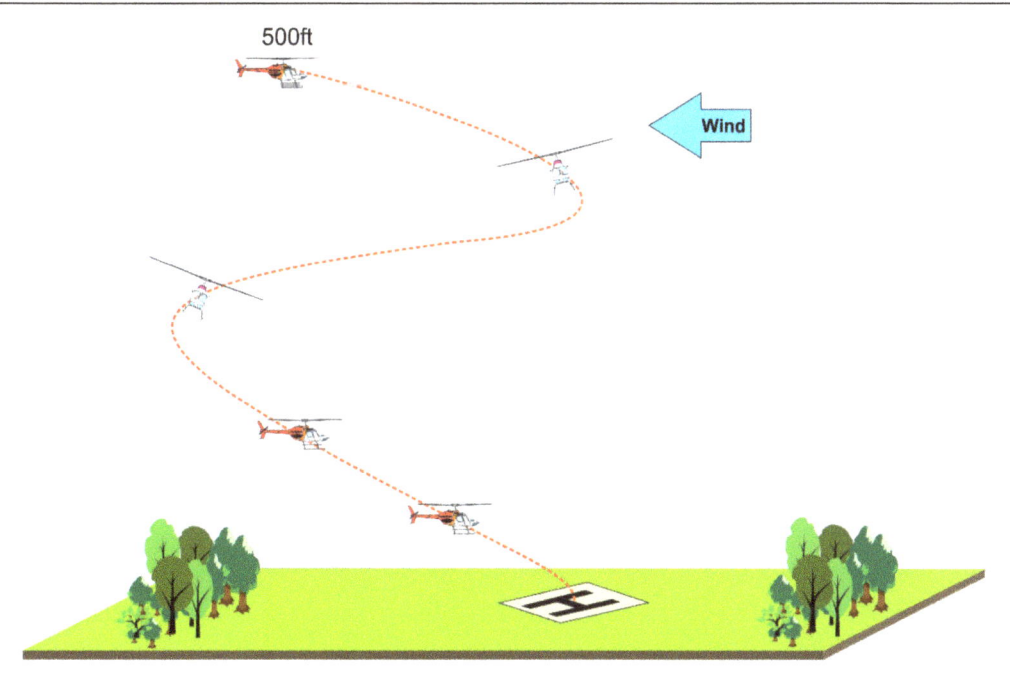

Low Level Emergencies

Introduction

When conducting low level operations, any emergency shall result in the pilot immediately landing the helicopter at the nearest suitable landing area.

The most important emergency to practice is engine failure. All other emergencies (apart from a failed or damaged component) will allow the helicopter to continue flying under power to at least a suitable landing site.

When low level and the engine fails, the pilot has very few options on where to land; instead, the pilot must accept a *controlled* crash straight ahead.

In general:

- the **lower** the helicopter, the more an engine failure is treated like a **quick stop to the ground**.
- the **higher** the helicopter, the more an engine failure is treated like a **normal autorotation to the ground**.

In either situation, the most important thing is to lower the collective, then manage the attitude and, if very low, the altitude with the cyclic while keeping straight with pedals.

There is no time or opportunity for checklists; there is usually no time to consult the crew; instead, the pilot must react instinctively and immediately.

There are three (3) engine failure exercises that may be practised:

1. Engine failure after takeoff (EFATO)
2. Engine failure on approach (EFOA), and
3. Engine failure while low level (EFLL below 150 ft AGL) at 60 kts IAS or greater.

Engine failure after takeoff (EFATO)

When conducting a takeoff and still below 500 ft AGL, if the engine fails, the pilot will have to enter autorotation immediately.

The lower the helicopter is, the less time the pilot has to respond and the fewer options available on where to go; therefore, plan on landing straight ahead as it can only be assumed the takeoff was into wind.

To practice an engine failure after takeoff, conduct a normal takeoff. At some point, the instructor will roll the throttle to IDLE and announce, "Engine Failure".

The pilot will immediately lower collective, maintain the attitude for the current speed of the helicopter, then conduct a flare and termination.

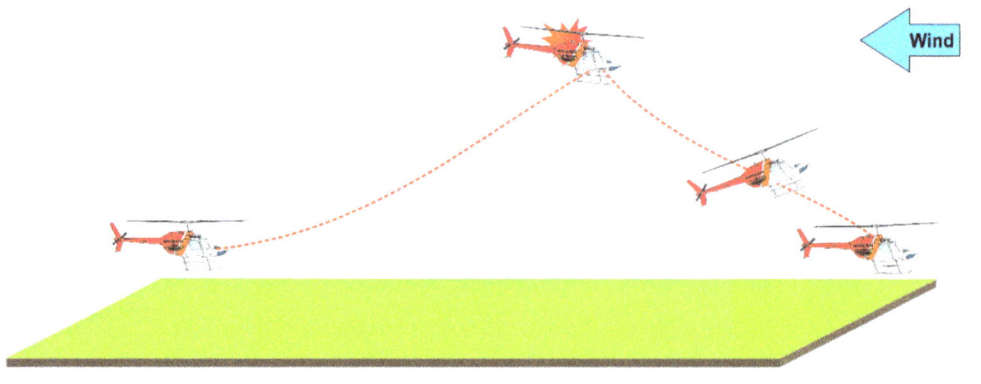

Engine failure on approach (EFOA)

When conducting a normal approach and below 500 ft AGL, if the engine fails, it can quite often go unnoticed for the first few seconds. This is because the collective is already down, and the engine does not have to produce a lot of power, so there may not be any big change in yaw or loss of Rotor RPM.

If a piston engine fails, there is an instant change in engine noise, and it is very obvious. If a turbine engine fails, it may not immediately be as obvious as the turbine can take some time to spool down, so there may not be an instant change in engine noise.

To practice an engine failure on approach, conduct a normal approach. At some point, the instructor will roll the throttle to IDLE and announce, "Engine Failure".

The pilot will immediately lower collective, maintain the attitude for the current speed of the helicopter, then conduct a flare and termination.

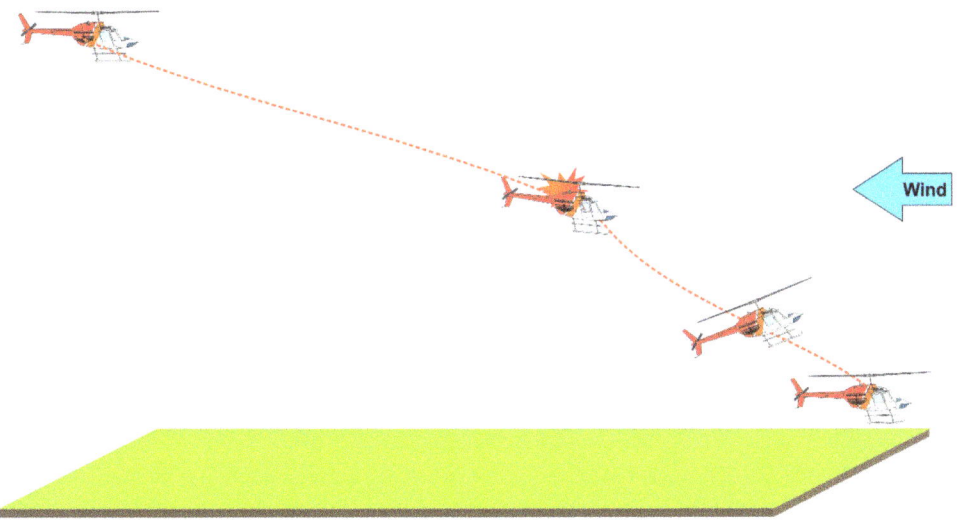

Engine failure low level high speed (EFLL)

The purpose of low flying is to be close to the ground. Often below 150 ft AGL.

Considering the HV Diagram, the pilot will understand the importance of having some airspeed whenever possible and limit the amount of time at very low speeds.

If the engine fails when the helicopter has a speed greater than 35 kts and below 150 ft AGL, the pilot shall immediately conduct a quick stop to the ground.

The higher the speed, the more inertia the pilot has to play with and may actually be able to climb over obstacles if required by sacrificing speed for height.

The lower the speed, the smaller the flare and the less inertia the pilot has to play with.

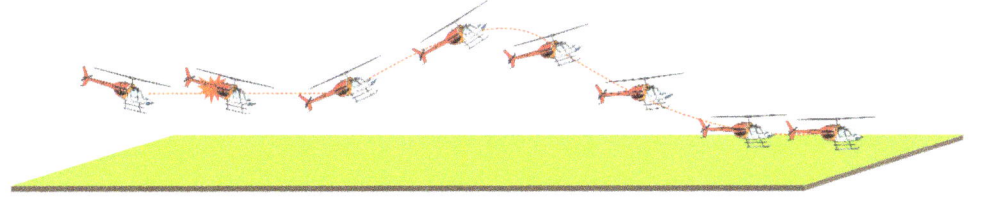

Air Exercises: Low Flying

Introduction

The Low Flying air exercises are not designed to teach specific low level operations; that becomes the responsibility of an operator who will check and train a pilot and crew for a specific task such as external loads, hoisting, sling loads, fire fighting, agricultural spraying, mustering, survey, Night Vision Goggles or anything else that requires the helicopter to be used low level.

Instead, these air exercises are designed to develop the underpinning knowledge, skills and attitudes required to coordinate the flight controls and manoeuvre the helicopter close to the ground.

The air exercises are divided into:

- Managing airspeed and groundspeed
- Turns
- Quick Stops
- Low Level Circuits, Maximum Performance Takeoffs and Zero-Zero landings
- Decelerating climb, high hover and rapid descent, and
- Engine failures low level.

Airmanship

- Conduct a risk analysis of the operation prior to departing, then constantly revise and be prepared to change it based on the PSWATP check and what is actually encountered on the day
- Minimise the time of exposure low level to only that needed
- Maintain a good lookout and communicate with the crew
- Maintain situational awareness

Common faults

Common faults to guard against during the air exercises:

- Poor lookout and no situational awareness of the surroundings
- Incorrect use of the collective and cyclic
- Not able to maintain an accurate height AGL, particularly in quick stops
- Not managing the airspeed
- Turning downwind
- Getting too close to obstacles
- Not making good command decisions if something changes

Prior Preparation   In preparation before a low flying operation and once airborne, complete the following steps:

Step	Action	Discussion
1	PREPARATION	**PRIOR PREPARATION & PLANNING PREVENT POOR PERFORMANCE** Complete a Risk Analysis. Conduct Flight Planning and a crew briefing. Prepare the helicopter for the operation: <ul><li>Fuel</li><li>Payload</li><li>Clean the windows</li><li>Secure the cabin</li><li>Doors on/off</li><li>Weight and balance</li></ul>
2	Enroute	Commence the PSWATP process as it relates to a low flying operation. Conduct a HEFFR check before arriving, adding: <ul><li>Heading home</li><li>Fuel calculation and nominate a time to depart the area, and</li><li>Radio calls before descending.</li></ul>
3	Recce	Conduct a high recce of the low flying area paying particular attention to identifying any unknown or unplanned for hazards or threats. Confirm the PSWATP as it relates to the low flying operation.
4	Plan	Based on all the relevant information, give a PWPTEM brief to/between the crew prior to entering the Low Flying area. *For example:* Power available based on the margin noted on lift-off from base is _____. Wind is from (direction) at (speed) kts. Plan is to conduct low level operations (describe the low level manoeuvres to be conducted or the operation). Threats. Advise and discuss any identified threats and mitigators. Ask the crew if there are any questions or concerns.
5	Manoeuvres	In the low flying training area, the instructor will demonstrate, direct, and monitor each of the exercises.

Chapter 12 Low Level Flying

Air Exercise 12-1: Introduction to Low Flying

Demonstration
The instructor will demonstrate:

- effect of wind on airspeed and ground speed, and
- S-Turns, Reversal Turns and Figure of 8s.

After the demonstration, the instructor will give the student the controls and position the helicopter in the low flying area.

Airspeed vs Groundspeed

Steps
To conduct Airspeed vs Groundspeed, complete the following steps:

Step	Action	Discussion
1	Conduct a Recce	On arrival into the Low Flying Area conduct a recce of the area at 500 ft AGL. This involves flying a large square circuit over the area to inspect it. This is an opportune time to also: - confirm the wind, and - demonstrate the airspeed vs groundspeed relationship to the student (steps 2 and 3).
2	Changes in ground speed with a constant airspeed	- Fly a straight leg at a constant airspeed and visually note the ground speed. - Turn through 90 degrees and then maintain another straight leg while maintaining a constant airspeed and observe the effect on ground speed. - Turn through 90 degrees with a straight leg two more times so that by the end, the helicopter has flown a square pattern, and the student has observed the changes in ground speed while at a constant airspeed. - Discuss where the wind is coming from and an estimated strength.
3	Changes in airspeed with a constant ground speed	- Fly a straight leg at a constant ground speed and visually note the airspeed on the ASI. - Turn through 90 degrees and then maintain another straight leg while maintaining a constant ground speed and observe the effect on airspeed on the ASI. - Turn through 90 degrees with a straight leg two more times so that by the end, the helicopter has flown a square pattern, and the student has observed the changes in airspeed while at a constant ground speed. - Discuss where the wind is coming from and an estimated strength.

S-Turns and Reversal Turns

Steps   To conduct S-Turns and Reversal Turns, repeat the following steps:

Step	Action	Discussion
1	S-Turns and Reversal Turns	Choose a straight-line feature that the S-turns can be flown along, and ensure it is approximately 90 degrees to the prevailing wind. Ensure there are no obstacles, hazards, or threats relative to the operation.
2	Manoeuvre the helicopter to a starting point	Manoeuvre the helicopter at 40-60 kts IAS at 50 ft AGL and cross the straight-line feature at a 45-degree angle.
3	Use cyclic to turn	Use the cyclic to turn left and right in a lazy S manoeuvre crossing the straight line feature each time. ■ Maintain height with collective ■ Balance with pedals ■ Ground track and airspeed with cyclic
4	Use throttle as required	If flying a piston engine helicopter without an electric governor, maintain RRPM with throttle.

5	180-degree turn	Coming to the end of the straight-line feature, conduct a 180-degree (Reversal Turn) turn into wind by:
		■ Initially, turn towards the downwind with cyclic
		■ Raise some collective to start the helicopter climbing
		■ Maintain balance with pedals
		■ As the helicopter starts a small climb, turn back into wind
		■ Passing halfway through the turn:
		    ■ Lower some collective
		    ■ Maintain balance with pedals
		■ Continue the turn through the 180 degrees
		■ Position the helicopter at 50 ft AGL and 40-60 kts IAS, and
		■ Continue to cross the straight-line feature at 45 degrees to continue the S-Turns back the other way.

6	Repeat	The exercise may be repeated at various nominated airspeeds and over multiple straight-line features.

Figure of 8 Turns

Steps To conduct Figure of 8 turns, repeat the following steps:

Step	Action	Discussion
1	Determine the wind	This may be known or have to be decided given the wind cues the pilot is experiencing in the low flying area on the day.
2	Select a central feature	Choose a feature that the Figure of 8 can be flown over, such as a tree, bush, fence, structure etc. Ensure there are no obstacles, hazards or threats relative to the operation.
3	Manoeuvre the helicopter to a starting point	Manoeuvre the helicopter to be at 40-60 kts IAS at 50 ft AGL and cross the central feature at a 45-degree angle relative to the wind.

4	Use cyclic to turn	Use the cyclic to turn left and right through 180 degrees crossing over the central feature each time by:
		
		- Initially, turn towards the downwind with cyclic
		- Raise some collective to start the helicopter climbing
		- Maintain balance with pedals
		- As the helicopter starts a small climb, turn back into wind
		- Passing halfway through the turn:
		    - Lower some collective
		    - Maintain balance with pedals
		- Continue the turn through the 180 degrees
		- Position the helicopter at 50 ft AGL and 40-60 kts IAS, and
		- Continue to cross the central feature at 45 degrees and then repeat the steps above to complete the second half of the Figure 8 turn.

5	Repeat	The exercise may be repeated at various nominated airspeeds and over multiple central features.

Air Exercise 12-2: Quick Stops

Demonstration The instructor will demonstrate:

- Straight ahead Quick Stops
- 90 degree Quick Stops
- 180 degree Quick Stops, and
- Turn and Flare and Flare and Turn methods.

After the demonstration, the instructor will give the student the controls and position the helicopter in the low flying area.

Straight ahead Quick Stops

Steps To conduct a Straight Ahead Quick Stop, complete the following steps:

Step	Action	Discussion
1	Determine the wind	This may be known or have to be decided given the wind cues the pilot is experiencing in the low flying area on the day.
2	Set up	Establish the helicopter at 50 ft AGL between 45-80 kts IAS into wind. (Height may vary at the discretion of the instructor.)
3	Lower collective	Commence the quick stop by lowering some collective and maintaining a constant heading with pedals.
4	Aft cyclic	At the same time the collective is being lowered, control height by moving the cyclic aft. (Remember the Control Effectiveness section.) This will "flare" the helicopter reducing speed while maintaining height AGL.
5	Coordinate the collective, cyclic and pedals	As the helicopter slows down, use the collective and cyclic together to maintain height while reducing speed. Maintain a constant heading with pedals.
6	Throttle	Because this can be a harsh manoeuvre, RRPM may want to increase. In a piston engine, the pilot may require some throttle to manage RRPM. In a turbine engine, there may be a transient overspeed of the rotor system. To avoid this, the Quick Stop will need to be entered with more finesse, and the collective may be required to help control RRPM (raise again), or the pilot may have to reduce the harshness of the flare so the governing system can keep up, as the throttle cannot be manipulated for this purpose in a turbine.
7	Recover	As the flare effects decrease, start raising the collective to maintain height. Before the helicopter comes to a complete stop, level the skids with the ground with some forward cyclic and return to the high hover. Keep straight with pedals.

90-degree Quick Stop

Steps   To conduct a 90-degree Quick Stop, complete the following steps:

Step	Action	Discussion
1	Determine the wind	This may be known or have to be decided given the wind cues the pilot is experiencing in the low flying area on the day.
2	Set up	Establish the helicopter at 50 ft AGL between 45-80 kts IAS 90 degrees to the wind.
3	Lower collective	Commence the quick stop by lowering some collective and maintaining a constant heading with pedals.
4	Aft cyclic	At the same time the collective is being lowered, control height by moving the cyclic aft. This will "flare" the helicopter reducing speed while maintaining height AGL.
5	Coordinate the collective, cyclic and pedals	As the helicopter slows down, use the collective and cyclic together to maintain height while reducing speed. Maintain a constant heading with pedals.
6	Throttle	Because this can be a harsh manoeuvre, and you will also be introducing a turn that will increase the disk loading, the RRPM may want to increase. In a piston engine, the pilot may require some throttle to manage RRPM. In a turbine engine, there may be a transient overspeed of the rotor system. To avoid this, the Quick Stop will need to be entered with more finesse, and the collective may be required to help control RRPM (raise again), or the pilot may have to reduce the harshness of the flare so the governing system can keep up, as the throttle cannot be manipulated for this purpose in a turbine.
7	Left or right cyclic	As the helicopter begins to slow, use left or right cyclic to turn the helicopter into the wind. As the helicopter slows down, use the collective and cyclic together to maintain height while reducing speed. Maintain balance in the turn with pedals.
8	Recover	As the flare effects decrease, start raising the collective to maintain height. Before the helicopter comes to a complete stop, level the skids with the ground with some forward cyclic and return to the high hover. Keep straight with pedals.

Mike Becker's Helicopter Handbook

180-Degree Quick Stop

Steps   To conduct a **Turn and Flare** 180-degree Quick Stop, complete the following steps:

Step	Action	Discussion
1	Determine the wind	This may be known or have to be decided given the wind cues the pilot is experiencing in the low flying area on the day.
2	Set up	Establish the helicopter at 50 ft AGL between 45-80 kts IAS downwind.
3	Cyclic	Commence a turn, left or right. The angle of bank does not have to be high. 10-15 degrees angle of bank is sufficient.
4	Aft cyclic	Once the turn has commenced, start using some aft cyclic to initiate a flare in the turn. It is the aft cyclic that will reduce the speed in the flare, not the angle of bank.
5	Collective	As the helicopter passes through the 90-degree point, start lowering some collective to maintain height. *If done correctly, the collective would be lowered in the turn, not raised.* Maintain balance with pedals and throttle, as required.
6	Coordinate	As the helicopter slows down, use the collective and cyclic together to maintain height while reducing speed. Maintain balance in the turn with pedals.
7	Recover	As the flare effects decrease, and the helicopter is now pointing into the wind, start raising the collective to maintain height. Before the helicopter comes to a complete stop, level the skids with the ground with some forward cyclic and return to the high hover. Keep straight with pedals.

Air Exercise 12-3: Reversal Turns

Demonstration The instructor will demonstrate some additional methods to complete a reversal turn.

After the demonstration, the instructor will give the student all of the controls and position the helicopter in the low flying area.

Flat 180-Degree Turn

Steps To conduct a flat 180-degree turn, complete the following steps:

Step	Action	Discussion
1	Determine the wind	This may be known or have to be decided given the wind cues the pilot is experiencing in the low flying area on the day.
2	Set up	Establish the helicopter at 50 ft AGL between 45-80 kts IAS. This can be done from any position relative to the wind, BUT it will be important to maintain airspeed throughout as wind and ground speed change.
3	Cyclic	Commence a turn, left or right, to approximately 15-30 degrees angle of bank.
4	Collective	Use collective as required to maintain height. Remember this is not a Quick Stop; the purpose is to do a smooth 180-degree turn at a constant airspeed. Maintain balance with pedals.
5	Cyclic	Fine turn altitude with fore or aft cyclic in the turn so the helicopter does not sink or climb. Roll out with opposite cyclic on reaching the 180-degree mark and fly away at a constant airspeed.

Cyclic 180-Degree Turn

Steps   To conduct a 180-degree cyclic turn, complete the following steps:

Step	Action	Discussion
1	Determine the wind	This may be known or have to be decided given the wind cues the pilot is experiencing in the low flying area on the day.
2	Set up	Establish the helicopter at 50 ft AGL between 45-80 kts IAS with the wind 90 degrees to the right or left of the current ground track.
3	Cyclic	If the wind is 90 degrees to the left, then the cyclic turn will be to the left: ■ Initially, turn right towards the downwind, then immediately turn left while pitching the nose up approximately 30 degrees. ■ Speed will wash off, and pedals will need to be used to assist the turn and maintain balance. ■ As the helicopter passes through 90 degrees, finish the turn with some forward cyclic returning to the entry speed and fly away. If the wind is 90 degrees to the right, then the cyclic turn will be to the right: ■ Initially, turn left towards the downwind, then immediately turn right while pitching the nose up approximately 30 degrees. ■ Speed will wash off, and pedals will need to be used to assist the turn and maintain balance. ■ As the helicopter passes through 90 degrees, finish the turn with some forward cyclic returning to the entry speed and fly away.
4	Collective	If done correctly, collective will not need to be used; however, use collective as required to fine-tune height if needed. Maintain balance with pedals.

Pedal and Torque 180 degree Turns

Steps   To conduct a pedal or torque turn, complete the following steps:

Step	Action	Discussion
1	Determine the wind	This may be known or have to be decided given the wind cues the pilot is experiencing in the low flying area on the day.
2	Set up	Establish the helicopter at 50 ft AGL between 45-80 kts IAS with the wind 90 degrees to the right or left of the current ground track.
3	Collective	Approaching the turn point, lower some collective. In a piston engine, reduce power by approximately 2" MAP. In a turbine engine, reduce power by approximately 10% Tq. Maintain balance. At this point, maintain height with cyclic.

Chapter 12 Low Level Flying

Step	Action	Discussion
4	Cyclic	At the same time collective is being lowered use some aft cyclic to pitch the nose up approximately 30 degrees, this will result in: ■ the helicopter climbing quickly ■ losing airspeed, and ■ a need to maintain the helicopter straight with pedals. If the wind is 90 degrees to the left, then the ***pedal turn*** will be to the left; commence this by: ■ As airspeed reduces to no less than translational lift (15-35 kts IAS). ■ Raise the collective back to the pre-entry amount (increase by 2" MAP or 10% Tq) and use left pedal to make the nose of the helicopter turn left 180 degrees about the main rotor mast. ■ As the helicopter now points down towards the ground, keep straight with pedals and use forward cyclic to allow the airspeed to again build up to the pre-entry speed. ■ Level out and fly away. If the wind is 90 degrees to the right, then the ***torque turn*** will be to the right; commence this by: ■ As airspeed reduces to no less than translational lift (15-35 kts IAS). ■ Raise the collective back to the pre-entry amount (increase by 2" MAP or 10% Tq) and allow ***torque effect*** to make the nose of the helicopter turn left 180 degrees about the main rotor mast. At times you may need some right pedal to help it on its way. ■ As the helicopter now points down towards the ground, keep straight with pedals and use forward cyclic to allow the airspeed to again build up to the pre-entry speed. ■ Level out and fly away.

Air Exercise 12-4: Low Level Circuits with a Maximum Performance Takeoff and a Zero Zero Landing

Demonstration	The instructor will demonstrate a maximum performance takeoff, a low level circuit and a zero-zero landing.
	After the demonstration, the instructor will give the student all of the controls and position the helicopter in the low flying training area over a helipad.

Maximum Performance Takeoffs

Steps	To conduct a Maximum Performance Takeoff, complete the following steps:

Step	Action	Discussion
1	Determine the wind	This may be known or have to be decided given the wind cues the pilot is experiencing in the low flying area on the day.
2	Set up	Establish the helicopter on the ground, ready for lift-off: The HEFFR check is completed, the RRPM is top of the green, and the pilot is ready to pull pitch (collective).
		Confirm there is a sufficient power margin (CAT4) to conduct a vertical takeoff. This may not be obvious as there is no ability to do a power check prior to the lift-off. Instead, the pilot will have to rely on previous experience and operations into and out of the same pad to make a guesstimation.
		Remember that if there is not sufficient power, the pilot can still land again before committing to forward flight.
3	Collective	Positively raise collective to lift the helicopter off the ground and continue with a vertical climb until clear of the rotor wash or obstacles.
		Use maximum power available or, at the very least, enough power to be moving away from the ground in a positive manner.
		This exercise requires a positive climb rate rather than a very slow measured climb rate.
4	Cyclic and pedal	Use cyclic and pedals to control ground position and direction.
5	Transition	Once into clear air and leaving the ground cushion or climbing clear of the obstacles and with a rate of climb, transition into forward flight to conduct a low level circuit.

Chapter 12 Low Level Flying

Low Level Circuit

Steps   To conduct a low level circuit, complete the following steps:

Step	Action	Discussion
1	Lift-off	Conduct the Maximum Performance Takeoff described above
2	Transition	Once clear of the obstacles, transition into forward flight in the normal manner.
3	Racetrack circuit not above 300 ft AGL	**Crosswind** Fly straight ahead, accelerating to 55 kts IAS and climb to half the nominated circuit height (in this case, 150 ft AGL) before conducting a climbing turn crosswind in a racetrack pattern. **Downwind** Level off downwind at 300 ft AGL and no more than 60-70 kts IAS. Abbreviate all checks and only consider those items that may have changed. There is no requirement for a HEFFR or PWP. The focus is on flying the helicopter, not on checks. **Base** Once abeam the nominated helipad, commence a descending, decelerating racetrack turn to establish on a high and steep finals leg. Do not go any lower than half the circuit height (in this case, 150 ft AGL) until rolling out on finals. **Finals** Establish a normal or high and steep finals leg (this can depend on the obstacles). Maintain some positive forward speed but do not be so absorbed in monitoring the ASI for the normal linear closure rate; instead, adjust speed as required to be fully in control early of the rate of descent.
4	Termination	Conduct a Zero Zero Landing.

Zero Zero landing

Steps To conduct a Zero Zero landing, complete the following steps:

Step	Action	Discussion
1	Set up	Establish the helicopter on a normal or steep final approach to the nominated helicopter landing site.
		On very short finals, visually look down the approach path to the touchdown area and determine if it is suitable before committing to the landing.
		You are checking:
		■ the landing area is big enough to accommodate the rotor disc plus a good margin (surrounds)
		■ where the tail will be placed in the area and that area is clear
		■ where the landing gear is going to be on the ground, and
		■ what is the condition of the surface and is there any obvious slope.
		If it is deemed suitable, continue. If it is deemed not suitable, go-around.
2	Zero Zero landing	Continue the approach and follow it through all the way to the ground.
		The goal is to try to stay slightly higher and always moving forward to stay ahead of the downwash until the last possible moment.
		Do not terminate at the hover; instead, terminate with the skids on the ground, reducing the time the downwash has to affect the surface.
		Upon landing, continue to lower the collective so that there is no longer any downwash.
3	Active scan	A zero-zero landing is usually for an operational reason, such as landing on sand, snow, or ash on a fire ground, etc. This means once the downwash disturbs the surface, the pilot can lose all visibility if looking out towards the horizon at a critical time.
		To avoid this, as the helicopter approaches the ground, the pilot should move their eyes vertically downwards to the ground and use that as the new hover reference point.
		With a loss of visibility, it is important to now commit to the ground.
		Do not hesitate and try to come to a hover as you cannot see the ground.
		A zero-zero landing is an advanced manoeuvre.

Air Exercise 12-5: Decelerating Climb, High Hover and Rapid Descents

Demonstration   The instructor will demonstrate a decelerating climb, high hover, and a rapid descent.

After the demonstration, the instructor will give the student all of the controls and position the helicopter in the low flying training area over a helipad.

Decelerating Climb, High Hover and Rapid Descent

Steps   To conduct a decelerating climb, high hover and rapid descent complete the following steps:

Step	Action	Discussion
1	Determine the wind	This may be known or have to be decided given the wind cues the pilot is experiencing in the low flying area on the day.
2	Normal takeoff and low level circuit	Conduct a normal or maximum performance takeoff and low level circuit.
3	Abeam the helipad	Once abeam the helipad: ■ Use some aft cyclic to commence a zoom climb using the effects of the flare. The helicopter will climb while losing speed. Use the collective as required to assist in the climb. ■ Maintain balance with pedals. ■ Passing through 30-40 kts IAS and 350 ft AGL, commence a left or right-hand turn and terminate in a high 400 ft hover into wind.
4	Hover	Maintain a high hover at 400 ft AGL (or higher) and conduct a left or right-hand pedal turn. Use ground reference points to maintain the ground position and the VSI, Altimeter and RAD ALT to maintain height. This can be a very tricky manoeuvre to get right.

Rapid Descent

Steps To conduct a rapid descent, complete the following steps:

Step	Action	Discussion
1	Determine the wind	This may be known or have to be decided given the wind cues the pilot is experiencing in the low flying area on the day.
2	From the high hover	Lower some collective to commence a descent and allow the nose to pitch down approximately 30 degrees.
3	Cyclic and collective	The helicopter will start to move forward and descend. As the IAS reaches 30 kts, start a left or right-hand turn with cyclic. Either continue the turn for a full 360 degrees or alternate left and right cyclic to conduct S-Turns to lose height but not move horizontally across the ground. It is important not to overshoot the landing area. Allow speed to increase to 45-60 kts IAS, but no more. Manage the ROD with collective. Maintain balance with pedals.
4	Rollout	Passing through 360 degrees or when low enough and the landing site is positioned in front, roll out with cyclic and conduct a steep approach. Terminate at the hover over the nominated helipad or continue for a zero-zero landing.

Chapter 12 Low Level Flying

Air Exercise 12-6: Low Level Emergencies

Demonstration The instructor will demonstrate the following low level emergencies in a runway environment where good touch down areas are assured.

- Engine failure after takeoff (EFATO)
- Engine failure on approach (EFOA)
- Engine failure low level high speed (EFLL)

These emergencies will not be practised within the low level flying area.

After the demonstration, the instructor will give the student all of the controls and position the helicopter inside a runway environment.

Engine Failure After Takeoff

Steps To conduct an engine failure after takeoff, complete the following steps:

Step	Action	Discussion
1	Determine the wind	This should be known and takeoff and approach would be expected to be into wind.
2	Set up	Conduct a normal takeoff and climb straight ahead.
3	Engine failure	The instructor will announce "Engine Failure".
4	Immediate actions	Lower Collective fully. Look outside. **Maintain the current attitude with cyclic.** The preferred speed is between 35-60 kts IAS. Keep straight with pedals. Entry to autorotation from a climb on takeoff will normally require a lot of **forward cyclic** as the helicopter aerodynamically changes from a climb to a descent. This can be very counterintuitive initially. The danger is the pilot conducts a flare at 150 ft or higher on entry which is way too early as the flare effects will not last, so 150 ft is a long way to fall.
5	Flare	The flare can seem different because the helicopter may still be accelerating in its descent, and the speed may be slow. The pilot will have to adjust based on the circumstances; however, if the exercise is done correctly, once reaching the normal flare height, conduct a flare and power termination to the hover (the instructor will manage the throttle).

Mike Becker's Helicopter Handbook

Engine Failure on Approach

Steps To conduct an engine failure on approach, complete the following steps:

Step	Action	Discussion
1	Determine the wind	This should be known and takeoff and approach would be expected to be into wind.
2	Set up	Conduct a normal approach to a helipad.
3	Engine failure	The instructor will announce "Engine Failure".
4	Immediate actions	Lower Collective fully. Look outside. **Maintain the current attitude with cyclic.** The preferred speed is between 35-60 kts IAS. Keep straight with pedals.
5	Flare	Conduct a flare and power termination to the hover. (Instructor will be managing the throttle). When on approach, the speed will be reducing, and when the engine fails and the pilot lowers the collective, the ROD will increase. The flare will vary depending on the current speed. The lower the speed, the less flare and the bigger the pitch pull.

Engine Failure Low Level High Speed

Steps To conduct an engine failure low level high speed, complete the following steps:

Step	Action	Discussion
1	Determine the wind	This may be known or have to be decided given the wind cues the pilot is experiencing in the low flying area on the day.
2	Set up	Establish the helicopter at 50-100 ft AGL at 80 kts IAS.
3	Engine failure	The instructor will announce "Engine Failure".
4	Immediate actions	Lower Collective as required. This may not be as quick as a normal autorotation as the flare effects from the aft cyclic will be causing the Rotor RPM to increase, so some collective may still be required to control the Rotor RPM. Treat the entry to autorotation in a similar manner to a Quick Stop. It is a coordinated effort. Look outside. Flare the helicopter with cyclic in the same manner as a quick stop to manage the height. Keep straight with pedals.
5	Termination	As the flare effects die out, conduct a power termination to the hover. (Instructor will be managing the throttle).

13

Weight and Balance

Aim
To ensure the helicopter remains within its weight and Centre of Gravity (CofG) limits before lift-off.

Objectives
On completion of this lesson, the student will be able to:

- list and explain some weight and balance terms
- explain a CofG Envelope
- state where to find a load data sheet for a particular helicopter, and
- complete a simple weight and balance calculation for a sample helicopter.

Motivation
Helicopters, like all aircraft, are limited by how much weight they can carry and where it is loaded within the fuselage.

If the Maximum lift-off weight is exceeded, the aircraft can experience stresses outside its design limits and not perform as expected.

If the Centre of Gravity (loading) is incorrect, the pilot may run out of cyclic control movement and be unable to control the helicopter.

Before every flight, the pilot must ensure the aircraft is within its weight and balance limits.

Preparation: Weight and Balance

Balance Terms

Terms

Terms described in this section include:

- Datum
- Buttline
- Arm
- Station
- Moment
- Index units, and
- Centre of Gravity (or balance point).

Datum

The *Datum* is an imaginary line from which all **longitudinal** (fore and aft) measurements of *Arms* are taken.

The reference *Datum* can be located at any position in the vicinity of the helicopter at the discretion of the helicopter manufacturer.

It is usually an imaginary line in front of the helicopter or running through the rotor mast.

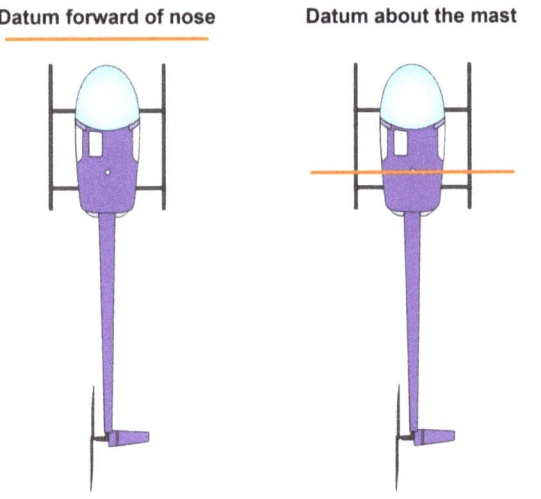

Buttline

The *Buttline* is an imaginary line from which all **lateral** (left and right) measurements of *Arms* are taken. The *Buttline* is normally drawn straight down the middle, dividing the helicopter into two equal parts.

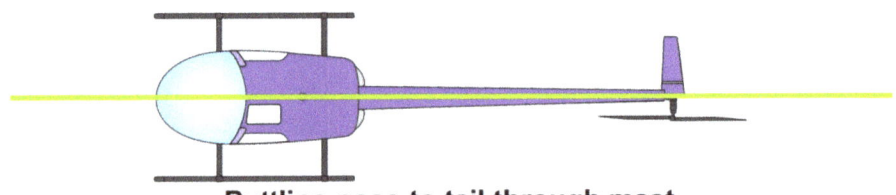

Buttline nose to tail through mast

Chapter 13 Weight and Balance

Arm

An *Arm* is the distance from a reference datum or Buttline to the Centre of Gravity of each item included in a weight and balance calculation.

The *Arm* is **positive** (+ve) if measured:

- aft of the Datum, or
- right of the Buttline.

The *Arm* is **negative** (-ve) if measured:

- forward of the Datum, or
- left of the Buttline.

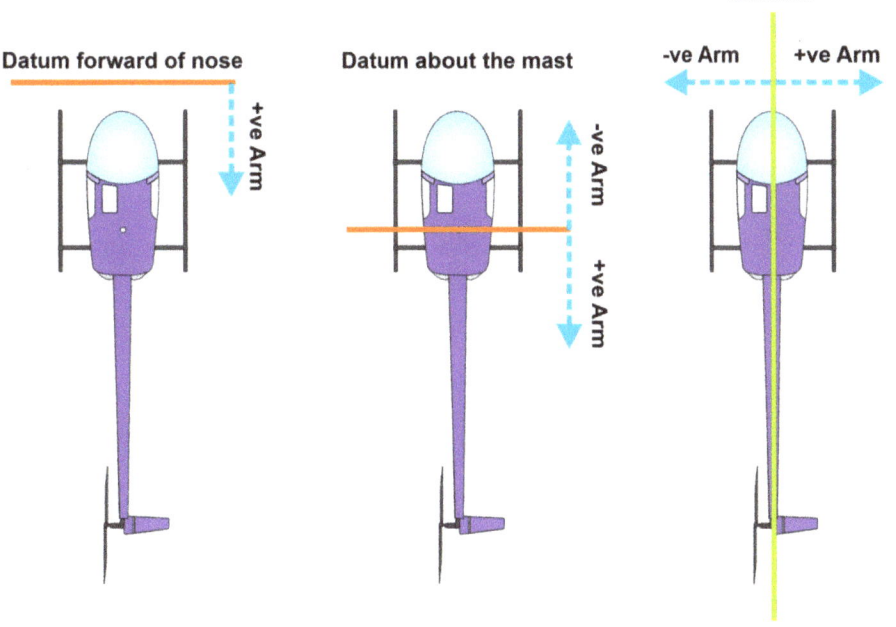

Station

The term *Station* has the same meaning as *Arm*.

Sta 32 means the item is placed 32 units aft of the *Datum*.

Sta -32 means the item is placed 32 units forward of the *Datum*.

Moment

A *Moment* (or turning moment) is the product of the force multiplied by the perpendicular (90 degrees) distance from the *Datum* and/or *Buttline* to the pivot point where the object will turn.

It can be calculated using the following formula.

$$\text{WEIGHT} \times \text{ARM} = \text{MOMENT}$$

All moments that are:

- **clockwise** are considered **positive**, and
- **anti-clockwise** considered **negative**.

The *Total Moment* is the total weight of the helicopter multiplied by the distance between the *Datum/Buttline* and the Centre of Gravity (balance point).

Index units

Some *Moments,* when calculated, are very large; therefore, instead of working with unmanageable numbers, the moment can be divided by a factor of 10, 100, 1000 or more. These moments are then referred to as index units.

For example, consider a Moment that is equal to 12936472. This would be a very large number to try to do calculations with.

If it were divided by 1000, then it would have an index unit of 12936.472.

CofG Range and Limits

The Centre of Gravity (CofG) is a calculated point where all the helicopter's weight is said to act. Its position is measured as a distance from the *Datum* and *Buttline*.

If moments are taken about this point, all the clockwise moments would equal the anti-clockwise moments.

It can be calculated by the following formula:

$$\frac{\textbf{TOTAL MOMENT}}{\textbf{TOTAL WEIGHT}} = \textbf{CENTRE OF GRAVITY POSITION}$$

There is no single CofG point for a helicopter, as its loading will change every flight. Instead, the manufacturer allows a CofG range. This means any CofG calculation must fall within the published range. This can then be graphically displayed as a CofG Envelope.

CofG Envelope

The CofG Envelope is a visual method of determining whether the CofG is within limits. Once calculations are made, the pilot can plot the result on the graph. If the calculated CofG point falls within the envelope or on the line, then the CofG is within limits.

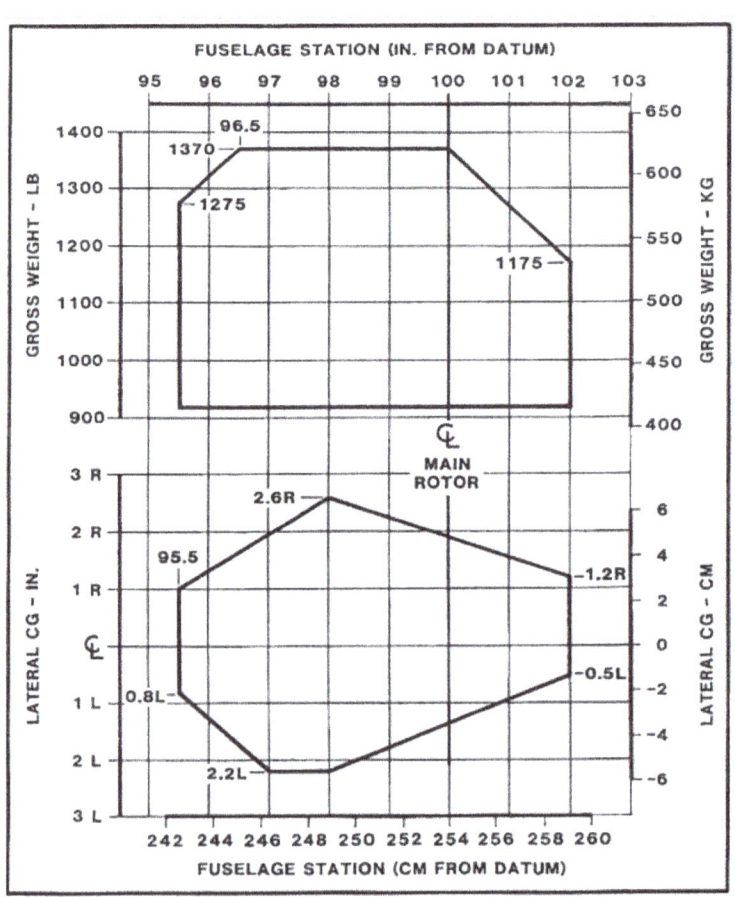

Chapter 13 Weight and Balance

Weight Terms

Terms

Terms described in this section include:

- Aircraft Empty Weight
- Basic Operating Weight
- Payload
- Zero Fuel Weight
- Ramp Weight
- Take-off Weight, and
- Landing Weight.

Aircraft Empty Weight

The *Aircraft Empty Weight* includes the:

- helicopter structure
- power plant
- equipment, as per the equipment list stated in the RFM, such as
  - fire extinguisher, and
  - first aid kit
- unusable fuel
- systems fluid (as in hydraulic fluid), and
- all engine oil.

Aircraft Empty CofG Position

The *Aircraft Empty Centre of Gravity Position* is the CofG of a helicopter in its *Empty Weight* condition. This is calculated by an engineer once every two years or after a major rebuild, repaint or installation of new equipment. The load data sheet is then added to the RFM (usually in the Weight and Balance Section) and remains available in the helicopter at all times, as the pilot will need it when making their weight and balance calculations.

EMPTY WEIGHT (kg)	LONGITUDINAL		LATERAL		CONFIGURATION (ALSO SEE EQUIPMENT LIST)
	ARM (mm)	INDEX MOMENT	ARM (mm)	INDEX MOMENT	
861.6	2938	2531397	9.9	8560.0	FIVE SEATS TOTAL
IMPERIAL					
EMPTY WEIGHT (lb)	ARM (in)	INDEX MOMENT	ARM (in)	INDEX MOMENT	FIVE SEATS TOTAL
1899.5	115.67	219713	0.39	743	

THE ABOVE WEIGHTS INCLUDE:
EMPTY WEIGHT: UNUSABLE FUEL & FULL ENGINE OILS

LOAD SYSTEM
1) REFER TO FLIGHT MANUAL & SUPPLEMENTS FOR LOAD LIMITATIONS.

WORK ORDER	DATUM	Lateral	Center Line of Aircraft
WB-6128		Horizontal	F.S. '0'

Basic Operating Weight

The *Basic Operating Weight* or Aircraft Prepared for Service Weight (APSW) includes the:

- helicopters empty weight
- crew
- crews' baggage, and
- water (toilet and drinking).

It is the aircraft's total weight, ready to fly, **excluding** the payload and fuel.

Payload	The *Payload* includes anything from which revenue is obtained. This can include passengers and baggage, freight or cargo.
Zero Fuel Weight	The *Zero Fuel Weight* is the *Basic Operating Weight* plus the *Payload*.
Ramp weight	The *Ramp Weight* is the *Zero Fuel Weight* plus the fuel loaded on board for the flight.
Take-off weight	*Take-off Weight* is the *Ramp Weight* less the fuel used for run-up and taxi.
	This is not usually applicable for a helicopter because any taxiing is done once the helicopter has lifted to the hover. However, with fixed wing and possibly even helicopters that have wheels, it is possible to be overweight when leaving the ramp due to excessive fuel, but the crew's intention is for the fuel to be consumed by the engines during the taxi and just before commencing the take-off.
	Think of a long-haul airliner needing maximum fuel but having a long taxi and hold at a big international airport.
Landing Weight	*Landing Weight* is the *Take-off Weight* less the fuel used during the flight.
Summary of weights	The different types of weights are summarised below.

Basic Operating Weight

    +   **Payload**

    =   **Zero Fuel Weight**

          +   **Fuel**

          =   **Ramp Weight**

                +   **Fuel for Taxi**

                =   **Take-off Weight**

                       -   **Flight fuel**

                       =   **Landing Weight**

Chapter 13 Weight and Balance

Converting Fluid Volume to Weight

Specific Gravity Specific Gravity (SG) is a term used to compare the weight of a certain fluid volume to the same volume of water (known as the reference fluid).

Because the reference fluid (water) has a specific gravity (SG) of one (1), any other fluid can be measured or compared against it.

If a volume of a substance is:

- **heavier** than that of the same volume of water, then the specific gravity will be **greater than one (1)**.
- **lighter** than that of the same volume of water, then the specific gravity will be **less than one (1)**.

Aviation fuel and oil are lighter than water; therefore, their SG is less than one (1). Avgas has a SG of 0.72. Jet A1 has a SG of 0.8.

Item	=	Weight	Specific Gravity
1 Imperial gal of water	=	10 lb	1
1 Imperial gal of AVGAS	=	7.2 lb	0.72
1 Imperial gal of oil	=	9.1 lb	0.91
1 litre of water	=	1 kg	1
1 litre of AVGAS	=	0.72 kg	0.72
1 litre of oil	=	0.91 kg	0.91

Example: Calculating the CofG Position

Introduction This section provides an example of calculating a CofG position given the relevant information of a sample R22 helicopter. This information is generic and not specific to any particular helicopter. For accurate information, refer to the RFM for the helicopter you are flying.

Weights The weight limits for the sample R22 are detailed below.

Empty weight	831.5 lb (377.16 kg)
Empty moment	87116 lb/in
Maximum All Up Weight (MAUW)	1370 lb (621.43 kg)
Minimum All Up Weight (Min AUW)	920 lb (417.31 kg)
Max per seat	240 lb (108.86 kg)
Max in BAX (BAX = Baggage)	50 lb (22.68 kg)

Centre of gravity limits

The CofG limits for the sample R22 are detailed below.

Forward CofG limit	95.5 inches aft of the Datum
Aft CofG limit	102 inches aft of the Datum
Left CofG limit	-2.2 inches left of the Buttline
Right CofG limit	+2.6 inches right of the Buttline

Lateral CofG arms

The **Lateral CofG** Arms for the sample R22 are detailed below.

Pilot and BAX under the right seat	+10.7 inches right of the Buttline
PAX and BAX under the left seat (PAX = Passengers)	- 9.3 inches left of the Buttline
Fuel	-11 inches left of the Buttline

Longitudinal CofG arms

The **Longitudinal CofG** Arms for the sample R22 are detailed below.

Pilot, PAX and BAX	+78 inches aft of the Datum
Fuel	+108 inches aft of the Datum

Loaded on board

The following is now loaded on board, requiring a calculation:

- Pilot weighing 173 lb
- Passenger weighing 160 lb
- 60 litres of fuel (Avgas with an SG of .72 = 95 lb)
- 20 lb of baggage under the pilot's seat

Note: It is common for pilots to receive information in a mix of imperial and metric measurements (e.g. lb and kg), so before commencing a CofG calculation, they convert the weights into a common system of measurement, that is, either imperial or metric. A calculation will not be accurate when mixing metric with imperial information and vice versa.

Calculating the CofG position

To calculate the new CofG position for the sample R22 above, the pilot needs to consider the CofG relative to the Datum and the Buttline.

Chapter 13 Weight and Balance

Buttline Calculating the CofG position relative to the Buttline is shown below.

Lateral

Item	Weight	X	Arm	=	Moment
Empty weight	831.5 lb	X	0	=	0
Pilot	173 lb	X	+10.7	=	+851.1
Passenger (PAX)	160 lb	X	-9.3	=	-1488
Baggage (BAX)	20 lb	X	+10.7	=	+214
Fuel (AVGAS)	95 lb	X	-11	=	-1045
Totals	1279.5 lb		-0.36		-467.9

Total moment of -467.9 ÷ Total Weight of 1279.5 lb = a **CofG of -0.36 inches left of the Buttline**.

The MAUW allowable is 1370 lb, and our calculation is **1279.5 lb**, so we are within the weight limit for lift-off.

The Lateral CofG limits are 2.2 inches left of the Buttline and 2.6 inches right of the Buttline. Our calculation is 0.36 inches left of the Buttline, so we are within the lateral CofG limits for lift-off.

Datum Calculating the new CofG position relative to the Datum is shown below.

Longitudinal

Item	Weight	X	Arm	=	Moment
Empty weight	831.5 lb	X	+104	=	+86476
Pilot	173 lb	X	+78	=	+13494
Passenger (PAX)	160 lb	X	+78	=	+12480
Baggage (BAX)	20 lb	X	+78	–	+1560
Fuel (AVGAS)	95 lb	X	+108	=	+10260
Totals	1279.5 lb		+97.12		+124270

Total moment of +124270 ÷ Total Weight of 1279.5 lb = **CofG of +97.12 inches aft of the Datum**

The MAUW allowable is 1370 lb, and our calculation is **1279.5 lb**, so we are within the weight limit for lift-off.

The Longitudinal CofG limits are 95.5 inches aft of the Datum to 102 inches aft of the Datum. Our calculation is 97.12 inches aft of the Datum, so we are within the Longitudinal CofG limits for lift-off.

Plotting the CofG Envelope   Once the calculations are done, the pilot has the option to then plot the information on the CofG Envelope as follows:

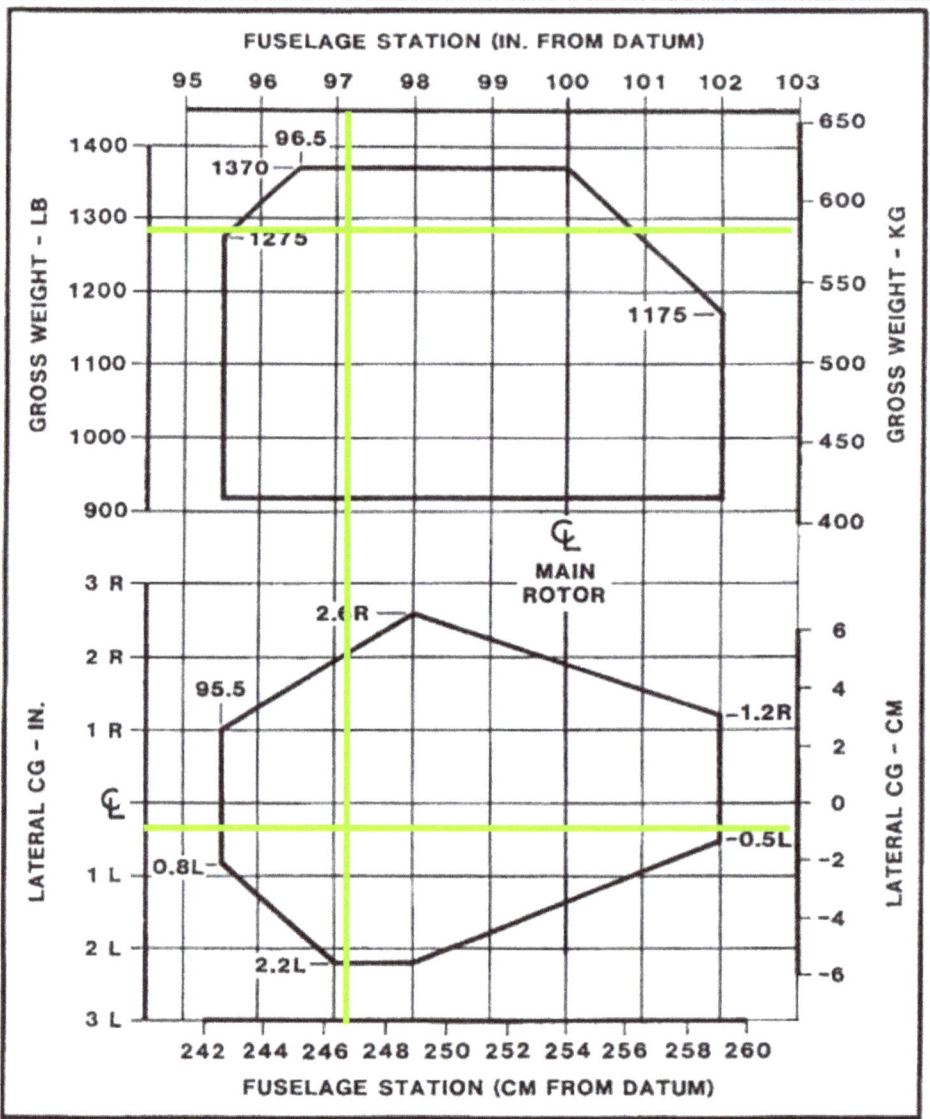

Where the two green lines intersect represents the current CofG Position, and this falls within the CofG Envelope.

www.ingramcontent.com/pod-product-compliance
Lightning Source LLC
Chambersburg PA
CBHW042020090526
44590CB00030B/4345